Mycobacterium MOLECULAR MICROBIOLOGY

Edited by:

Tanya Parish
Institute of Cell and Molecular Science, Barts and The London, Queen Mary's School of Medicine and Dentistry, London

horizon bioscience

Horizon Bioscience
32 Hewitts Lane
Wymondham
Norfolk NR18 0JA
U.K.

www.horizonbioscience.com

British Library Cataloguing-in-Publication Data

A catalogue record for this book is available from the British Library

ISBN: 1-904933-14-9

Printed and bound in Great Britain by Antony Rowe Ltd, Chippenham, Wiltshire

Contents

Contributors

Miriam Braunstein*
Department of Microbiology and Immunology
University of North Carolina
School of Medicine
Chapel Hill
NC 27599-7290
USA

B.M. Buddle
AgResearch
Wallaceville Animal Research Centre
Upper Hutt
New Zealand

D.M. Collins*
AgResearch
Wallaceville Animal Research Centre
Upper Hutt
New Zealand

Rabia Johnson
US/MRC Centre of Molecular and Cellular Biology
Department of Medical Biochemistry
Faculty of Health Sciences
Stellenbosch University
South Africa

Arseny S. Kaprelyants
Bakh Institute of Biochemistry
Russian Academy of Sciences
Leninski pr.33
119071 Moscow
Russia

Sharon L. Kendall
Department of Pathology and Infectious Diseases
The Royal Veterinary College
Royal College Street
London NW1 0TU
UK

Sherry Kurtz
Department of Microbiology and Immunology
University of North Carolina
School of Medicine
Chapel Hill
NC 27599-7290
USA

G.W. de Lisle
AgResearch
Wallaceville Animal Research Centre
Upper Hutt
New Zealand

Gail E. Louw
US/MRC Centre of Molecular and Cellular Biology
Department of Medical Biochemistry
Faculty of Health Sciences
Stellenbosch University
South Africa

Murty V. Madiraju
Biomedical Research
The University of Texas Health Center at Tyler
Tyler
TX 75708-3154
USA

Farahnaz Movahedzadeh
Department of Pathology and Infectious Diseases
The Royal Veterinary College
Royal College Street
London NW1 0TU
UK

Galina V. Mukamolova
Institute of Biological Sciences
University of Wales
Aberystwyth
Ceredigion SY23 3DD
UK

I. Papatheodorou
Department of Computing
Imperial College
London SW7 2AZ
UK

Tanya Parish
Institute of Cell and Molecular Science
Barts and The London
Queen Mary's School of Medicine and Dentistry
London E1 2AD
UK

Malini Rajagopalan*
Biomedical Research
The University of Texas Health Center at Tyler
Tyler
TX 75708-3154
USA

Kyu Y. Rhee*
Division of International Medicine and Infectious Diseases
Weill Medical College of Cornell University
1300 York Avenue A-421
New York
NY 10021
USA

Stuart C. G. Rison
Department of Pathology and Infectious Diseases
The Royal Veterinary College
Royal College Street
London NW1 0TU
UK

P.L.C. Small*
University of Tennessee
Knoxville
USA

G. R Stewart*
School of Biomedical and Molecular Sciences
University of Surrey
Guildford
Surrey GU2 7XH
UK

Neil G. Stoker*
Department of Pathology and Infectious Diseases
The Royal Veterinary College
Royal College Street
London NW1 0TU
UK

Elizabeth M. Streicher
US/MRC Centre of Molecular and Cellular Biology
Department of Medical Biochemistry
Faculty of Health Sciences
Stellenbosch University
South Africa

Paul D. van Helden
US/MRC Centre of Molecular and Cellular biology
Department of Medical Biochemistry
Faculty of Health Sciences
Stellenbosch University
South Africa

Thomas C. Victor*
US/MRC Centre of Molecular and Cellular Biology
Department of Medical Biochemistry
Faculty of Health Sciences
Stellenbosch University
South Africa

Robin M. Warren
US/MRC Centre of Molecular and Cellular Biology
Department of Medical Biochemistry
Faculty of Health Sciences
Stellenbosch University
South Africa

D.B. Young
Department of Infectious Diseases and Microbiology
Imperial College
London SW7 2AZ
UK

Michael Young*
Institute of Biological Sciences
University of Wales
Aberystwyth
Ceredigion SY23 3DD
UK

* Corresponding author

Preface

The mycobacteria are a fascinating group of bacteria as well as being of extreme medical importance. *Mycobacterium tuberculosis* is the biggest bacterial killer, responsible for approximately two million deaths every year. *Mycobacterium leprae,* which causes leprosy, still poses severe problems in parts of the world and environmental mycobacteria are causing increasing numbers of serious infections, mainly in immunocompromised people. For these reasons research on mycobacteria has focussed on pathogenic mechanisms, drug resistance mechanisms and ways of improving diagnosis, treatment and prevention.

Mycobacteria are difficult organisms to work with; *M. leprae* cannot be cultured *in vitro* and the other mycobacteria have relatively slow growth rates. The cells have a tough lipid-rich cell wall which poses both a permeability barrier, particularly to antibiotics, and an obstacle to lysis techniques. Coupled with the practical restrictions that the pathogenic nature of some species poses, this has hampered the progress of the biological study of these bacteria.

However, despite, or perhaps because of, these limitations, mycobacterial researchers have been quick to utilise the tools of genomics and proteomics and *M. tuberculosis* was one of the first bacteria to be completely sequenced. Several other mycobacterial genomes are now fully sequenced and annotated or near completion and this has enabled the wide use of techniques such as microarrays or 2D gel electrophoresis for proteins. The ability to construct both defined mutants and random libraries now provide a battery of tools to answer some of the key biological questions. Thus research into the basic biology of these organisms has bloomed.

The topics covered in this volume are wide ranging and intend to provide a snapshot of current areas of research. Some of these subjects are not unique to the mycobacteria, but the particular ways in which mycobacteria deal with these problems are different from the classical *Escherichia coli* model system. Basic cell functions such as DNA replication and secretion of proteins are covered revealing significant differences. The response to the environment which may play a large role in defining the pathogenic potential of mycobacteria is covered in several chapters dealing with two component regulatory systems, the heat shock response and defence against reactive nitrogen and oxygen compounds. A particular type of response – the persistent/latent/dormancy response is also covered since much recent work has attempted to determine the basis of this phenomenon. The practical aspects of developing new vaccines are covered, as well as methods for detecting drug resistance and an overview of the molecular mechanisms involved. Finally the role of environmental mycobacteria in virulence and the consideration of emerging mycobacterial pathogenic species are covered.

Although this book reviews the current state of play in several promising areas of research, it cannot cover all topics of interest. However, I hope that this selection will stimulate the reader as well was provide an insight into ongoing research and raise new questions for future fruitful avenues.

Tanya Parish
July 2005

Books of Related Interest

Hepatitis C Viruses: Genomes and Molecular Biology	2006
Microbial Bionanotechnology	2006
Molecular Diagnostics: Current Technology and Applications	2006
DNA Microarrays: Current Applications	2006
Computational Biology: Current Methods	2006
Lactobacillus Molecular Microbiology	2006
Probiotics and Prebiotics: Scientific Aspects	2005
Cancer Therapy: Molecular Targets in Tumour-Host Interactions	2005
Biodefense: Principles and Pathogens	2005
Mycobacterium: Molecular Microbiology	2005
Dictyostelium Genomics	2005
Epstein Barr Virus	2005
Cytomegaloviruses: Molecular Biology and Immunology	2005
Papillomavirus Research: From Natural History To Vaccines	2005
HIV Chemotherapy: A Critical Review	2005
Food Borne Pathogens: Microbiology and Molecular Biology	2005
SAGE: Current Technologies and Applications	2005
Microbial Toxins: Molecular and Cellular Biology	2005
Vaccines: Frontiers in Design and Development	2005
Antimicrobial Peptides in Human Health and Disease	2005
Campylobacter: Molecular and Cellular Biology	2005
The Microbe-Host Interface in Respiratory Tract Infections	2005
Malaria Parasites: Genomes and Molecular Biology	2004
Pathogenic Fungi: Structural Biology and Taxonomy	2004
Pathogenic Fungi: Host Interactions and Emerging Strategies for Control	2004
Strict and Facultative Anaerobes: Medical and Environmental Aspects	2004
Brucella: Molecular and Cellular Biology	2004
Yersinia: Molecular and Cellular Biology	2004
Bacterial Spore Formers: Probiotics and Emerging Applications	2004
Foot and Mouth Disease: Current Perspectives	2004
Sumoylation: Molecular Biology and Biochemistry	2004
DNA Amplification: Current Technologies and Applications	2004
Prions and Prion Diseases: Current Perspectives	2004
Real-Time PCR: An Essential Guide	2004
Protein Expression Technologies: Current Status and Future Trends	2004
Computational Genomics: Theory and Application	2004
The Internet for Cell and Molecular Biologists (2nd Edition)	2004
Tuberculosis: The Microbe Host Interface	2004
Metabolic Engineering in the Post Genomic Era	2004
Peptide Nucleic Acids: Protocols and Applications (2nd Edition)	2004
Ebola and Marburg Viruses: Molecular and Cellular Biology	2004

Full details of all these books at: www.horizonpress.com

Chapter 1

DNA Replication and Cell Division

Murty V. Madiraju and Malini Rajagopalan

ABSTRACT

The eubacterial cell cycle consists of two important aspects; one is the DNA replication cycle and the other is the cell division cycle. Information on how the critical events involved in the *Mycobacterium tuberculosis* cell cycle are initiated, regulated and coordinated with each other is critical for understanding the state of the bacterium during active and persistent growth *in vivo*. Although we have a long way to go in understanding the molecular mechanisms responsible for the initiation and the regulation of replication and cell division processes in mycobacteria, we are gaining ground in achieving this goal. Studies on the replication origin(s) and the DnaA initiator protein and their interactions *in vitro* and *in vivo* are providing clues to the roles of the initiator protein and the regulation of *ori*C replication. Epidemiological studies on clinical isolates suggest that *M. tuberculosis* strains can use alternate origins for replication initiation. The involvement of the inducible DnaE2 polymerase in generating spontaneous mutations following DNA damage has been established. Investigations into activities of DNA chain elongation enzymes and the inducible DNA damage response are actively being pursued. Fluorescent microscopy techniques to visualize the FtsZ protein catalyzed septal rings at the mid-cell location and genetic and biochemical characterization of other cell division protein components that either associate with and/or affect the FtsZ-ring assembly, are being carried out.

INTRODUCTION

This review focuses on the *M. tuberculosis* cell cycle. Consequently, issues related to DNA replication and cell division are discussed in detail. Bacterial replication is believed to be regulated at the level of initiation and once initiated, DNA chain elongation continues at a uniform pace, unless impeded by replication blocking lesions (Kornberg and Baker, 1991). Hence, we discuss in depth what we know about the replication initiation process in *M. tuberculosis* and mechanisms and concepts, if any, used for the regulation of DNA replication in *M. tuberculosis*.

The genus *Mycobacterium* includes both rapid-growing members, *e.g. M. smegmatis, M. fortuitum* (doubling time is ~ 2 to 3 h) and slow-growing members *e.g. M. avium-intracellulare complex* (doubling time ~ 10 to 12 h), the *M. tuberculosis* complex (doubling time ~ 22 to 24h) and *M. leprae* (doubling time ~ 14 days *in vivo*). The genetic and biochemical factors responsible for growth rate differences and their correlations, if any, with replication in different mycobacterial species are largely unknown. Nonetheless, where appropriate, we discuss information known on this subject in other members of mycobacteria. Replication blocking lesions often impede DNA synthesis by polymerases. These replication blocks can be repaired by various error-free pathways or bypassed at the expense of generating mutations - referred to as trans-lesion synthesis (TLS) (Walker, 1984; Walker, 1985). We will discuss recent advances in unraveling the DNA damage response and replication through DNA lesions in *M. tuberculosis*. DNA repair in *M. tuberculosis* was the subject of a recent review and is not discussed here (Mizrahi et al., 2000; Mizrahi and Andersen, 1998). Cell division constitutes an essential part of cell cycle without which separation of daughter cells does not occur. Current studies on this subject in mycobacterial members are limited to the initiation aspect of the cell division and these are summarized. Finally, conclusions focusing on gaps and relevant questions in the subject are presented at the end of replication and cell division topics.

REPLICATION

The molecular details that dictate how *M. tuberculosis* replicates its genome, and how the mechanism of replication is coordinated to other critical events in cell cycle such as cell division, is an enigma as compared to the recent advances made in mycobacterial pathogenesis research in identifying virulence factors, secreted proteins, cell wall biosynthesis, vaccines (see other chapters in this book). Replication of genomic DNA and cell duplication processes resulting in the multiplication of the pathogenic organisms is a necessary critical step for establishing infections. Thus, detailed studies on the replication and cell division processes in *M. tuberculosis* offer two-fold significance. First, they will help us to understand how the *M. tuberculosis* pathogen prepares its genome for replication and subsequent cell division both during intracellular and extracellular growth conditions. Second, they will help us to identify novel drug targets against which new generation of antimycobacterial agents can be developed.

Replication is divided into three steps—initiation, elongation and termination, each of which involves the action of multiple proteins (Kornberg and Baker, 1991). Much of our understanding of the replication process and its regulation comes from detailed studies carried out on the key players involved (*i.e. ori*C and DnaA) and the process itself in the Gram-negative bacterium *Escherichia coli*. We will present our current understanding of the process in *M. tuberculosis* and draw attention to both common and unique features with the well-characterized *E. coli* replication process.

MYCOBACTERIAL OriC

Replication in eubacteria is initiated at a unique site on the chromosome called the origin of replication or *ori*C (Jacob et al., 1963). The *ori*C of *M. tuberculosis*

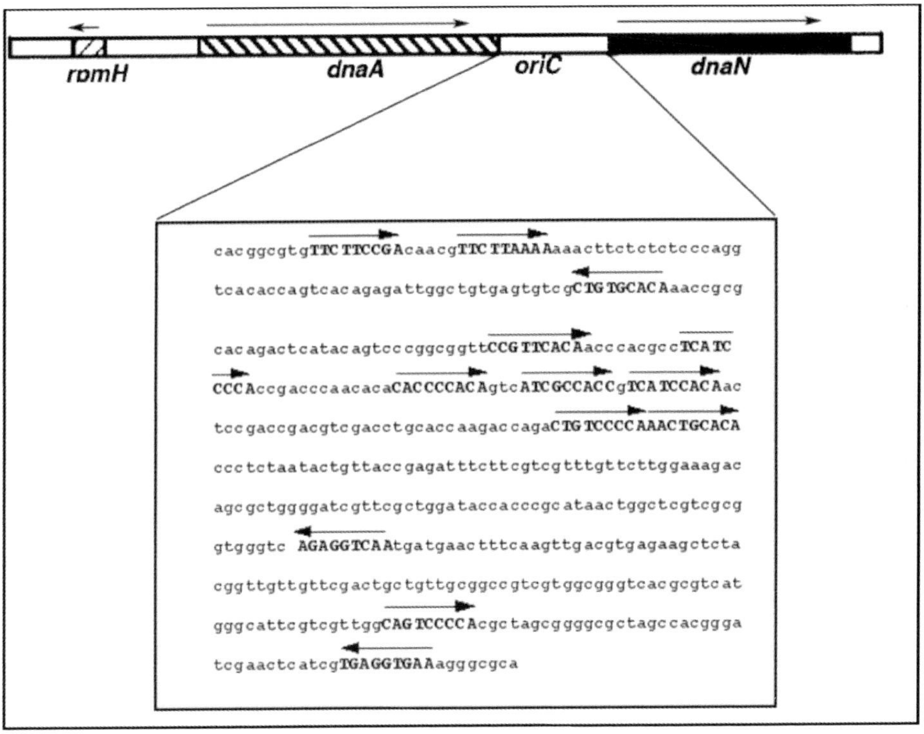

Fig. 1. The *dna*A region of *M. tuberculosis*. A 5-kb DNA fragment bearing the *rpmH, dna*A and *dna*N genes is shown (the region is not drawn to scale). Direction of arrowheads above the *rpmH* and *dna*A genes indicate that they are divergently transcribed. The intergenic region between *dna*A and *dna*N shown as a white box is designated as *ori*C. The 550 bp nucleotide sequence of this region is shown below. Arrows above the individual nucleotide sequence indicate the presumptive multiple DnaA boxes and the direction of the arrowheads indicates the orientation of the DnaA boxes.

was identified prior to the completion of genome sequence by both comparative sequence analysis of the *dna*A flanking regions to identify characteristic features of replication origins (Fig. 1), and plasmid transformation assays to identify mycobacterial sequences that support autonomous replication activity in mycobacteria (Qin et al., 1999; Qin et al., 1997; Rajagopalan et al., 1995; Salazar et al., 1996). Based on sequence comparisons, a nine base pair (bp) mycobacterial DnaA box (TTGTGCACA) has been defined and several sequences with one to two mismatches to this consensus sequence are identified in the 5' (*rpmH-dna*A) and the 3' (*dna*A-*dna*N) flanking regions of *dna*A. While these may be authentic mycobacterial DnaA-box sequences, they are often referred to as "non-perfect DnaA box sequences", because they bear little or no sequence similarity to the DnaA box sequences of *E. coli ori*C (McGarry et al., 2004; Messer et al., 2001). DNA fragments bearing the *dna*A-*dna*N intergenic region, but not those of the *rpmH-dnaA* intergenic region of *M. tuberculosis, M. avium* and *M. smegmatis*, exhibited autonomous replication activity in their respective mycobacterial hosts (Madiraju et al., 1999; Qin et al., 1999; Qin et al., 1997; Rajagopalan et al., 1995; Salazar et al., 1996), hence the *dna*A-*dna*N intergenic region is referred to as *ori*C. Point mutations in the DnaA box sequences severely decreased plasmid *ori*C replication activity whereas insertions in the DnaA box sequences abolished *ori*C activity indicating that integrity of DnaA box sequence is essential for *ori*C plasmid replication of mycobacteria (Dziadek et al., 2002b; Qin et al., 1999; Qin et al., 1997). These results are in stark contrast with the situation seen with *E. coli ori*C where single, double and triple mutant combinations in the DnaA boxes do not affect plasmid *ori*C activity (Langer et al., 1996). The replication origins of *E.coli* chromosomes and plasmids contain A-T rich sequence regions which serve as sites for the coordinated actions of the DnaA initiator protein and DnaB replicative helicase and thereby enable DNA replication to begin (Bramhill and Kornberg, 1988). *M. tuberculosis ori*C lacks these distinct A-T rich repeat motifs. Another notable feature of mycobacterial origins is that they are species specific *i.e. ori*C from *M. tuberculosis* does not function in *M. smegmatis* or *M. avium* and vice versa (Madiraju et al., 1999; Qin et al., 1999). In contrast *ori*C from distantly related enterobacterial members, *e.g. Vibrio cholera* and *Salmonella typhimurium* does function in *E. coli* and vice versa (Skovgaard and Hansen, 1987; Zyskind and Smith, 1986). Similarly, *ori*C of *Pseudomonas putida* functions in *P. aeruginosa;* and *Streptomyces coelicolor ori*C functions in *S. lividans* (Yee and Smith, 1990; Zakrzewska-Czerwinska and Schrempf, 1992). It is unknown why mycobacterial *ori*C are species-specific. We speculate that there may be species-specific regulators acting on *ori*C, which could be responsible for this phenomenon.

Two transcripts, designated as leftward and rightward, are mapped in the *ori*C of *M. bovis* BCG, a slow grower, but no corresponding transcripts are found in the *ori*C of *M. smegmatis* (Salazar et al., 2003). Since the *ori*C sequences of *M. bovis* BCG and *M. tuberculosis* are identical (Qin et al., 1999), it is expected that such transcriptional activity will also be detected in the *M. tuberculosis ori*C. Transcriptional activation of *ori*C replication in *E. coli* and the regulatory role of the *ori*C flanking *gid*A and *mio*C genes transcription on *ori*C activation are well recognized (Asai et al., 1992; Nozaki et al., 1988). It remains to be seen whether the observed transcriptional activity in the *ori*C of slow-growing mycobacteria is relevant for their *ori*C replication.

CRYPTIC ORIGINS?

The *rpmH-dna*A intergenic regions of *M. avium* and *M. tuberculosis* show significant sequence identity, although they do not exhibit *ori*C activity. Interestingly, the *M. avium rpmH-dnaA* intergenic region exhibits *ori*C activity in *M. tuberculosis*. These rather surprising results led to an interesting hypothesis that the *rpmH-dnaA* is a cryptic origin whose activity might be unmasked under specific growth conditions, *e.g.* under the conditions where *ori*C activity originating at the *dna*A-

*dna*N intergenic region is prevented (Madiraju et al., 1999). It is noted that the *ori*C transformation assays with *M. tuberculosis* genomic libraries demonstrated that only the *dnaA-dna*N intergenic sequences exhibited autonomous replication activity. We have speculated that functional *ori*C activity originating at the *dna*A-*dna*N intergenic region suppresses possible replication activity originating elsewhere on the genome (Madiraju et al., 1999).

SOME *M. TUBERCULOSIS* CLINICAL STRAINS ARE NATURAL ORIC MUTANTS

A variable number of the IS*6110* insertion element, a member of IS*3* transposable elements, are found distributed around the chromosome of *M. tuberculosis* strains (Cole et al., 1998). Molecular epidemiological studies using a probe specific to IS*6110* sequence revealed a non-random distribution of IS*6110* sequences in the *ori*C region of *M. tuberculosis* strains. The majority of these insertions are located outside of the DnaA box regions, although one insertion, designated as A4, disrupted the CCGTTCACA DnaA box (Fig. 2) (Kurepina et al., 1998). Surface plasmon resonance technique confirmed that CCGTTCACA is an authentic DnaA box (Dziadek et al., 2002b). Insertions outside the DnaA boxes allowed *ori*C activity on plasmids, whereas those that interrupted the DnaA box sequences did not (Dziadek et al., 2002b). The *ori*C of some clinical strains also contain two copies of IS*6110* sequences which are often associated with up to 330-bp deletions in the *ori*C region (Krieswirth, B. et al, unpublished). Replacement of the *M. tuberculosis* H37Rv *ori*C sequence with that of the corresponding sequence of clinical strains by homologous recombination revealed that the *ori*C of *M. tuberculosis* H37Rv *strain* can tolerate

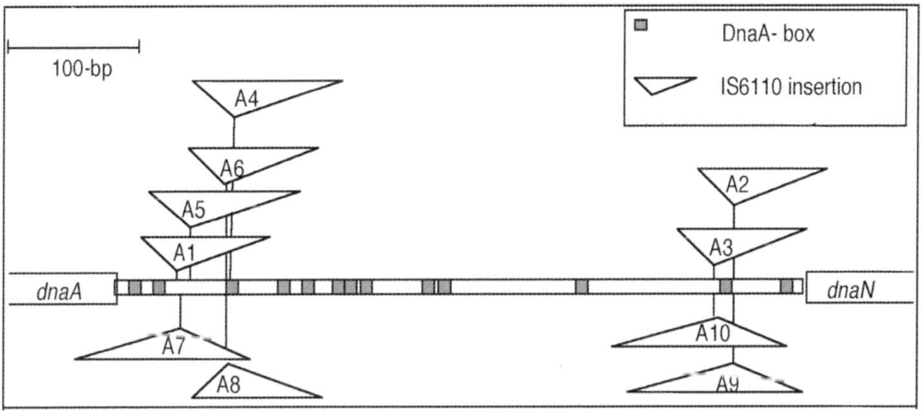

Fig. 2. The non-random distribution of IS*6110* sequences in the *oriC* region of clinical strains. IS*6110* sequences are shown as triangles and the DnaA boxes as shaded boxes. Note the DnaA boxes and individual insertions are not marked to the scale. The cartoon shows individual IS*6110* insertions found in clinical isolates, designated as A1 to A10. Although not shown, some clinical strains have two IS*6110* insertions within the 550-bp region.

IS*6110* insertions and associated deletions of up to 60-bp *ori*C sequence (Dziadek et al., 2002b; Madiraju et al., unpublished). These data raise important questions about replication initiation mechanisms operational in strains with mutant origins. Does the *rpmH-dnaA* intergenic region or another region serve as an alternate *ori*C or do clinical strains with deletions in *ori*C carry out *ori*C independent replication? Clearly further studies are required to answer these questions.

DnaA: *IN VITRO* STUDIES

DnaA is the initiator of DNA replication. *M. tuberculosis* DnaA protein is associated with the membrane fractions in non-overproducing cells of *M. tuberculosis*. Characterization of recombinant *M. tuberculosis* DnaA ($DnaA_{TB}$) purified under denaturing conditions indicated that $DnaA_{TB}$ binds *ori*C, ATP and ADP with high affinity and exhibits weak ATPase activity (Yamamoto et al., 2002a). *M. tuberculosis* cells producing ATPase defective DnaA protein are nonviable indicating that the ATPase activity is biologically relevant (Madiraju et al, unpublished data). Like its *E. coli* counterpart, $DnaA_{TB}$ retains ADP following the hydrolysis of ATP. Acidic phospholipids such as cardiolipin, phosphatidylinositol and phosphotidylglycerol, but not basic phospholipids such as phosphatidylethanolamine, promote dissociation of ATP or ADP from $DnaA_{TB}$ (Yamamoto et al., 2002a; Yamamoto et al., 2002b). In the presence of *ori*C, dissociation of ATP from $DnaA_{TB}$ is decreased whereas that of the ADP is stimulated. This latter result is different from that reported for *E. coli* where in the presence of *ori*C, phospholipid-mediated dissociation of ATP or ADP from DnaA is prevented (Crooke, 2001; Crooke et al., 1992). $DnaA_{TB}$ binds to some presumptive DnaA boxes including the CCGTGCACA (Dziadek et al., 2002b), but not to all (Madiraju et al., unpublished). Presumably, the *ori*C region contains both high and low affinity DnaA boxes, and the binding of $DnaA_{TB}$ to high affinity boxes promotes contact with other low affinity boxes thereby organizing the formation of an efficient *ori*C- $DnaA_{TB}$ initiation complex. Interestingly, $DnaA_{TB}$ purified under native conditions failed to bind to a near perfect DnaA box sequence [TTGTCCACA], but binding was promoted when two DnaA boxes were present [TTGTCCACA, TTGTCCCA and with one mismatch] (Zawilak et al., 2004). DNase I foot printing as well as electron microscopy showed that $DnaA_{TB}$ binds four specific regions in *ori*C, corresponding to 11 out of the 13 presumptive DnaA boxes (Zawilak et al., 2004). It is hypothesized that DnaA binding to *ori*C involves co-operative binding to several non-perfect DnaA boxes resulting in a massive nucleoprotein complex. It remains to be tested whether differences in the activities of $DnaA_{TB}$ are due to the differences in the purification procedures such that the $DnaA_{TB}$ purified under native conditions is associated with bound nucleotides, phospholipids or both.

DnaA *IN VIVO* STUDIES

DnaA protein is present at approximately 800 molecules per cell and the intracellular levels of DnaA are stable during all growth phases examined. Optimal levels of

DnaA appear to be essential for cell cycle progression as elevated intracellular levels of DnaA either in *M. smegmatis* or in *M. tuberculosis* delayed growth, but did not affect cell viability (Greendyke et al., 2002). This is in contrast to the filamentation and reduced viability seen with overproduction of DnaA in *Bacillus subtilis* and *E. coli* (Grigorian et al., 2003; Kawakami et al., 2001; Ogura et al., 2001). The *M. smegmatis dna*A gene is expressed from single promoter ($P1_{dnaA}$) whereas that of the *M. bovis* BCG is expressed from two promoters ($P1_{dnaA}$ and $P2_{dnaA}$). Presumably, regulation of *dna*A transcription in slow-growing mycobacteria is more complex (Salazar et al., 2003). Overproduction or depletion of DnaA in *M. smegmatis* results in filamentation and a multinucleoidal state indicating a close correlation between DnaA-mediated DNA replication and cell division processes (Greendyke et al., 2002). Mycobacterial DnaA proteins show significant sequence identity with differences in the amino-terminal region. The latter is believed to be involved in interactions with the DnaB helicase. Mycobacterial DnaAs, unlike their *ori*C, are interchangeable, *i.e. M. tuberculosis dna*A can replace the function of its *M. smegmatis* counterpart and vice versa. $DnaA_{TB}$ binds to the *M. smegmatis ori*C with the same kinetics as it does to its own *ori*C (Madiraju et al., unpublished). Presumably, steps subsequent to the binding of DnaA to *ori*C in cell cycle progression limit interspecies *ori*C replication.

DnaB HELICASE

E. coli DnaA-*ori*C complex serves as a launch pad for interactions with a host of other proteins that include the loading of the DnaB helicase that ultimately marks the formation of bidirectional replication forks. The locus coding for *dna*B in *M. tuberculosis* (Cole et al., 1998), *M. leprae* (Cole et al., 1998), *M. avium* and *M. smegmatis* (http://www.tigr.org/tdb/ufmg) also codes for an *intein*. Inteins are protein splicing elements that are produced following self-splicing of the precursor protein (reviewed in Perler, 1998; Perler et al., 1994). The intein coding sequence is usually inserted in-frame within a protein coding sequence, hence protein-splicing is essential to regenerate functional host protein. DnaB precursor protein splicing has been demonstrated for mycobacterial proteins (Yamamoto et al., 2001). The DnaB inteins of *M. leprae* (123 aa) and *M. avium* (330 aa) are inserted at the amino-terminal region of DnaB and contain alanine at the amino-terminal splice junction, unlike other inteins that contain either cysteine, serine or threonine (Perler, 1998; Perler et al., 1994). The DnaB intein of *M. tuberculosis* is large (433 aa) and is located at the carboxy terminal end of the DnaB precursor (Cole et al., 1998). Interestingly, the *M. smegmatis* DnaB precursor protein contains two inteins, one at the amino-terminus like *M. avium* and the other at the carboxy-terminus like *M. tuberculosis* (http://www.tigr.org/tdb/ufmg).

The *M. tuberculosis* genome lacks *dna*C, whose gene product acts as a molecular matchmaker that helps load the DnaB helicase onto the DnaA-*ori*C complex (Cole et al., 1998). Consequently the mechanism by which the DnaB helicase makes contact with the DnaA-*oriC* complex is predicted to differ from that established for *E. coli* (Messer et al., 2001; Sutton et al., 1998; Sutton and Kaguni, 1997). Global gene expression profiling showed that *dna*B expression is altered in several genetic backgrounds including *regX-senX* and *dna*E2 deletion strains (Boshoff et al., 2003; Parish et al., 2003). Since DnaB is essential for survival and since DnaB precursor protein must undergo post-translational protein splicing for generation of the intein and the DnaB extein, the DnaB protein splicing reaction is considered as an excellent drug target. A positive genetic selection system for inhibition of DnaB intein splicing is being developed with the potential for developing splicing inhibitors that prevent mycobacterial infections (Adam and Perler, 2002).

OTHER PROTEINS

The primase encoded by *dna*G, is essential for priming of DNA synthesis and replicative polymerase cannot start chain elongation without the action of DnaG. *M. tuberculosis dna*G (Rv2343c) is an essential replication gene (Sassetti et al., 2001; Sassetti et al., 2003). A temperature sensitive *M. smegmatis dna*G mutant strain has been characterized and showed a filamentation phenotype at the restrictive temperature indicating a co-ordination between replication and cell division (Klann et al., 1998). Other proteins involved in replication include histone like proteins *hup*B (Rv2986c) (Prabhakar et al., 2004); RNase H (Rv2228c) which appears to be a bifunctional protein that shows homology to N-terminal ribonuclease H and a C-terminal phosphoglyceratemutase, single stranded DNA binding protein, gyrase and topoisomerase. The last three are also involved in other cellular processes such as repair, recombination and transcription.

SINGLE STRANDED DNA BINDING (SSB) PROTEIN

As the name implies, Ssb binds non-specifically to single stranded DNA produced during replication, recombination and repair and protects it from subsequent degradation by nucleases. Ssb proteins of *M. tuberculosis* (Rv0054) and *M. smegmatis* have been purified and characterized (Acharya and Varshney, 2002; Handa et al., 2000; Purnapatre and Varshney, 1999). The recombinant proteins exist as stable tetramers in solution and share properties with their *E. coli* counterpart. Ssb_{TB} interacts with single stranded DNA (Purnapatre and Varshney, 1999; Reddy et al., 2001) and influences RecA and topoisomerase activities (Reddy et al., 2001; Sikder et al., 2001). The carboxy terminal domain of Ssb_{TB} makes stable associations with RecA (Reddy et al., 2001) and uracil-DNA-glycosylases (Acharya and Varshney, 2002; Purnapatre and Varshney, 1998; Sikder et al., 2001). Presumably stable interactions between these proteins are essential for processing damaged DNA bases for repair. In stark contrast, functional but *not* physical interactions between Ssb and RecA proteins of *E. coli* have been demonstrated (Kowalczykowski et al., 1994). RecO, a protein involved in DNA recombination, repair and replication, is the only protein that has been shown to interact with *E. coli* Ssb physically and functionally (Umezu et al., 1993; Umezu and Kolodner, 1994). Presumably, the mechanistic aspects of DNA repair, replication and recombination mediated by Ssb_{TB} are different from those established for *E. coli*. Ssb_{TB} has been crystallized and the tetrameric Ssb_{TB} crystals show variations in the quaternary structure that are implicated to provide greater stability to the oligomeric protein during dormant growth or stress (Saikrishnan et al., 2002; Saikrishnan et al., 2003).

TOPOISOMERASES

Topoisomerases (Topo I) are a class of enzymes that alter the topology of DNA by concerted breaking and rejoining of the phosphodiester backbone of DNA and thereby affect the superhelical density of chromosomal DNA. The activities of these enzymes are essential for DNA replication, repair and recombination. Recombinant Topo I from *M. smegmatis* has been purified and characterized. A notable feature of this protein is that, unlike other eubacterial Topo I enzymes, it lacks the characteristic zinc-binding motif and its activity has been shown to be stimulated by Ssb (Nagaraja et al., 2002; Sikder and Nagaraja, 2001; Yang et al., 1996).

GYRASE

Gyrases are ATP-dependent enzymes that catalyze negative supercoiling of DNA and are crucial for DNA replication, recombination and transcription. Fluoroquinolones are potential antimycobacterial agents that act on DNA gyrase and topoisomerase IV enzymes (Drlica et al., 1996; Xu et al., 1996). *M. tuberculosis* genome lacks *par*C and *par*E, the genes for Topo IV enzyme (Cole et al., 1998) making DNA gyrase the sole quinolone target. Consequently, significant efforts on characterization of *M. tuberculosis* gyrase and analysis of quinolone action on gyrase activities have been carried out. The contiguous *gyr*A and *gyr*B genes encode DNA gyrase, a tetrameric A2B2 protein. The A subunit is required for strand breakage and rejoining activities whereas the B subunit is required for ATP hydrolysis providing the energy needed for catalysis (Aubry et al., 2004; Manjunatha et al., 2002; Manjunatha et al., 2001a; Manjunatha et al., 2001b). The genome of *M. smegmatis* contains a second orphan *gyr*B gene of unknown function (Jain and Nagaraja, 2002). Native gyrase protein from extracts of *M. smegmatis, M. avium,* and *M. fortuitum* as well as recombinant fusion proteins have been purified and characterized. Investigation of the interactions of gyrase with quinolones showed a correlation between inhibition of DNA supercoiling and *M. tuberculosis* growth (reviewed in Aubry et al., 2004).

ELONGATION, DNA SYNTHESIS AND POLYMERASES

Replicative DNA polymerases are the class of enzymes that carry out DNA synthesis accurately without errors. Classical studies from late Dr. Ramakrishnan's group showed the DNA synthesis rate (replication time) of *M. tuberculosis* to be 1200 nucleotides/min which is approximately 11 times slower than that of *M. smegmatis* and 13 to 18 times slower than that of *E. coli* (Hiriyanna and Ramakrishnan, 1986). The reported slow rates are consistent with the slow growth rate of *M. tuberculosis*. Both

PolI and PolIII holoenzyme alpha subunit coding genes have been identified in the genomes of *M. tuberculosis* and *M. smegmatis* (reviewed in Mizrahi et al., 2000).

POLYMERASE I (PolI)

DNA PolI encoded by *pol*A gene (Rv1629), is one of the best-characterized polymerases of *M. tuberculosis*. The *M. tuberculosis* PolI enzyme possess both polymerase and 5'—>3' exonuclease functions and is believed to be required for the processing of Okazaki fragments (Huberts and Mizrahi, 1995; Mizrahi and Huberts, 1996). An *M. smegmatis pol*A mutant strain shows increased sensitivity to UV and hydrogen peroxide consistent with the prediction that PolI is a DNA repair polymerase (Gordhan et al., 1996). Recombinant *M. tuberculosis* PolI does not discriminate between deoxynucleotide-triphosphates and dideoxynucleotide-triphosphates during DNA synthesis, thus making it amenable to inhibition by chain-terminating nucleotide analogs (Mizrahi and Huberts, 1996). A putative gene with 5'—>3' exonuclease domain (Rv2090) with unknown function has also been identified in the *M. tuberculosis* genome.

POLYMERASE III HOLOENZYME (PolIII)

PolIII is the principal replicative polymerase in bacteria and is a multi-subunit protein. Putative genes coding for the subunits of PolIII holoenzyme complex have been identified in the *M. tuberculosis* genome sequence (Cole et al., 1998) and are recently summarized (Mizrahi et al., 2000). Homologues of these genes are also present in the unannotated genome sequence of *M. smegmatis* (http://www.tigr.org/tdb/ufmg). The alpha subunit of *M. tuberculosis* polymerase III is encoded by two distinct genes designated as *dna*E1 (Rv1547) and *dna*E2 (Rv3370c). A recent study shows that the former encodes the essential subunit for DNA synthesis whereas the latter encodes for the non-essential error-prone polymerase and is required for SOS-induced (DNA damage induced) DNA mutagenesis (Boshoff et al., 2003).

DnaN

The β-subunit of Pol III, also called the beta-clamp loader, is encoded by *dna*N (Rv0002). DnaN is a processivity factor that helps tether the polymerase to the DNA and hence plays an important role in DNA chain elongation. Recent studies in *E. coli* showed that DnaN is involved in the negative regulation of *ori*C replication. In the presence of Hda protein, DnaN following the replisome assembly, activates the intrinsic ATPase activity of DnaA and converts it to the ADP-form ($DnaA_{ADP}$) that is refractory to subsequent reinitiations of *ori*C replication (Katayama, 2001; Katayama et al., 1998; Katayama and Sekimizu, 1999; Kato and Katayama, 2001). Functional *in vivo* interactions between DnaA- and DnaN proteins have also been identified in *E. coli* and *B. subtilis* (Ogura et al., 2001; Ortenberg et al., 2004). It is unknown if negative regulation of *ori*C replication, as reported for *E. coli,* works in *M. tuberculosis*. Detailed studies on DnaA-*ori*C interactions could provide clues, if any, to regulation of *M. tuberculosis ori*C replication.

OTHER SUBUNITS

The proof-reading function during DNA synthesis is carried out by the epsilon subunit (*dna*Q) encoded by Rv3711c. The coding regions for other polymerase subunits such as "γ, δ, τ" have also been identified in genome sequence but have not been functionally characterized.

CONCLUSIONS AND MODELS FOR REPLICATION INITIATION AND REGULATION

Although we have identified the key players involved in the replication process in *M. tuberculosis,* the key questions pertinent to replication remain largely unexplored. For example, how is the *ori*C-DnaA complex organized and *ori*C replication initiated in *M. tuberculosis*? How does the replisome complex coordinate its assembly and action with the DnaA mediated *ori*C complex? Molecular epidemiological studies indicate that the *ori*C of clinical strains is a hot-spot for IS*6110* insertions (Kurepina et al., 1998), but how do clinical strains with major deletions in *ori*C initiate replication? A model for replication initiation and the role(s) for DnaA in the cell cycle is shown in Fig. 3. According to this model, genomic DNA replication in *M. tuberculosis* proceeds from *ori*C, *i.e.* the *dna*A-*dna*N intergenic region or in the absence of its function from alternate sites elsewhere on the genome in an *ori*C-independent manner. The ability to utilize alternate replication origins could

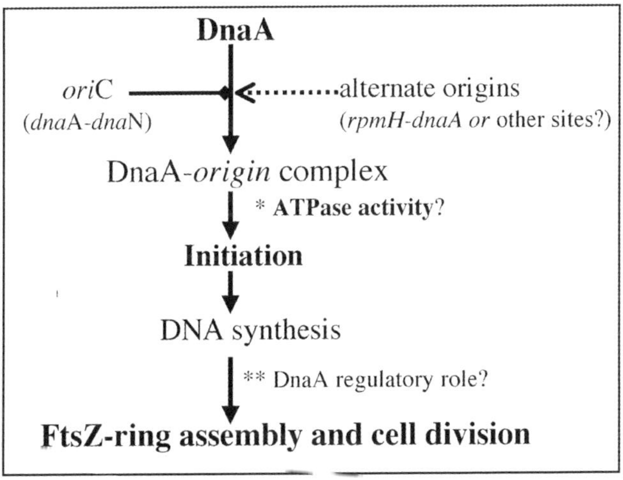

Fig. 3. A model for *M. tuberculosis* replication initiation mediated by DnaA. It is shown that DnaA protein interacts with either the *ori*C (i.e. *dnaA-dna*N intergenic region) or alternate origins shown as dashed arrow. * indicates a hypothetical role for ATPase activity of DnaA in organization of DnaA-initiation complex. It is suggested that the ATPase activity promotes rapid oligomerization in the organization of the initiation complex. ** indicates a role for DnaA in the septation process. Since cells producing altered levels of DnaA show delayed growth and filamentation, it is hypothesized that the DnaA protein acts as regulator in this process.

be an effective survival mechanism employed by the pathogen. The *rpmH-dna*A intergenic region containing multiple DnaA box like sequences may serve as an alternate origin and initiate replication in a DnaA dependent manner. Since ATPase defective *M. tuberculosis dna*A cells are nonviable, it is speculated that the intrinsic ATPase activity of DnaA plays an important role in the organization of the *M. tuberculosis ori*C initiation complex. Once organized, the DnaA-*ori*C initiation complex could serve as a launch pad for interactions with host of other proteins including direct contact with the DnaB helicase. In addition to the initiation of DNA synthesis, DnaA protein may act as a transcriptional regulator and control the expression of genes involved in septation and cytokinesis, although specific details as to how this process is accomplished are unknown.

We speculate, based on the available data, that *M. tuberculosis* DNA replication is subject to regulation by at least two different mechanisms. First, the *rpmH-dna*A intergenic region may serve as a sink to trap excess DnaA that is not engaged in *ori*C replication and thereby prevent reinitiation events. This kind of reasoning is consistent with the *dat*A locus concept proposed for *E. coli*. The *dat*A locus is located at 94.6 min, contains clusters of DnaA boxes and is implicated in sequestration of DnaA following initiation of *ori*C replication (Katayama and Sekimizu, 1999; Kelley et al., 1982; Kitagawa et al., 1998; Morigen et al., 2003; Ogawa et al., 2002). It remains to be seen whether the *rpm*H-*dna*A region, the only region besides *ori*C with multiple DnaA-boxes, serves as a mycobacterial *dat*A counterpart. Second, *M. tuberculosis ori*C replication could be subject to regulation by membrane lipids, as in *E. coli* (Crooke, 1995; Crooke, 2001). This is an appealing concept because *M. tuberculosis* DnaA in non-overproducing cells is membrane associated, and that membrane lipids modulate the interactions of DnaA with adenine nucleotide (Yamamoto et al., 2002a). Presumably, the ratio of acidic to basic phospholipids and their interaction with DnaA will determine whether or not the nucleotide bound form of DnaA is available for *ori*C binding and replication initiation. Such interactions could in turn determine whether the organism is in non-replicative persistent state or active replication state.

REPLICATION BLOCKING LESIONS AND REPAIR

Precise genome duplication requires accurate copying of DNA. The so-called high-fidelity replicative polymerases, in the absence of any proof-reading function, generate replication errors in the range of 10^{-4} to 10^{-5} per base pair. However, in the presence of proof reading functions and the associated MutSL-methyl directed mismatch repair (MMR) function, the spontaneous mutation frequency can be reduced to 10^{-5} to 10^{-7} (reviewed in Tippin et al., 2004). The *M. tuberculosis* genome lacks homologs of the MutSL-dependent MMR system (Cole et al., 1998), yet the spontaneous mutation frequency is not very high (recently reviewed Springer et al., 2004). Presumably, there are hitherto unidentified processes that regulate the spontaneous mutation frequency or increase the fidelity of DNA synthesis.

According to a recent hypothesis put forwarded by Radman the fidelity of DNA polymerase may be increased by decreasing the rate of DNA synthesis in the absence of MMR function (Radman, 1998) and the reported slow replication rates in mycobacteria are consistent with this notion (Hiriyanna and Ramakrishnan, 1986). Spontaneous deamination of cytosine to uracil, exposure of cells to reactive oxygen intermediates, nitrosative stress, hydroxyl radicals, endogenous alkylating agents and chemicals could all produce replication blocking lesions that hinder the action of high-fidelity DNA polymerases (Walker, 1984; Walker, 1985) and *M. tuberculosis* appears to have the capacity to deal with various kinds of DNA damage (reviewed in Mizrahi and Andersen, 1998). The subject of interest pertinent to replication is trans-lesion synthesis, which requires the action of low-fidelity DNA replicative polymerases.

DnaE2- A DNA DAMAGE-RESPONSIVE MUTATOR GENE

*Dna*E2 is one of several *M. tuberculosis* genes whose expression increases upon exposure of cells to DNA damaging agents such as UV irradiation and mitomycin C. (Boshoff et al., 2003). Expression of *dna*E2 is under the control of *lexA* regulon and DnaE2, along with other unidentified DNA damage inducible proteins, is required for generation of mutations upon DNA damage. An *M. tuberculosis dna*E2 mutant shows reduced virulence in mice and is not proficient in generating resistance to antibiotics which can be directly correlated to a reduction in DnaE2 dependent SOS-induced mutagenesis (Boshoff et al., 2003). These *in vivo* studies suggest that the DnaE2 protein is a low-fidelity error-prone DNA polymerase. DnaE2 of *M. tuberculosis* does not contain the β-clamp-binding motif necessary for interaction with DnaN (Dalrymple et al., 2001) raising an interesting question as to how this error-prone DNA polymerase carries out TLS. Presumably, the interactions between DnaE2 and DnaN are facilitated by hitherto unidentified and possibly DNA damage inducible protein(s). Clearly, more studies are required in this exciting area.

RecA AND DNA DAMAGE RESPONSE IN MYCOBACTERIA

The *M. tuberculosis rec*A locus also codes for an *intein* and protein splicing of the RecA precursor protein is essential for generation of functional RecA (Davis et al., 1992; Davis et al., 1991) although RecA-Intein does not interfere with recombination or survival normally (Papavinasasundaram et al., 1998; Sander et al., 2003; Sander et al., 2001). $RecA_{TB}$ protein has been characterized extensively with respect to ATP, DNA binding characteristics and ability to catalyze heteroduplex formation (Ganesh and Muniyappa, 2003a; Ganesh and Muniyappa, 2003b). The *M. tuberculosis rec*A is cotranscribed with *rec*X, a potential regulatory gene (Papavinasasundaram et al., 1997). Purified RecX protein binds to and interferes with the RecA protein catalyzed ATPase and heteroduplex formation processes (Ganesh and Muniyappa, 2003a; Ganesh and Muniyappa, 2003b; Venkatesh et al., 2002). These studies led to a hypothesis that RecX acts as an antirecombinase and regulates *M. tuberculosis* RecA function.

The *M. tuberculosis recA* is also a DNA damage inducible gene, but is induced slowly upon DNA damage. *recA* is required for the induction of genes controlled by *lexA* repressor following DNA damage (Brooks et al., 2001; Davis et al., 2002; Papavinasasundaram et al., 2001). Two promoters for *recA* expression have been identified, and interestingly only one of which is under the control of LexA (Gopaul et al., 2003; Rand et al., 2003). Presumably, a DNA damage independent expression of *recA* is required for hitherto unidentified DNA repair processes in *M. tuberculosis*. Global gene expression studies revealed that RecA protein is not required for the expression of several genes including those involved in the nucleotide excision DNA repair pathway (Rand et al., 2003). A *M. bovis* BCG *recA* mutant strain showed increased sensitivity to DNA damaging agents such as MMS, EMS and metroniazide, but had no effect on virulence in mice (Sander et al., 2003; Sander et al., 2001). While *recA* is required for survival of cells following DNA damage and expression of several unlinked genes involved in DNA repair including *dnaE2* (Rand et al., 2003), it is unknown whether *recA* has any role in TLS mediated by DnaE2. Analysis of spontaneous mutation frequency and development of antibiotic resistance in *recA* and *recAdnaE2* mutant background could provide clues to the survival mechanisms used by *M. tuberculosis* during *in vivo* growth.

CELL DIVISION

Cell division blockage leads to filamentation and eventual cell death. Thus, understanding cell division and its regulation in *M. tuberculosis* could help to identify potential molecular drug targets against which novel antimycobacterial agents can be developed. Although the annotated mycobacterial sequences identified several genes that are involved in the cell division process, to date only *ftsZ, ftsW* and *ftsH* gene products have been characterized (see below). In addition, a novel gene *whmD* with a role in cell division, unrelated to any of the genes involved in *E. coli* cell division machinery, has been identified. In the following section, we will briefly summarize what is known about these proteins and identify major questions and gaps in this important area of investigation.

Extensive genetic, molecular biological and cell biological studies with *E. coli, B. subtilis* and *C. cresentus* have focused on how the septation process is initiated and regulated, how the mid-cell site relative to the polar sites is chosen for cell division protein assembly, how multiple proteins are assembled in an orderly manner and importantly, how the cell division process is coordinated with the replication cycle (reviewed in Errington et al., 2003). Whilst the rules and themes developed from these studies can be extended to other organisms including *M. tuberculosis*, genome data reveals the absence of several genes that are believed to be crucial for the septation process. For example, the *M. tuberculosis* genome contains *ftsZ, ftsQ, ftsK, ftsW* and *ftsI*, but lacks identifiable analogs of *ftsA, ftsL, ftsN* and *ZipA* (Cole et al., 1998). All of these gene products in *E. coli* are shown to be localized to the Z-ring (for recent reviews Buddelmeijer and Beckwith, 2004; Errington et al., 2003). These data suggest that either the cell division process in *M. tuberculosis* does not require these missing gene products or could involve the action of hitherto unidentified proteins. Furthermore, *sul*A, *ezr*A, *zap*A and *min*C, the putative regulators of *ftsZ*, the central player for the assembly of the septal Z-ring (Errington et al., 2003), are absent indicating that the mechanism of regulation of cell division in *M. tuberculosis* could be different from that reported in other bacteria.

FtsZ *IN VIVO* STUDIES FILAMENTATION

FtsZ, the eubacterial homologue of tubulin, is an essential cell division protein and is critical for initiation of the cell division process. This process is believed to be initiated when FtsZ anchors to the membrane at the mid-cell location and organizes in the form of rings called the "Z-rings"(Errington et al., 2003; Margolin, 2000a; Margolin, 2000b). *In vitro*, FtsZ polymerizes in a GTP-dependent manner. Ordered assembly of several other proteins at the Z-ring takes place ultimately leading to septation. *M. tuberculosis* FtsZ accounts for approximately 20, 000 to 30, 000 molecules per cell during exponential growth and 7, 000 molecules per cell during stationary growth (Dziadek et al., 2002a). Increasing the intracellular FtsZ levels in *M. tuberculosis* by elevating gene expression leads to inviability (Dziadek et al., 2002a). Similarly, FtsZ deficiency or its overproduction in *M. smegmatis* also leads to filamentation and lethality (Dziadek et al., 2003). Together, these results indicate that optimal levels of FtsZ are required to sustain cell division in mycobacteria. *M. tuberculosis* FtsZ-Green fluorescent protein (GFP), has been visualized by fluorescent microscopy in the form of bands or rings in filamentous *M. smegmatis* cells (Dziadek et al., 2002a). FtsZ protein of *M. smegmatis* can be exchanged with that of the *M. tuberculosis* counterpart indicating that the FtsZ-catalyzed cell division, if not its regulation, is similar.

FtsZ I*N VITRO* STUDIES

Recombinant FtsZ ($FtsZ_{TB}$) catalyzes GTP dependent polymerization and hydrolysis activities at rates much lower than those reported for other eubacterial FtsZ proteins (White et al., 2000). The rate limiting step in the $FtsZ_{TB}$ catalyzed GTPase activity has not been established although FtsZ polymerization approach that of the *E. coli* FtsZ if the carboxy terminus of $FtsZ_{TB}$ is deleted (Anand et al., 2004). The slow GTP-dependent polymerization and hydrolysis activities of $FtsZ_{TB}$ make it an appealing system to identify compounds that inhibit these activities and eventual proliferation of *M. tuberculosis*. The GTPase activity of FtsZ is inhibited by putative tubulin inhibitors such as SRI-3072 and SRI-7614 and these compounds also inhibit growth of *M. tuberculosis* (White et al., 2002). However, it is not known whether the inhibitory effects are due directly to interference with FtsZ ring formation. *M. smegmatis ftsZ* mutant strains that are defective for hydrolysis and/or binding to GTP are nonviable indicating that these activities are biologically relevant (Rajagopalan et al., unpublished). FtsZ protein is a substrate for FtsH, a AAA family protease *in vitro* suggesting a regulatory role for FtsH (Anilkumar et al., 2004). $FtsZ_{TB}$, unlike its *E. coli* counterpart, interacts with FtsW, a putative membrane protein involved in cell division (Datta et al., 2002). Very recently, $FtsZ_{TB}$ structures associated with citrate, GDP and GTPγS have been have been resolved at 1. 8, 2. 60 and 2. 08 Å resolution (Leung et al., 2004). $FtsZ_{TB}$ is a laterally oriented dimer, unlike α–β tubulin, which is a longitudinal polymer (Leung et al., 2004). The FtsZ dimer interface is believed to be important for formation of FtsZ protofilaments, Z-ring assembly and function (Lowe, 1998). Crystal structure indicates that an α to β secondary structure conformation switch at the dimer interface of $FtsZ_{TB}$ is spatially analogous to and has many hallmarks of the conformational changes exhibited by the G- proteins upon activation (Leung et al., 2004). Although magnesium is required for GTP binding and hydrolysis activities, $FtsZ_{TB}$ crystals with magnesium ions were not obtained in this study. Genetic and biochemical studies are required to validate these models, nonetheless, these studies mark the beginning of an exciting area of $FtsZ_{TB}$ biochemistry.

FtsH AND FILAMENTATION

FtsH is a AAA (ATPases Associated with a variety of cellular Activities) family protease believed to play an important role in the septation process. FtsH mutants show a filamentous phenotype. $FtsH_{TB}$ has been purified and shown to degrade $FtsZ_{TB}$ in an ATP and zinc- dependent manner. $FtsH_{TB}$ can also digest $FtsZ_{Ecoli}$ protein. Expression of *M. smegmatis ftsH* in *E. coli* leads to filamentation and subsequent growth arrest but cannot complement *E. coli ftsH* mutants (Anilkumar et al., 2004). Since FtsH digests FtsZ, and since FtsZ levels are growth phase-dependent, it is important to address whether reduction in FtsZ levels in non-dividing cells of *M. tuberculosis* can be correlated with FtsH activity.

FtsW

FtsW, a highly conserved membrane protein, is present in all bacteria that have a peptidoglycan cell wall. FtsW protein is hypothesized to have two roles in the septation process (reviewed in Errington et al., 2003). First, FtsW is required for stabilization of the "Z-rings" at the mid-cell site. Second, FtsW, facilitates cell wall synthesis by recruitment of FtsI, a penicillin binding protein that exhibits transpeptidase activity. *M. tuberculosis* FtsW (Rv2154c), contains a cluster of hydrophilic arginine residues that are absent in the *E. coli* protein. Using protein-blot overlay and pull down assays, physical interactions between $FtsW_{TB}$ and $FtsZ_{TB}$ have been demonstrated (Datta et al., 2002). It is interesting to note that genetic and biochemical studies did not reveal any interactions between the corresponding proteins of *E. coli*. It has been suggested that $FtsW_{TB}$ helps $FtsZ_{TB}$ anchor to membrane and link septum formation to peptidoglycan synthesis. Using yeast two hybrid systems, the interactions between these two proteins have been confirmed. Furthermore, FtsZ-enhanced Cyan Fluorescent Protein and enhanced Yellow Fluorescent Protein-FtsW fusions have been shown to co-localize to the septum in *M. smegmatis* (Rajagopalan et al., unpublished).

OTHER PROTEINS AND FILAMENTATION

Although *fts*Z and *fts*H are listed as bonafide cell division genes, recent studies have identified other genes in the cell division process. Included in this class is the *whm*D gene. WhmD, the gene product of *whm*D, is homologous to a *Streptomyces coelicolor* protein (WhiB) required for the maturation of aerial hyphae during sporulation. The *whm*D gene of *M. smegmatis* is an essential cell division gene (Gomez and Bishai, 2000). Characterization of a *whm*D conditionally expressing strain revealed that depletion of WhmD levels resulted in irreversible filamentation with aberrant septa whereas overproduction of WhmD led to hyperseptation and growth retardation, indicating that WhmD is required for cell division (Gomez and Bishai, 2000). Intracellular levels of FtsZ remained unchanged under conditions of WhmD deficiency. It remains to be tested how FtsZ catalyzed Z-rings are affected in cells producing different levels of WhmD. Nonetheless, these results suggested that WhmD is required for septum formation and cell division.

CONCLUSIONS

Assembly of the FtsZ-ring at the putative mid-cell site is a crucial event in cell cycle and marks the beginning of the cell division process. Good progress has been made with respect to the characterization of some of the players involved in the cell division process in mycobacteria. While biochemical studies have been carried out with *M. tuberculosis* cell division proteins, much of the *in vivo* studies have focused on the fast-grower *M. smegmatis*. These studies demonstrate that optimal levels of FtsZ and WhmD proteins are required to sustain cell division and that FtsZ is localized to the mid-cell site as bands. Although both *M. smegmatis* and *M. tuberculosis* exhibit clearly different life styles and doubling times, their *fts*Z genes are exchangeable (Dziadek et al., 2003; Rajagopalan et al., unpublished). Cell division is a complex multi-protein act and one effective way to understand this process is to construct defined mutant strains, visualize Z-rings and carry out co-localization experiments with variant GFP fusion proteins. Such studies would reveal if FtsW targets FtsZ to the presumptive cell division site, and what if any, is the order of assembly of proteins associated with the septal Z-ring. Information from *M. tuberculosis* FtsZ crystal structure can be used to define the structural attributes associated with the slow GTP dependent polymerization and GTPase activities of the protein.

Merodiploid strains of *M. tuberculosis* expressing *ftsZ* or *ftsZ-gfp* from the inducible amidase promoter are non viable indicating that *M. tuberculosis* is exquisitely sensitive to the levels and possibly activities of FtsZ. This necessitates the development of tightly regulated promoter system(s) to regulate *M. tuberculosis ftsZ* expression and possibly its activity *in vivo*. Independently, *fts*Z promoters and their activities in different genetic backgrounds need to be determined in an effort to understand how the expression of *M. tuberculosis fts*Z is regulated. Since elevated levels of FtsZ lead to inviability, it is desirable to explore whether *ftsZ-gfp* can serve as a sole source of *ftsZ* in *M. tuberculosis*. Such a strain, if developed, could provide a screening tool for identifying compounds that selectively target the Z-ring assembly and the septation process in *M. tuberculosis*.

The study of cell division is incomplete without the involvement of the regulators of Z-ring assembly. The regulators are those that help target FtsZ to the mid-cell site, that prevent Z-ring assembly elsewhere or those that stabilize the FtsZ-ring assembly at mid-cell site, *e.g*. MinCDE, EzrA, ZapA, SulA etc (Errington et al., 2003; Margolin, 2000b). The *M. tuberculosis* genome lacks any identifiable analogs of these proteins, but it has ParA, ParB and a related family of proteins, functions of which are unknown. Understanding how the Z-ring in *M. tuberculosis* is placed and targeted to the mid-cell site will not only help to understand the cell division process *per se* but will also help to define novel drug targets for preventing multiplication.

PERSPECTIVES AND FUTURE TRENDS

In this review we have attempted to summarize what is known about the key players involved in the mycobacterial replication and cell division processes. These processes are already elucidated in detail in *E. coli, B. subtilis* and *C. cresentus*. However, the absence of many identifiable homologues of key genes and regulators of these processes in the genome of *M. tuberculosis* questions how widely these models will hold. The identification and characterization of mycobacterial specific genes that are involved in the various steps of the cell cycle and their regulation is much awaited. It is recognized that many of the genes involved in replication and cell division processes in *M. tuberculosis* are expected to be essential *in vitro*, *in vivo* or both. Currently, only the *M. smegmatis* derived acetamidase promoter system is being used to regulate the expression of essential genes in *M. tuberculosis* and so the use of alternate tightly regulated systems will be critical. Much of the knowledge on the roles catalyzed by the key proteins involved in cell cycle of *E. coli* and *C. cresentus* has been obtained using well characterized synchronous cultures. Currently no synchronous cultures available although the hypoxia-induced dormant cultures of the Wayne model may be of use (Wayne, 1977; Wayne and Hayes, 1996). Alternatively, the construction and characterization of temperature sensitive mutants may be explored.

Successful survival of *M. tuberculosis in vivo* depends in part on its ability to adapt to complex hostile environments, and consequently, the activities of the cell replication and division machinery would be expected to be adjusted accordingly. It is well recognized that growth *in vivo* includes an active replicative state (intracellular replication) and a latent or persistent state (extracellular replication). It is important to understand how *M. tuberculosis* cell cycle parameters and activities of the key players involved are altered under these growth conditions.

Finally, recent studies on the characterization of the non-essential *dna*E2 gene function lay a foundation for studying the trans-lesion synthesis, an inducible mutagenesis pathway in *M. tuberculosis*. The other genetic components involved in the *dna*E2 dependent trans-lesion synthesis pathway may have to be identified. Biochemical experiments with purified proteins may be explored to define how trans-lesion synthesis occurs in *M. tuberculosis* pathogens. Clearly, exciting times are ahead in mycobacterial cell cycle research.

WEB RESOURCES

The Institute for Genome Research (TIGR) for sequences of *M. smegmatis, M. avium, M. tuberculosis* CDC1551:
http://tigr.org/tdb/ufmg

Tuberculist web server:
http://genolist.pasteur.fr/TubercuList/

Expasy (Expert Protein Analysis System) proteomics server:
http://au.expasy.org/

ACKNOWLEDGEMENTS

Work presented from our laboratory is supported in part from the public health service grants AI41406 (MM) and AI48417 (MR). We thank Drs. Julia Grimwade, Zafer Hatahet, Barry Krieswirth and Tanya Parish, for sharing unpublished data, helpful suggestions and interest during the course of work, and Ms. Hava Lofton for critically reading the manuscript.

REFERENCES

Acharya, N., and Varshney, U. (2002). Biochemical properties of single-stranded DNA-binding protein from *Mycobacterium smegmatis*, a fast-growing *Mycobacterium* and its physical and functional interaction with uracil DNA glycosylases. J. Mol. Biol. *318*, 1251-1264.

Adam, E., and Perler, F.B. (2002). Development of a positive genetic selection system for inhibition of protein splicing using mycobacterial inteins in *Escherichia coli* DNA gyrase subunit A. J. Mol. Microbiol. Biotechnol. *4*, 479-487.

Anand, S.P., Rajeswari, H., Gupta, P., Srinivasan, R., Indi, S., and Ajitkumar, P. (2004). A C-terminal deletion mutant of *Mycobacterium tuberculosis* FtsZ shows fast polymerization *in vitro*. Microbiology *150*, 1119-1121.

Anilkumar, G., Srinivasan, R., and Ajitkumar, P (2004). Genomic organization and *in vivo* characterization of proteolytic activity of FtsH of *Mycobacterium smegmatis* SN2. Microbiology *150*, 2629-2639.

Asai, T., Chen, C.P., Nagata, T., Takanami, M., and Imai, M. (1992). Transcription *in vivo* within the replication origin of the *Escherichia coli* chromosome: a mechanism for activating initiation of replication. Mol. Gen. Genet. *231*, 169-178.

Aubry, A., Pan, X.S., Fisher, L.M., Jarlier, V., and Cambau, E. (2004). *Mycobacterium tuberculosis* DNA gyrase: interaction with quinolones and correlation with antimycobacterial drug activity. Antimicrob. Agents Chemother. *48*, 1281-1288.

Boshoff, H.I., Reed, M.B., Barry, C.E.,3rd, and Mizrahi, V. (2003). DnaE2 polymerase contributes to *in vivo* survival and the emergence of drug resistance in *Mycobacterium tuberculosis*. Cell *113*, 183-193.

Bramhill, D., and Kornberg, A. (1988). Duplex opening by *dna*A protein at novel sequences in initiation of replication at the origin of the *E. coli* chromosome. Cell *52*, 743-755.

Brooks, P.C., Movahedzadeh, F., and Davis, E.O. (2001). Identification of some DNA damage-inducible genes of *Mycobacterium tuberculosis*: apparent lack of correlation with LexA binding. J. Bacteriol. *183*, 4459-4467.

Buddelmeijer, N., and Beckwith, J. (2004). A complex of the *Escherichia coli* cell division proteins FtsL, FtsB and FtsQ forms independently of its localization to the septal region. Mol. Microbiol. *52*, 1315-1327.

Cole, S.T., Brosch, R., Parkhill, J., Garnier, T., Churcher, C., and 30 authors. (1998). Deciphering the biology of *Mycobacterium tuberculosis* from the complete genome sequence. Nature *393*, 537-544.

Crooke, E. (1995). Regulation of replication in *E. coli:* sequestration and beyond. Cell *82*, 877-880.

Crooke, E. (2001). *Escherichia coli* DnaA protein--phospholipid interactions: *in vitro* and *in vivo*. Biochimie. *83*, 19-23.

Crooke, E., Castuma, C.E., and Kornberg, A. (1992). The chromosome origin of *Escherichia coli* stabilizes DnaA protein during rejuvenation by phospholipids. J. Biol. Chem. *267*, 16779-16782.

Dalrymple, B.P., Kongsuwan, K., Wijffels, G., Dixon, N.E., and Jennings, P.A. (2001). A universal protein-protein interaction motif in the eubacterial DNA replication and repair systems. Proc. Natl. Acad. Sci. USA *98*, 11627-11632.

Datta, P., Dasgupta, A., Bhakta, S., and Basu, J. (2002). Interaction between FtsZ and FtsW of *Mycobacterium tuberculosis*. J. Biol. Chem. *277*, 24983-24987.

Davis, E.O., Dullaghan, E.M., and Rand, L. (2002). Definition of the mycobacterial SOS box and use to identify LexA- regulated genes in *Mycobacterium tuberculosis*. J. Bacteriol. *184*, 3287-3295.

Davis, E.O., Jenner, P.J., Brooks, P.C., Colston, M.J., and Sedgwick, S.G. (1992). Protein splicing in the maturation of *M. tuberculosis rec*A protein: a mechanism for tolerating a novel class of intervening sequence. Cell *71*, 201-210.

Davis, E.O., Sedgwick, S.G., and Colston, M.J. (1991). Novel structure of the *rec*A locus of *Mycobacterium tuberculosis* implies processing of the gene product. J. Bacteriol. *173*, 5653-5662.

Drlica, K., Xu, C., Wang, J.Y., Burger, R.M., and Malik, M. (1996). Fluoroquinolone action in mycobacteria: similarity with effects in *Escherichia coli* and detection by cell lysate viscosity. Antimicrob. Agents Chemother. *40*, 1594-1599.

Dziadek, J., Madiraju, M.V., Rutherford, S.A., Atkinson, M.A., and Rajagopalan, M. (2002a). Physiological consequences associated with overproduction of *Mycobacterium tuberculosis* FtsZ in mycobacterial hosts. Microbiology *148*, 961-971.

Dziadek, J., Rajagopalan, M., Parish, T., Kurepina, N., Greendyke, R., Kreiswirth, B.N., and Madiraju, M.V. (2002b). Mutations in the CCGTTCACA DnaA Box of *Mycobacterium tuberculosis ori*C That Abolish Replication of *ori*C Plasmids Are Tolerated on the Chromosome. J. Bacteriol. *184*, 3848-3855.

Dziadek, J., Rutherford, S.A., Madiraju, M.V., Atkinson, M.A., and Rajagopalan, M. (2003). Conditional expression of *Mycobacterium smegmatis ftsZ*, an essential cell division gene. Microbiology *149*, 1593-1603.

Errington, J., Daniel, A.S., and Scheffers, D.J. (2003). Cytokinesis in bacteria. Micro. Mol. Biol. Rev. *67*, 52-65.

Ganesh, N., and Muniyappa, K. (2003a). Characterization of DNA strand transfer promoted by *Mycobacterium* smegmatis RecA reveals functional diversity with *Mycobacterium tuberculosis* RecA. Biochemistry *42*, 7216-7225.

Ganesh, N., and Muniyappa, K. (2003b). *Mycobacterium smegmatis* RecA protein is structurally similar to but functionally distinct from *Mycobacterium tuberculosis* RecA. Proteins *53*, 6-17.

Gomez, J.E., and Bishai, W.R. (2000). *whm*D is an essential mycobacterial gene required for proper septation and cell division. Proc. Natl. Acad. Sci. USA *97*, 8554-8559.

Gopaul, K.K., Brooks, P.C., Prost, J.F., and Davis, E.O. (2003). Characterization of the two *Mycobacterium tuberculosis rec*A promoters. J. Bacteriol. *185*, 6005-6015.

Gordhan, B.G., Andersen, S.J., De Meyer, A.R., and Mizrahi, V. (1996). Construction by homologous recombination and phenotypic characterization of a DNA polymerase domain *pol*A mutant of *Mycobacterium smegmatis*. Gene *178*, 125-130.

Greendyke, R., Rajagopalan, M., Parish, T., and Madiraju, M.V. (2002). Conditional expression of *Mycobacterium smegmatis dna*A, an essential DNA replication gene. Microbiology *148*, 3887-3900.

Grigorian, A.V., Lustig, R.B., Guzman, E.C., Mahaffy, J.M., and Zyskind, J.W. (2003). *Escherichia coli* cells with increased levels of DnaA and deficient in recombinational repair have decreased viability. J. Bacteriol. *185*, 630-644.

Handa, P., Acharya, N., Thanedar, S., Purnapatre, K., and Varshney, U. (2000). Distinct properties of *Mycobacterium tuberculosis* single-stranded DNA binding protein and its functional characterization in *Escherichia coli*. Nucleic. Acids Res. *28*, 3823-3829.

Hiriyanna, K.T., and Ramakrishnan, T. (1986). Deoxyribonucleic acid replication time in *Mycobacterium tuberculosis* H37Rv. Arch. Microbiol. *144*, 105-109.

Huberts, P., and Mizrahi, V. (1995). Cloning and sequence analysis of the gene encoding the DNA polymerase I from *Mycobacterium tuberculosis*. Gene *164*, 133-136.

Jacob, F., Brenner, S., and Cuzin, F. (1963). On the regulation of replication in bacteria. Cold Spring Harbor Symp. Quant. Biol. *28*, 329-348.

Jain, P., and Nagaraja, V. (2002). An orphan *gyr*B in the *Mycobacterium smegmatis* genome uncovered by comparative genomics. J. Genet. *81*, 105-110.

Katayama, T. (2001). Feedback controls restrain the initiation of *Escherichia coli* chromosomal replication. Mol. Microbiol. *41*, 9-17.

Katayama, T., Kubota, T., Kurokawa, K., Crooke, E., and Sekimizu, K. (1998). The initiator function of DnaA protein is negatively regulated by the sliding clamp of the *E. coli* chromosomal replicase. Cell *94*, 61-71.

Katayama, T., and Sekimizu, K. (1999). Inactivation of *Escherichia coli* DnaA protein by DNA polymerase III and negative regulations for initiation of chromosomal replication. Biochimie. *81*, 835-840.

Kato, J., and Katayama, T. (2001). Hda, a novel DnaA-related protein, regulates the replication cycle in *Escherichia coli*. EMBO J. *20*, 4253-4262.

Kawakami, H., Iwura, T., Takata, M., Sekimizu, K., Hiraga, S., and Katayama, T. (2001). Arrest of cell division and nucleoid partition by genetic alterations in the sliding clamp of the replicase and in DnaA. Mol. Genet. Genomics *266*, 167-179.

Kelley, K.W., Greenfield, R.E., Evermann, J.F., Parish, S.M., and Perryman, L.E. (1982). Delayed-type hypersensitivity, contact sensitivity, and phytohemagglutinin skin-test responses of heat- and cold-stressed calves. Am. J. Vet. Res. *43*, 775-779.

Kitagawa, R., Ozaki, T., Moriya, S., and Ogawa, T. (1998). Negative control of replication initiation by a novel chromosomal locus exhibiting exceptional affinity for *Escherichia coli* DnaA protein. Genes Dev. *12*, 3032-3043.

Klann, A.G., Belanger, A.E., Abanes-de Mello, A., Lee, J.L., and Hatful, G.H. (1998). Chracterization of the *dna*G locus in *Mycobacterium smegmatis* reveals linkage of DNA replication and cell division. J. Bacteriol. *180*, 65-72.

Kornberg, A., and Baker, T. (1991). DNA replication (New York, W.H. Freeman and Company).

Kowalczykowski, S.C., Dixon, D.A., Eggleston, A.K., Lauder, S.D., and Rehrauer, W.M. (1994). Biochemistry of Homologous recombination in *Escherichia coli*. Microbiol. Rev. *58*, 401-465.

Kurepina, N.E., Sreevatsan, S., Plikaytis, B.B., Bifani, P.J., Connell, N.D., Donnelly, R.J., van Sooligen, D., Musser, J.M., and Kreiswirth, B.N. (1998). Characterization of the phylogenetic distribution and chromosomal insertion sites of five *IS*6110 elements in *Mycobacterium tuberculosis*: non-random integration in the *dnaA-dnaN* region. Tuber. Lung Dis. *79*, 31-42.

Langer, U., Richter, S., Roth, A., Weigel, C., and Messer, W. (1996). A comprehensive set of DnaA-box mutations in the replication origin, *ori*C, of *Escherichia coli*. Mol. Microbiol. *21*, 301-311.

Leung, A.K., White, L.E., Ross, L.J., Reynolds, R.C., DeVito, J.A., and Borhani, D.W. (2004). Structure of *Mycobacterium tuberculosis* FtsZ reveals unexpected, G protein-like conformational switches. J. Mol. Biol. *342*, 953-970.

Lowe, J. (1998). Crystal structure determination of FtsZ from *Methanococcus jannaschii*. J. Struct. Biol. *124*, 235-243.

Madiraju, M.V., Qin, M.H., Yamamoto, K., Atkinson, M.A., and Rajagopalan, M. (1999). The *dna*A gene region of *Mycobacterium avium* and the autonomous replication activities of its 5' and 3' flanking regions. Microbiology *145*, 2913-2921.

Manjunatha, U.H., Dalal, M., Chatterji, M., Radha, D.R., Visweswariah, S.S., and Nagaraja, V. (2002). Functional characterisation of mycobacterial DNA gyrase: an efficient decatenase. Nucleic Acids Res. *30*, 2144-2153.

Manjunatha, U.H., Mahadevan, S., Visweswariah, S.S., and Nagaraja, V. (2001a). Monoclonal antibodies to mycobacterial DNA gyrase A inhibit DNA supercoiling activity. Eur. J. Biochem. *268*, 2038-2046.

Manjunatha, U.H., Somesh, B.P., Nagaraja, V., and Visweswariah, S.S. (2001b). A *Mycobacterium smegmatis* gyrase B specific monoclonal antibody reveals association of gyrase A and B subunits in the cell. FEMS Microbiol. Lett. *194*, 87-92.

Margolin, W. (2000a). Organelle division: Self-assembling GTPase caught in the middle. Curr. Biol. *10*, R328-330.

Margolin, W. (2000b). Themes and variations in prokaryotic cell division. FEMS Microbiol Rev. *24*, 531-548.

McGarry, K.C., Ryan, V.T., Grimwade, J.E., and Leonard, A.C. (2004). Two discriminatory binding sites in the *Escherichia coli* replication origin are required for DNA strand opening by initiator DnaA-ATP. Proc. Natl. Acad. Sci. USA *101*, 2811-2816.

Messer, W., Blaesing, F., Jakimowicz, D., Krause, M., Majka, J., Nardmann, J., Schaper, S., Seitz, H., Speck, C., Weigel, C. (2001). Bacterial replication initiator DnaA. Rules for DnaA binding and rolesof DnaA in origin unwinding and helicase loading. Biochimie. *83*, 5-12.

Mizrahi, M., Dawes, S., and Rubin, H. (2000). DNA replication. In Molecular Genetics of Mycobacteria, G.F. Hatfull, and W.R. Jacobs, eds. (Washington D.C., ASM Press), pp. 159-172.

Mizrahi, V., and Andersen, S.J. (1998). DNA repair in *Mycobacterium tuberculosis*. What have we learnt from the genome sequence? Mol. Microbiol. *29*, 1331-1339.

Mizrahi, V., and Huberts, P. (1996). Deoxy- and dideoxynucleotide discrimination and identification of critical 5' nuclease domain residues of the DNA polymerase I from *Mycobacterium tuberculosis*. Nucleic Acids Res. *24*, 4845-4852.

Morigen, A., Lobner-Olesen, A., and Skarstad, K. (2003). Titration of the *Escherichia coli* DnaA protein to excess datA sites causes destabilization of replication forks, delayed replication initiation and delayed cell division. Mol. Microbiol. *50*, 349-362.

Nagaraja, V., Sikder, D., and Jain, P. (2002). DNA topoisomerase I from mycobacteria--a potential drug target. Curr. Pharm. Des. *8*, 1995-2007.

Nozaki, N., Okazaki, T., and Ogawa, T. (1988). *In vitro* transcription of the origin region of replication of the *Escherichia coli* chromosome. J. Biol. Chem. *263*, 14176-14183.

Ogawa, T., Yamada, Y., Kuroda, T., Kishi, T., and Moriya, S. (2002). The *dat*A locus predominantly contributes to the initiator titration mechanism in the control of replication initiation in *Escherichia coli*. Mol. Microbiol. *44*, 1367-1375.

Ogura, Y., Imai, Y., Ogasawara, N., and Moriya, S. (2001). Autoregulation of the *dna*A-*dna*N operon and effects of DnaA protein levels on replication initiation in *Bacillus subtilis*. J. Bacteriol. *183*, 3833-3841.

Ortenberg, R., Gon, S., and Beckwith, J. (2004). interactions of glutaredoxins, ribonucleotide reductase, and components of DNA replication systems of *Escherichia coli*. Proc. Natl. Acad. Sci. USA *101*, 7439-7444.

Papavinasasundaram, K.G., Anderson, C., Brooks, P.C., Thomas, N.A., Movahedzadeh, F., Jenner, P.J., Colston, M.J., and Davis, E.O. (2001). Slow induction of RecA by DNA damage in *Mycobacterium tuberculosis*. Microbiology *147*, 3271-3279.

Papavinasasundaram, K.G., Colston, M.J., and Davis, E.O. (1998). Construction and complementation of a *rec*A deletion mutant of *Mycobacterium smegmatis* reveals that the intein in *Mycobacterium tuberculosis rec*A does not affect RecA function. Mol. Microbiol. *30*, 525-534.

Papavinasasundaram, K.G., Movahedzadeh, F., Keer, J.T., Stoker, N.G., Colston, M.J., and Davis, E.O. (1997). Mycobacterial *rec*A is cotranscribed with a potential regulatory gene called *rec*X. Mol. Microbiol. *24*, 141-153.

Parish, T., Smith, D.A., Roberts, G., Betts, J., and Stoker, N.G. (2003). The *sen*X3-*reg*X3 two-component regulatory system of *Mycobacterium tuberculosis* is required for virulence. Microbiology *149*, 1423-1435.

Perler, F.B. (1998). Protein splicing of inteins and hedgehog autoproteolysis: structure, function, and evolution. Cell *92*, 1-4.

Perler, F.B., Davis, E.O., Dean, G.E., Gimble, F.S., Jack, W.E., Neff, N., Noren, C.J., Thorner, J., and Belfort, M. (1994). Protein splicing elements: inteins and exteins--a definition of terms and recommended nomenclature. Nucleic Acids Res. *22*, 1125-1127.

Prabhakar, S., Mishra, A., Singhal, A., Katoch, V.M., Thakral, S.S., Tyagi, J.S., and Prasad, H.K. (2004). Use of the *hup*B gene encoding a histone-like protein of *Mycobacterium tuberculosis* as a target for detection and differentiation of *M. tuberculosis* and *M. bovis*. J. Clin. Microbiol. *42*, 2724-2732.

Purnapatre, K., and Varshney, U. (1998). Uracil DNA glycosylase from *Mycobacterium smegmatis* and its distinct biochemical properties. Eur. J. Biochem. *256*, 580-588.

Purnapatre, K., and Varshney, U. (1999). Cloning, over-expression and biochemical characterization of the single-stranded DNA binding protein from *Mycobacterium tuberculosis*. Eur. J. Biochem. *264*, 591-598.

Qin, M.H., Madiraju, M.V., and Rajagopalan, M. (1999). Characterization of the functional replication origin of *Mycobacterium tuberculosis*. Gene *233*, 121-130.

Qin, M.H., Madiraju, M.V., Zachariah, S., and Rajagopalan, M. (1997). Characterization of the *ori*C region of *Mycobacterium smegmatis*. J. Bacteriol. *179*, 6311-6317.

Radman, M. (1998). DNA replication: one strad may be more equal. Proc. Natl. Acad. Sci. USA *95*, 9718-9719.

Rajagopalan, M., Qin, M.H., Nash, D.R., and Madiraju, M.V. (1995). *Mycobacterium smegmatis dna*A region and autonomous replication activity. J. Bacteriol. *177*, 6527-6535.

Rand, L., Hinds, J., Springer, B., Sander, P., Buxton, R.S., and Davis, E.O. (2003). The majority of inducible DNA repair genes in *Mycobacterium tuberculosis* are induced independently of RecA. Mol. Microbiol. *50*, 1031-1042.

Reddy, M.S., Guhan, N., and Muniyappa, K. (2001). Characterization of single-stranded DNA-binding proteins from Mycobacteria. The carboxyl-terminal of domain of SSB is essential for stable association with its cognate RecA protein. J. Biol. Chem. *276*, 45959-45968.

Saikrishnan, K., Jeyakanthan, J., Venkatesh, J., Acharya, N., Purnapatre, K., Sekar, K., Varshney, U., and Vijayan, M. (2002). Crystallization and preliminary X-ray studies of the single-stranded DNA-binding protein from *Mycobacterium tuberculosis*. Acta Crystallogr. D. Biol. Crystallogr. *58*, 327-329.

Saikrishnan, K., Jeyakanthan, J., Venkatesh, J., Acharya, N., Sekar, K., Varshney, U., and Vijayan, M. (2003). Structure of *Mycobacterium tuberculosis* single-stranded DNA-binding protein. Variability in quaternary structure and its implications. J. Mol. Biol. *331*, 385-393.

Salazar, L., Fsihi, H., de Rossi, E., Riccardi, G., Rios, C., Cole, S.T., and Takiff, H.E. (1996). Organization of the origins of replication of the chromosomes of *Mycobacterium smegmatis*, *Mycobacterium leprae* and *Mycobacterium tuberculosis* and isolation of a functional origin from *M. smegmatis*. Mol. Microbiol. *20*, 283-293.

Salazar, L., Guerrero, E., Casart, Y., Turcios, L., and Bartoli, F. (2003). Transcription analysis of the dnaA gene and oriC region of the chromosome of *Mycobacterium smegmatis* and *Mycobacterium bovis* BCG, and its regulation by the DnaA protein. Microbiology *149*, 773-784.

Sander, P., Bottger, E.C., Springer, B., Steinmann, B., Rezwan, M., Stavropoulos, E., and Colston, M.J. (2003). A *rec*A deletion mutant of *Mycobacterium bovis*

BCG confers protection equivalent to that of wild-type BCG but shows increased genetic stability. Vaccine *21*, 4124-4127.

Sander, P., Papavinasasundaram, K.G., Dick, T., Stavropoulos, E., Ellrott, K., Springer, B., Colston, M.J., and Bottger, E.C. (2001). *Mycobacterium bovis* BCG *rec*A deletion mutant shows increased susceptibility to DNA-damaging agents but wild-type survival in a mouse infection model. Infect. Immun. *69*, 3562-3568.

Sassetti, C.M., Boyd, D.H., and Rubin, E.J. (2001). Comprehensive identification of conditionally essential genes in mycobacteria. Proc. Natl. Acad. Sci. USA *98*, 12712-12717.

Sassetti, C.M., Boyd, D.H., and Rubin, E.J. (2003). Genes required for mycobacterial growth defined by high density mutagenesis. Mol. Microbiol. *48*, 77-84.

Sikder, D., and Nagaraja, V. (2001). A novel bipartite mode of binding of *M. smegmatis* topoisomerase I to its recognition sequence. J. Mol. Biol. *312*, 347-357.

Sikder, D., Unniraman, S., Bhaduri, T., and Nagaraja, V. (2001). Functional cooperation between topoisomerase I and single strand DNA-binding protein. J. Mol. Biol. *306*, 669-679.

Skovgaard, O., and Hansen, F.G. (1987). Comparison of *dna*A nucleotide sequences of *Escherichia coli, Salmonella typhimurium*, and *Serratia marcescens*. J. Bacteriol. *169*, 3976-3981.

Springer, B., Sander, P., Sedlacek, L., Hardt, W.D., Mizrahi, V., Schar, P., and Bottger, E.C. (2004). Lack of mismatch correction facilitates genome evolution in mycobacteria. Mol. Microbiol. *53*, 1601-1609.

Sutton, M.D., Carr, K.M., Vicente, M., and Kaguni, J.M. (1998). *Escherichia coli* DnaA protein. The N-terminal domain and loading of DnaB helicase at the *E. coli* chromosomal origin. J. Biol. Chem. *273*, 34255-34262.

Sutton, M.D., and Kaguni, J.M. (1997). The *Escherichia coli dna*A gene: four functional domains. J. Mol. Biol. *274*, 546-561.

Tippin, B., Pham, P., and Goodman, M.F. (2004). Error-prone replication for better or worse. Trends Microbiol. *12*, 288-295.

Umezu, K., Chi, N.W., and Kolodner, R.D. (1993). Biochemical interaction of the *Escherichia coli* RecF, RecO and RecR proteins with RecA protein and single stranded DNA binding proteinProc. Natl. Acad. Sci. USA *90*, 3875-3879.

Umezu, K., and Kolodner, R.D. (1994). Protein interactions in genetic recombination in *Escherichia coli*. J. Biol. Chem. *269*, 30005-30013.

Venkatesh, R., Ganesh, N., Guhan, N., Reddy, M.S., Chandrasekhar, T., and Muniyappa, K. (2002). RecX protein abrogates ATP hydrolysis and strand exchange promoted by RecA: insights into negative regulation of homologous recombination. Proc. Natl. Acad. Sci. USA *99*, 12091-12096.

Walker, G.C. (1984). Mutagenesis and inducible responses to deoxyribonucleic acid damage in *Escherichia coli*. Microbiol. Rev. *48*, 60-93.

Walker, G.C. (1985). Inducible DNA repair systems. Ann. Rev. Biochem. *54*, 425-457.

Wayne, L.G. (1977). Synchronized replication of *Mycobacterium tuberculosis*. Infect. Immu. *17*, 528-530.

Wayne, L.G., and Hayes, L.G. (1996). An *in vitro* model for sequential study of shift down of *Mycobacterium tuberculosis* through two stages of nonreplicating presistence. Infect. Immun. *64*, 2062-2069.

White, E.L., Ross, L.J., Reynolds, R.C., Seitz, L.E., Moore, G.D., and Borhani, D.W. (2000). Slow polymerization of *Mycobacterium tuberculosis* FtsZ. J. Bacteriol. *182*, 4028-4034.

White, E.L., Suling, W.J., Ross, L.J., Seitz, L.E., and Reynolds, R.C. (2002). 2-Alkoxycarbonylaminopyridines: inhibitors of *Mycobacterium tuberculosis* FtsZ. J. Antimicrob. Chemother. *50*, 111-114.

Xu, C., Kreiswirth, B.N., Sreevatsan, S., Musser, J.M., and Drlica, K. (1996). Fluoroquinolone resistance associated with specific gyrase mutations in clinical isolates of multidrug-resistant *Mycobacterium tuberculosis*. J. Infect. Dis. *174*, 1127-1130.

Yamamoto, K., Low, B., Rutherford, S.A., Rajagopalan, M., and Madiraju, M.V. (2001). The *Mycobacterium avium-intracellulare* complex *dna*B locus and protein intein splicing. Biochem. Biophys. Res. Commun. *280*, 898-903.

Yamamoto, K., Muniruzzaman, S., Rajagopalan, M., and Madiraju, M.V. (2002a). Modulation of *Mycobacterium tuberculosis* DnaA protein-adenine- nucleotide interactions by acidic phospholipids. Biochem. J. *363*, 305-311.

Yamamoto, K., Rajagopalan, M., and Madiraju, M.V. (2002b). Phospholipids Promote Dissociation of ADP from the *Mycobacterium avium* DnaA Protein. J. Biochem. (Tokyo) *131*, 219-224.

Yang, F., Lu, G., and Rubin, H. (1996). Cloning, expression, purification and characterization of DNA topoisomerase I of *Mycobacterium tuberculosis*. Gene *178*, 63-69.

Yee, T.W., and Smith, D.W. (1990). *Pseudomonas* chromosomal replication origins: a bacterial class distinct from *Escherichia coli*-type origins. Proc. Natl. Acad. Sci. USA *87*, 1278-1282.

Zakrzewska-Czerwinska, J., and Schrempf, H. (1992). Characterization of an autonomously replicating region from the *Streptomyces lividans* chromosome. J. Bacteriol. *174*, 2688-2693.

Zawilak, A., Kois, A., Konopa, G., Smulczyk-Krawczyszyn, A., and Zakrzewska-Czerwinska, J. (2004). *Mycobacterium tuberculosis* DnaA initiator protein: purification and DNA-binding requirements. Biochem. J. *382*, 247-252.

Zyskind, J.W., and Smith, D.W. (1986). The bacterial origin of replication. Cell *46*, 489-490.

Chapter 2

The Mycobacterial Two-Component Regulatory Systems

Stuart C. G. Rison, Sharon L. Kendall, Farahnaz Movahedzadeh and Neil G. Stoker

ABSTRACT

The ability to respond to different environments is critical for bacterial survival. Two-component regulatory systems are widely used to sense such changes. These systems usually involve a membrane-bound sensor that detects an environmental change, and activates a cytoplasmic response-regulator that is able to bind to relevant promoters, and to switch gene expression on and off as appropriate. In pathogens, virulence genes are often controlled in this way, and this proves to be the case also for *Mycobacterium tuberculosis*, which has 11 such paired systems and eight orphan genes (six response regulators and two sensors). In this chapter we review the current state of knowledge of these gene systems in *M. tuberculosis* (and their orthologues in related bacteria), including functional information from genetic studies, and expression data from microarray studies.

INTRODUCTION

Bacteria occupy many different environments, and they need to adapt to changes in their surroundings in order to protect themselves from external stresses, or to make the most of the available nutrients. Much of this adaptation is carried out by altering transcription levels of genes. There are many highly specific regulators (such as the much-studied *lac* repressor of *Escherichia coli* which controls expression of the *lac* operon and is allosterically regulated by the presence or absence of lactose). However, more complex regulation involving many genes in different parts of the genome is generally accomplished through the use of alternative sigma factors, or of two-component regulatory systems (2CRs).

2CRs have been found in most bacteria, and may be present in very small or large numbers. Thus *Chlamydia trachomatis* has only one paired 2CR (Stephens et al., 1998), while *Streptomyces coelicolor* has 53 paired systems and 58 orphan genes (Bentley et al., 2002). Presumably the number bears some relation to the number of different environments an organism is likely to encounter.

2CRs are composed of two proteins: a signal-sensing autophosphorylating histidine kinase and a cognate response regulator which often acts as a transcription factor. Fig. 1 shows the 2CR paradigm, where the sensor (often membrane-bound)

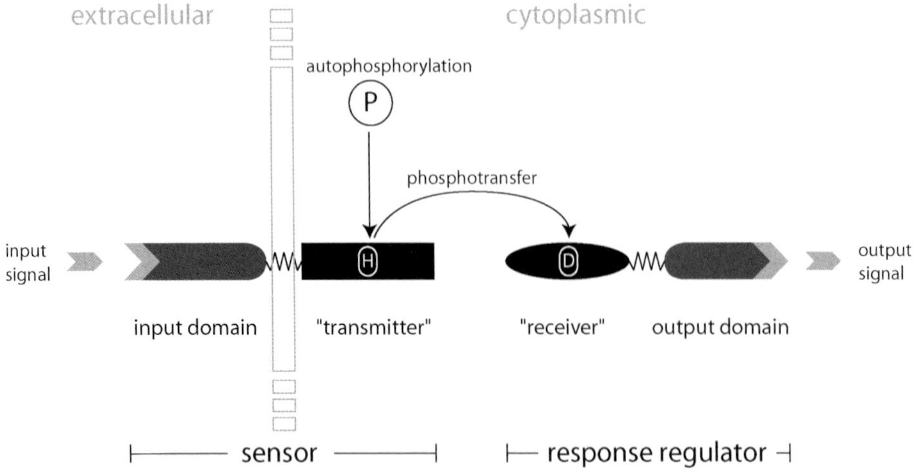

Fig. 1. The two-component regulatory system (2CR) paradigm. The canonical 2CR consists of two proteins and four domains. The sensor protein (often membrane bound), with its input and transmitter domains, is responsible for environmental signal detection, whilst the response regulator, with its receiver and output domains, effects the appropriate response to the signal following activation by the sensor. Following signal detection by the input domain, the sensor auto-phosphorylates a conserved histidine residue in the transmitter domain. The activated transmitter domain of the sensor then interacts with its cognate response regulator and activates the latter by transfer of its histidine-bound phosphate group to a conserved aspartate residue in the receiver domain. In many cases the activated response regulator acts as a transcription factor with a DNA-binding output domain regulating the expression of signal-related genes.

detects an environmental change through an N-terminal sensing domain, and autophosphorylates a conserved histidine residue in the C-terminal (transmitter) domain. The activated sensor then phosphorylates a conserved aspartate residue in the cognate response regulator, which binds to specific promoter regions, causing transcription to be enhanced or repressed, depending on the situation.

The general biology of 2CRs has been extensively reviewed by others, see for example Parkinson and Kofoid (1992) and Parkinson (1993), or the monograph on the subject (Hoch and Silhavy, 1995), and the whole sets of 2CRs for certain organisms have been the subject of publications, *e.g.* the cyanobacterium *Synechocystis* sp. (Mizuno et al., 1996), the actinomycete *S. coelicolor*, and eukaryotic filamentous ascomycetes (Catlett et al., 2003). In this chapter, we consider the 2CRs of mycobacteria in general, and of *Mycobacterium tuberculosis* in particular, focusing only on 2CR genes that are present in *M. tuberculosis* and their orthologues in other species, because there is most information about these. The *M. tuberculosis* 2CRs have also been reviewed by Tyagi and Sharma (2004); this excellent article has its own focus, and is in many respects complementary to this chapter.

MYCOBACTERIAL 2CRS

In most bacteria studied, the genes encoding the cognate sensor and response regulator of a 2CR are contiguous on a bacterium's chromosome, with occasional components (sensor or response regulator) found alone in the genome. We refer to the latter components (lacking a contiguous cognate partner) as orphan components. In *M. tuberculosis* there are 11 paired 2CRs, two orphan sensors and six orphan response regulators (Table 1). The output domain of a response regulator can have different functions, but all those in *M. tuberculosis* are predicted to bind nucleic acids. All are OmpR-like, with the exception of three (NarL, DosR and Rv0195) that have LuxR-type domains. The OmpR-like response regulators from other bacteria classified within this family regulate σ^{70} promoters and contain a C-terminal helix-turn-helix motif responsible for DNA binding (Pao and Saier, 1995). The *M. tuberculosis* sensor proteins vary between 410 and 860 aa in length, and the response regulators between 133 and 381 aa, demonstrating a considerable degree of variability within the families.

The origins of the 2CR gene names in *M. tuberculosis* are somewhat arcane, but are useful to know both because they can give a clue –sometimes misleading– to their function, and because there are some situations where more than one name is used. They are summarised in Table 1, and expanded upon in the relevant sections below. Some genes have yet to be named, and are still referred to by their unique identifier (which for *M. tuberculosis* H37Rv is the Rv number). To be consistent, 2CRs are identified in this chapter with the response regulator listed first; for example *trcRS* is a 2CR where *trcR* codes for the response regulator, and *trcS* codes for the sensor.

The genomic contexts of the genes are summarised in Fig. 2A for the paired 2CRs, and in Fig. 2B for the orphan sensors and response regulators. This is of interest because it can be seen that the paired systems are generally transcribed in the same direction and are likely to be in operons. However, there is no consistency as to which order they lie in; nine have the response regulator first, and two have the sensor first, while in one case (*narLS*), the genes are transcribed divergently. The second reason for showing gene context is that 2CRs often regulate genes close by (as well as elsewhere on the chromosome); thus this information can give clues to the genes under the control of that system. This is known to be true for the *kdpED* and *dosRS* systems (see below).

We can also use gene context as a way to identify true orthologues in other species. Two genes are considered homologous if they have a common evolutionary ancestor; such homology is usually assessed on the basis of sequence similarity. Homologous genes can be orthologues: similar genes in *different* species, usually with conserved function, *e.g. M. tuberculosis Rv1099c (glpX)* is an orthologue of *E. coli* minor fructose 1,6-bisphosphatase (FBPase) gene *glpX* and is the *de facto* FBPase in *M. tuberculosis* (Movahedzadeh et al., 2004). Alternatively, genes can be paralogues: homologous genes in the same species, the product of

Table 1. The two-component regulatory systems of *Mycobacterium tuberculosis*.

Paired systems: Sensor	Paired systems: Regulator	Rv Number	Arrangement[1]	Length[2]	Acronym	Reference
senX3		Rv0490	S+R+	410	SENsor X3 and REGulator X3	Wren et al. (1992)
	regX3	Rv0491		227		
tcrB2[3]		Rv0600c	R-S1-S2-	168	TcrA's cognate sensor "broken" N-term fragment (see text)	This chapter
tcrB1		Rv0601c		156	TcrA's cognate sensor "broken" C-term fragment (see text)	This chapter
	tcrA	Rv0602c		253	Two-Component Regulator	Buchmeier (1998)
	phoP	Rv0757	R+S+	247	By homology to *Bacillus subtilis* PhoR (for alkaline PHOsphatase regulation)	Cole et al. (1998)
phoR		Rv0758		485		
	narL	Rv0844c	R-S+ (Div)[4]	216	By homology to NarL proteins; NitrAte Reductase (from Stewart et al., 1982); *narS* coined by Parish et al. (2003) to distinguish *Rv0845* from *Rv1736c*, a probable nitrate reductase gene, also known as *narX*	Parish et al. (2003)
narS [narX][5]		Rv0845		425		
prrB		Rv0902c	R-S-	446	Phagocytosis Response Regulator	Graham and Clark-Curtiss (1999)
	prrA	Rv0903c		236		
	mprA	Rv0981	R+S+	230	Mycobacterial Persistence Regulator	Zahrt et al. (2001)
mprB		Rv0982		504		
	kdpE	Rv1027c	S-R-	226	By homology to *E. coli kdpED* (from Potassium (K) DePendence, Epstein and Davies (1970)).	Cole et al. (1998)
kdpD		Rv1028c		860		
trcS		Rv1032c	R-S-	509	Tuberculosis Regulatory Component	Haydel et al. (1999)
	trcR	Rv1033c		257		
dosS [devS]		Rv3132c	R-S-	578	dev: Differentially Expressed in the Virulent strain (H37Rv vs H37Ra)	Kinger and Tyagi (1993)
	dosR [devR]	Rv3133c		217	dos: DOrmancy Survival **or** Dependent on Oxygen Status	Boon and Dick (2002) This chapter
mtrB		Rv3245c	R-S-	567	*Mycobacterium tuberculosis* Response regulator	Via et al. (1996)
	mtrA	Rv3246c		228		

tcrY	*tcrX*	Rv3764c Rv3765c	R-S-	475 234	Two-Component Regulator	Parish et al. (2003)

Unpaired components		Rv Number	Strand	Length	Acronym	Reference
Sensor	Regulator					
dosT		Rv2027c	-	573	cf. *dosRS*	Roberts et al. (2004)
Rv3220c		Rv3220c	-	501		
	Rv0195[6]	Rv0195	+	211		
	Rv0260c	Rv0260c	-	381		
	Rv0818	Rv0818	+	255		
	Rv1626	Rv1626	+	205		
	Rv2884	Rv2884	+	252		
	Rv3143	Rv3143	+	133		

[1] For the paired 2CRs the arrangement column, genes are listed in their order of transcription, S: sensor, R: response regulator. For both paired and 2CRs and orphan 2CRs components, the + and - indicated the gene is on the forward and reverse strand respectively.
[2] Length of encoded protein in amino acids.
[3] For TcrA(B), the sensor is "split" due to a frameshift mutation (see main text)
[4] The *narLS* genes are divergent.
[5] Alternative names are listed in square brackets. The gene name without brackets is the one we recommend and use in this chapter
[6] Not listed in Cole et al. (1998), but listed in Haydel & Clark-Curtiss (2004), and annotated as "possible two-component transcriptional regulatory protein (probably LuxR-family)" in Tuberculist (http://genolist.pasteur.fr/TubercuList/).

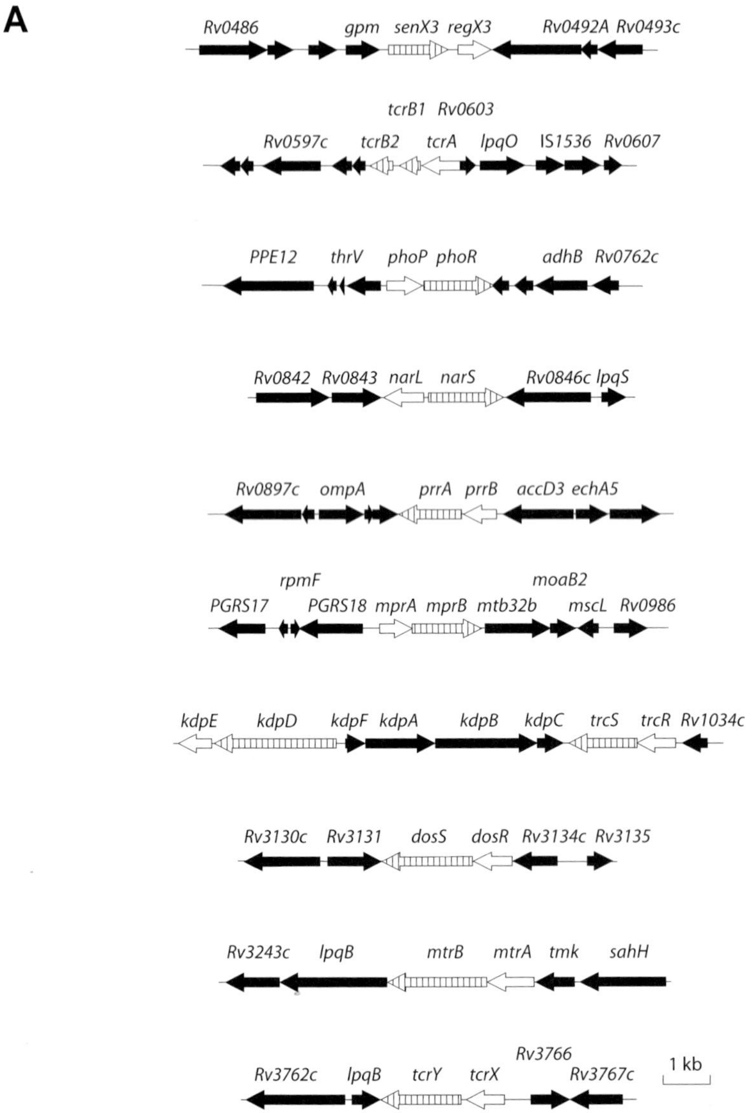

Fig. 2. Genomic contexts for (A) the 11 paired 2CRs and (B) orphan 2CR components of *M. tuberculosis* H37Rv. Hatched arrows, sensor genes; white arrows, response regulator genes; filled arrows, other genes. Space permitting, gene names (where available), or Rv numbers are shown above the genes.

gene or chromosomal duplication, which may have different functions following independent mutation (*e.g.* the *dosS* sensor of *M. tuberculosis* 2CR *dosRS* and *dosT* its orphan paralogue, Roberts et al., 2004).

Identification of orthologues allows us to integrate information about a system collated from different species. For example, *dosR* mutants have been isolated in

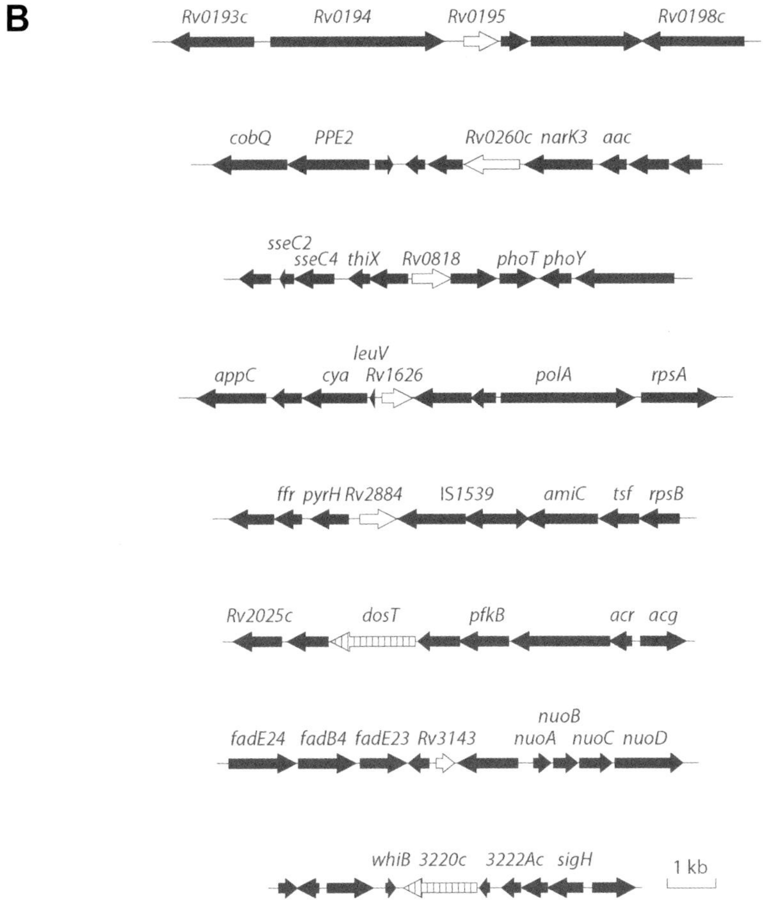

both *M. tuberculosis* and *M. smegmatis*, and a great deal is known about the *kdpED* system of *E. coli*. It is also interesting to know which species do or do not have an orthologue, as this can give clues to function and suggest ways of addressing an unknown function. With gene families where there is a great deal of similarity between all the members, homology is not necessarily definitive proof of orthology. In 2CRs this problem is compounded by the domain nature of the system (Fig. 1), in particular the transmitter and receiver domains that are homologous in all sensors and response regulators respectively. In a recent review, Tyagi and Sharma (2004) use a high cut-off (>50% identity over 90% of the length of the protein) as a way of defining orthologues. We have used a different criterion, where our primary consideration is the conservation of genomic context around a gene that shows homology. This is possible because several other genomes from the mycobacteria and the closely related corynebacteria have been completed or are near completion. Table 2 shows our analyses of eight genomes compared to *M. tuberculosis*. Even e 2. Orthologues of *M. tuberculosis* two-component genes in related genomes, determined by a combination of homology and conservation of context. Organisms are: *Mycobacterium bovis (*Mbo*)*, *Mycobacterium leprae* (Mlp), *Mycobacterium marinum* (Mma), *Mycobacterium avium* (Mav), *Mycobacterium smegmatis* (Msm), *Corynebacterium glutamicum* (Cgl) and *Corynebacterium diphtheriae* (Cdi).

	Mbo	**Mlp**	**Mma[1]**	**Mav[2]**	**Mpa**	**Msm[3]**	**Cgl**	**Cdi**
Paired systems								
regX3-senX3	✓	✓	✓	✓	✓	✓	✓	✓
tcrA(B)	✓[4]	-	-	(✓)[5]	(✓)[6]	-	-	
phoPR	✓	P	✓	✓	✓	✓	✓	✓
narLS	✓	-	✓[7]	✓	✓	(✓)[8]	-	-
prrAB	✓	✓	✓	✓	✓	✓	-	-
mprAB	✓	✓	✓	✓	✓	✓	✓	✓
kdpDE	✓	-	✓	✓	✓	✓[9]	-	-
trcRS	✓	P[10]	✓	✓	✓	(✓)	-	-
dosRS	✓	-	✓	✓	✓	✓	-	-
mtrAB	✓	✓	✓	✓	✓	✓	✓	✓
tcrXY	✓	(P)	✓	✓	✓	(✓)	-	-
Orphan sensors								
dosT	✓	-	✓[11]	[✓]	[✓][12]	✓[13]	-	-
Rv3220c	✓	✓	✓	✓	✓	✓	-	-
Orphan response regulators								
Rv0195	✓	-	-	-	-	-	-	-
Rv0260c	✓	-	✓	✓	✓	✓	-	-
Rv0818	✓	P	✓	✓	✓	✓	-	-
Rv1626	✓	✓	✓	✓	✓	✓	-	-
Rv2884	✓	P	✓	✓	✓	-	-	-
Rv3143	✓	-	✓	✓	✓	✓	-	-

[1] The *M. marinum* sequence was obtained from the Sanger centre.
[2] The *M. avium* sequence was obtained from TIGR.
[3] The *M. smegmatis* sequence was obtained from TIGR.
[4] A tick suggests detection of convincing homologues, and conservation of context (*i.e.* the presence of genes homologous to the *M. tuberculosis* genes surrounding the listed 2CR).
[5] When the tick is bracketed, convincing homologues were identified, but context conservation was lacking.
[6] There is a gene annotated as *tcrA* in the initial submission to EMBL, but no *tcrB* homologue.
[7] *M. marinum* appears to have a second *narS* orthologue which is contiguous, but not in an operon (see text and Fig. 3C).
[8] In *M. smegmatis*, the putative orthologues are in an operon (as opposed to the divergent arrangement in *M. tuberculosis*).
[9] *M. smegmatis* has two *kdpD* homologues; the second is isolated and context is not conserved (see text and Fig. 3D).
[10] P = pseudogene.
[11] In *M. marinum*, the *dosRS* region is duplicated; each *M. marinum* region (*dosR/S* homologues and *dosR2/T*) conserves context (see text and Figs. 3A and 3B).
[12] Square brackets indicate reasonable synteny, but unconvincing homology.
[13] As for *M.* marinum, *M. smegmatis* also appears to have a *dosR2* gene in the *dosT* region (see main text and Figs. 3A and 3B).

using a system that includes conserved synteny, there are situations where we cannot be certain of orthology, and we have indicated those. The homology between paralogues makes it difficult to extend searches beyond the actinomycetes. Thus it is difficult to directly apply much of the knowledge about the specific functions of systems in *Bacillus subtilis* and *E. coli* that have been extensively researched. A case beyond homology has to be made to link 2CR genes from such widely divergent species.

Only three systems from *M. tuberculosis* (*regX3senX3*, *mprAB* and *mtrAB*) are present and apparently functional in all species we looked at (*Mycobacterium bovis*, *Mycobacterium leprae*, *Mycobacterium marinum*, *Mycobacterium avium*, *Mycobacterium smegmatis*, *Corynebacterium glutamicum* and *Corynebacterium diphtheriae*). This suggests these 2CRs have important and highly conserved functions. The complement of genes in *M. leprae* is particularly interesting because of the huge gene loss that has occurred. Orthologues of eight of the genes we discuss are present as pseudogenes (three paired 2CRs and two orphan response regulators; see Table 2). Thus *M. leprae* only has four functional paired 2CR systems: *regX3senX3*, *mprAB*, *mtrAB* and the *prrAB* system, as well as one isolated sensor and one isolated regulator. These isolated genes may work with one of the paired systems; an attractive alternative hypothesis is that *Rv1626* and *Rv3220c* work together. One paired system (*phoPR*) is present in all species except *M. leprae*, while most of the rest are present in all mycobacteria (except *M. leprae*) but not in the corynebacteria (although not all orthologues are completely convincing). *Rv0195* has only been seen in the *M. tuberculosis* complex.

Our rationale for initially studying the *M. tuberculosis* 2CRs was their importance in pathogenesis of other bacteria. This has been borne out by observations that virulence is affected following mutagenesis of most of the paired systems, although there are some fascinating complexities already being revealed. The effects of 2CR gene mutagenesis are summarised in Table 3. Both mutagenesis and virulence experiments are extremely time-consuming with this pathogen, so it is important to take as much from each report as is possible. In some cases, different phenotypes have been observed by different groups working on the same system. The most likely causes for this are the nature of the mutation and the model used to assay virulence. We have therefore listed both these factors to allow inconsistencies and possible reasons for such inconsistencies to be assessed.

We have included data in Table 3 on experiments using the TraSH methodology pioneered by Sassetti et al. (2001), and it is important to understand the basis of this technology. TraSH (transposon site hybridization) combines the concept of signature-tagged mutagenesis (Mei et al., 1997), where a transposon library of a bacterium is grown and relative loss of mutants in individual genes assayed to indicate the requirement for that gene in that condition, with the ability of microarrays to scan every gene. DNA is extracted from bacteria before and after culture, and regions adjacent to each transposon are labeled using T7 promoters

Table 3. Phenotypes of 2CR mutants.

2CR system	Mutation[1]	Phenotype				
		Axenic culture	Macrophages		*In vivo*	
			Human	Mouse	Immuno-compromised (SCID) mice	Immunocompetent
regX3/senX3	*regX3/senX3* deletion (Parish et al., 2003b)	Unreproducible defect in aerobic culture	Attenuated[2]	Attenuated	Attenuated	Attenuated in DBA mice
	Hygromycin cassette in *senX3*(Rickman et al., 2004)	No phenotype	-	-	-	Attenuated in BalB/c mice
	*regX3*deletion (Rickman et al., 2004)[3]	lower O.D. in stationary phase	-	-	Attenuated in nude mice	Attenuated in BALB/c mice
	Transposon in 5' region (after nt 57) of *regX3* [4] (Ewann et al., 2002)	-	-	No phenotype	-	No phenotype in C57BL/6J or BALB/c mice
	TraSH (Sassetti et al., 2003; Sassetti and Rubin, 2003)	No phenotype	-	-	-	*senX3* required for survival in C57/BL6J mice
tcrA/tcrB1/tcrB2	TraSH	No phenotype	-	-	-	No phenotype
phoPR	Kanamycin cassette in *phoP* (Perez et al., 2001)	Smaller bacilli in log phase; smaller colonies; defective cording	-	Attenuated	-	Attenuated in BALB/c mice
	TraSH	*phoP* minor growth defect	-	-	-	No phenotype
narLS	Δ*narL* (Parish et al., 2003a)	-	-	No phenotype	No phenotype	-
	TraSH	No phenotype	-	-	-	No phenotype
prrAB	Transposon in 5' region of *prrA* [5](Ewann et al., 2002)	-	-	Attenuated in early stages	-	No phenotype in C57BL/6J or BALB/c mice
	TraSH	No phenotype	-	-	-	-
mprAB	Kanamycin cassette in *mprA* (Zahrt and Deretic, 2001)	-	-	Enhanced growth[6]	-	Attenuated in a low dose model of infection of BALB/c mice
	mprA D48A mutation in *M. bovis BCG* (Zahrt et al., 2003)	-	-	Attenuated	-	-
	mprB H249Q mutation in *M. bovis BCG* (Zahrt et al., 2003)	-	-	Attenuated	-	-
	TraSH	*mprB* required for optimal growth; *mprA* no phenotype	-	-	-	No phenotype for *mprA;* no data for *mprB*

kdpED	*kdpDE* deletion (Parish et al., 2003b)	No phenotype	-	-	Hypervirulent in SCID mice	-
	TraSH	*kdpE* required for optimal growth; *kdpD* no phenotype	-	-	-	*kdpD* required for optimal growth; *kdpE* no phenotype
trcRS	*trcS* deletion (Parish et al., 2003b)	No phenotype	-	-	Hypervirulent in SCID mice	-
	Transposon in 5' region of *trcS*[7] (Ewann et al., 2002)	-	-	No phenotype	-	No phenotype in C57BL/6J or BALB/c mice
	TraSH	No phenotype	-	-	-	No phenotype
dosRS	*dosR* deletion (Parish et al., 2003b)	No phenotype	-	Increased growth and survival	Hypervirulent in SCID mice	Increased growth in lung, liver and spleen of DBA/2 mice.
	Kanamycin cassette in *dosR* (Malhotra et al., 2004)	Reduced cell-cell aggregation	No phenotype	-	-	Attenuated in guinea pigs
	Kanamycin cassette in *M. bovis* BCG *Rv3134c* (with polar effects of *dosR*) (Sherman et al., 2001)	Reduced survival in stationary phase	-	-	-	-
	Kanamycin cassette in *M. bovis* BCG *dosR* (Boon and Dick, 2002)	Marked survival reduction in the Wayne model	-	-	-	-
	Kanamycin cassette in *dosS* in *M. bovis* BCG (Boon and Dick, 2002)	Moderate survival reduction the Wayne model	-	-	-	-
	Hygromycin cassette in *dosR* in *M. smegmatis* (O'Toole et al., 2003)	Greater reduction in survival in hypoxia	-	-	-	-
	TraSH	*dosS*, but not *dosR, is* required for optimal growth	-	-	-	*dosR* no phenotype; *dosS* no data
mtrAB	Attempted Δ*mtrA* deletion (Zahrt and Deretic, 2000)	-	-	-	-	-
	Attempted Δ*mtrB* and Δ*mtrAB* deletion (Parish et al., 2003a)	-	-	-	-	-
	TraSH	*mtrB*, but not *mtrA*, required for optimal growth	-	-	-	*mtrA* required for optimal growth; *mtrB* no data

tcrXY	Hygromycin cassette in *tcrXY* (Parish et al., 2003b)	No phenotype	-	-	Hypervirulent in SCID mice	-
	TraSH	No phenotype	-	-	-	No phenotype
Unpaired sensors						
Rv2027c[8]	TraSH	No phenotype	-	-	-	No phenotype
Rv3220c	*Rv3220c* deletion (Parish et al., 2003b)	No phenotype	-	-	No phenotype in SCID mice	-
	TraSH	-				-
Unpaired regulators						
Rv0195	TraSH	No phenotype	-	-	-	No phenotype
Rv0260c	TraSH	No phenotype	-	-	-	No phenotype
Rv0818	TraSH	No phenotype	-	-	-	No phenotype
Rv1626	TraSH	Required for optimal growth	-	-	-	-
Rv2884	TraSH	No phenotype	-	-	-	No phenotype
Rv3143	TraSH	No phenotype	-	-	-	No phenotype

[1] In *M. tuberculosis* H37Rv unless otherwise stated
[2] THP1 cell line
[3] Deletion also removes 78 residues from the adjacent gene Rv0492c
[4] *M. tuberculosis* MT103
[5] *M. tuberculosis* MT103
[6] Enhanced growth of mutant abrogated upon macrophage activation
[7] *M. tuberculosis* MT103

within the transposon. This labeled DNA is hybridized to a microarray, and presence of a signal for a gene indicates that a transposon lies in or near that gene. Statistical comparisons determine where loss of a signal has occurred during growth; this statistical element means that results are indicative, but need confirmation. TraSH technology has been used to identify 614 *M. tuberculosis* genes required for optimal growth *in vitro* (Sassetti et al., 2003), and a further 194 genes required for growth following infection of C57BL/6J mice (Sassetti and Rubin, 2003). Of course mutations in genes absolutely required for growth *in vitro* will not be present in the *in vivo* data, so data cannot be complete.

The 2CRs are means for bacteria to respond to changes in the environment by altering gene expression of a set of genes (the regulon). Activation of the system should therefore be measurable by assaying gene expression in those genes. This will allow the nature of the signal to be addressed. However, in most cases the regulon has not been defined. Fortunately in many cases (though by no means all), 2CRs autoregulate; thus activation leads to an increase in their own expression, allowing a rapid amplification of the response. We can therefore use changes in expression of the 2CR genes themselves as an assay for conditions in which they are expressed. Not only does this help define the signal (and thus the possible reason for the response), but it also helps the design of experiments to characterize the regulons as the genes must be activated in order to see a difference when comparing wild type and mutant strains.

Several whole genome microarray experiments have now been published for *M. tuberculosis*. One advantage of these experiments is that results for all genes are available, even if the publication focused on a subset, although these data tend to be very heterogeneous, and their post-publication analysis by no means trivial (Kendall et al., 2004b). In Table 4 we have extracted data from published and supplementary data on microarray experiments for *M. tuberculosis* grown in different *in vitro* and *in vivo* conditions, and highlight situations where a 2CR gene appears to be strongly up- or down-regulated. We do not have the information to carry out statistical analyses of these, and they are presented merely as hypothesis-generating information. The most striking down-regulation occurs in most systems in late stages of growth in stationary phase experiments, but this is most likely to be due to a general reduction in transcription rates. The most striking pattern of up-regulation is seen with the *dosRS* system, where several *in vitro* and *in vivo* conditions cause induction of expression. This and other specific cases are discussed in more detail below.

BIOLOGY OF INDIVIDUAL MYCOBACTERIAL 2CRS

In this section we review the available information on the 30 proteins of the 2CRs listed in Table 1. Each paired 2CR is considered in turn, and the orphan 2CR proteins are then briefly discussed.

Table 4A. 2CR gene expression data (*in vitro*). 2CR gene expression data as measured *in vitro*. These data were collected from various papers and their supplementary data. When numbers are given, these represent fold-induction in microarray experiments. Light shading represents repression (fold change 0.5 or below), dark shading induction (fold change above 1.5), no shading indicates no clear change or absence of data. Two other types of data are considered, GFP-fusions, and SCOTS data. For these, only gene up-regulation can be measured, and level of expression is indicated by pluses (+: low expression to ++++: high expression). 'no sig.' means no signal detected, '-' means no data available. References are: (1) Rodriguez et al. (2002); (2) Manganelli et al. (2001); (3) Schnappinger et al. (2003); (4) Bacon et al. (2004); (5) Muttucumaru et al. (2004); (6) Voskuil et al. (2004); (7) Zahrt and Deretic (2001); (8) Haydel and Clark-Curtiss (2004).

	Condition:	Low iron	SDS	H_2O_2	Palmitic acid	Hypoxia	Wayne model		Wayne model									Stationary phase					
	Reference:	(1)	(2)	(3)	(3)	(4)	(5)		(6)									(6)					
	Data type:	µarray	µarray	µarray	µarray	µarray	microarray		microarray									microarray					
Rv number	**Gene name**						**NRP1**	**NRP2**	**4h**	**6h**	**8h**	**10h**	**12h**	**14h**	**20h**	**30h**	**80h**	**0h**	**6h**	**8h**	**14h**	**24h**	**60h**
Rv0490	senX3	-	-	1.1	1.0	1.085	1.16	1.84	0.79	0.81	0.95	1.01	0.89	0.71	0.67	0.27	0.05	0.99	1.14	1.05	0.68	0.47	0.22
Rv0491	regX3	-	-	1.0	1.0	0.968	0.86	1.24	0.76	0.72	0.82	0.81	0.75	0.70	0.77	0.28	0.08	0.96	0.84	0.82	0.53	0.32	0.25
Rv0600	Rv0600c	-	-	0.8	1.2	0.461	0.37	0.85	0.75	0.57	0.51	0.44	0.50	0.53	0.99	0.55	0.09	1.02	0.52	0.62	-	0.40	0.44
Rv0601	Rv0601c	-	-	0.8	0.8	0.701	0.31	0.45	1.10	0.77	0.74	0.68	0.72	0.68	0.73	0.79	0.25	1.06	0.91	0.90	0.55	0.71	1.14
Rv0602	tcrA	-	-	0.8	0.8	0.748	0.43	0.60	0.79	0.79	0.65	0.64	0.73	0.69	0.86	0.56	0.18	1.13	0.81	0.82	0.72	0.62	0.72
Rv0757	phoP	-	1.7	1.7	1.2	0.968	0.77	1.86	0.65	0.84	0.91	0.66	0.86	0.67	0.76	0.13	0.06	0.96	0.78	0.76	0.56	0.25	0.17
Rv0758	phoR	-	-	-	-	0.921	0.70	1.40	0.83	0.91	0.97	0.91	1.01	0.83	1.06	0.27	0.05	1.16	0.76	0.72	0.56	0.33	0.30
Rv0844	narL	-	-	0.7	1.3	0.939	1.19	1.45	0.69	0.72	0.59	0.82	0.77	0.64	0.79	0.13	0.04	1.07	0.69	0.73	0.40	0.25	0.30
Rv0845	narS	-	-	0.7	1.1	0.9	1.46	2.76	0.82	0.80	0.71	0.59	0.86	0.75	0.69	0.24	0.15	0.93	0.60	0.61	0.44	0.27	0.48
Rv0902	prrB	-	-	1.0	0.9	1.109	0.82	0.84	0.79	0.63	0.74	0.64	0.69	0.57	0.51	0.16	0.07	0.93	0.84	0.80	0.65	0.42	0.35
Rv0903	prrA	-	-	0.9	0.9	0.269	0.81	0.80	0.68	0.65	0.78	0.74	0.80	0.71	0.57	0.17	0.08	0.92	0.78	0.77	0.57	0.64	0.51
Rv0981	mprA	1.7	3.5	1.5	0.7	1.068	0.57	0.51	1.07	0.82	0.70	0.72	0.59	0.91	1.00	0.96	0.24	1.17	0.96	1.08	1.03	1.19	0.82
Rv0982	mprB	-	3.5	-	-	1.344	0.61	0.76	-	-	-	-	-	-	-	-	0.08	-	-	-	-	-	-
Rv1027	kdpE	-	-	0.7	1.1	0.926	0.97	1.13	0.72	0.81	0.87	0.78	0.73	0.68	0.48	0.12	0.12	0.98	0.80	0.86	0.59	0.45	0.26
Rv1028	kdpD	-	-	-	-	1.464	1.05	1.46	0.77	0.87	0.85	0.95	1.02	1.02	1.21	0.48	0.12	0.98	0.73	0.94	0.72	0.69	0.38
Rv1032	trcS	-	-	0.8	1.0	0.531	0.62	0.76	0.56	0.59	0.66	0.58	0.68	0.50	0.65	0.26	0.09	0.99	0.82	0.73	0.53	0.29	0.41
Rv1033	trcR	-	0.67	0.7	1.0	0.412	0.61	0.72	0.57	0.47	0.53	0.54	0.54	0.38	0.41	0.16	0.09	1.04	0.79	0.72	0.42	0.32	0.50
Rv3132	dosS			0.8	0.8	4.05	11.55	14.87	10.24	11.29	6.32	4.04	3.27	3.13	1.65	0.31	0.08	0.99	0.63	0.68	1.07	0.44	0.20
Rv3133	dosR	-	-	0.8	1.0	3.029	10.28	5.52	10.01	9.82	13.23	8.59	4.89	5.32	1.93	0.41	0.27	1.14	0.69	0.82	2.31	1.28	0.39
Rv3245	mtrB	-	-	0.9	0.8	1.281	0.79	0.70	0.71	0.69	0.84	0.74	0.68	0.65	0.65	0.22	0.06	0.99	0.75	0.79	0.53	0.35	0.20
Rv3246	mtrA	0.5	-	0.6	0.7	0.83	1.57	1.40	0.41	0.32	0.26	0.23	0.20	0.26	0.13	0.03	0.01	1.04	0.50	0.54	0.23	0.14	0.06
Rv3764	tcrY	1.5	0.52	1.0	0.7	0.734	0.49	0.38	0.72	0.72	0.83	0.68	0.78	0.58	0.80	0.60	0.20	-	0.72	0.76	0.56	0.46	0.44
Rv3765	tcrX	-	2.5	0.6	0.9	1.289	1.83	1.49	0.57	0.72	0.69	0.61	0.54	0.67	0.70	0.16	0.07	0.94	0.66	0.76	0.54	0.43	0.39
Unpaired sensors																							
Rv2027	Rv2027c	-	-	0.8	1.7	1.303	0.75	0.94	0.93	1.03	0.95	0.75	0.77	0.56	-	0.24	0.01	1.00	1.17	1.11	0.90	0.77	0.70
Rv3220	Rv3220c	-	-	1.1	1.0	0.902	0.35	0.30	0.88	0.80	0.81	0.65	0.81	0.49	0.31	0.16	0.17	1.05	0.56	0.55	0.32	0.21	0.23
Unpaired regulators																							
Rv0195	Rv0195	-	-	0.9	0.9	1.448	1.56	3.30	0.81	0.82	0.71	1.12	0.90	2.17	8.03	4.11	0.33	0.79	0.50	0.63	0.47	0.36	0.28
Rv0260	Rv0260c	-	-	1.6	0.8	1.096	0.77	1.21	1.00	1.30	1.77	0.95	1.59	1.27	1.44	0.74	0.16	1.01	0.99	1.05	0.93	0.95	0.83
Rv0818	Rv0818	-	-	1.4	1.4	0.561	0.77	1.21	0.81	0.97	0.94	1.15	1.08	0.90	0.86	0.27	0.05	0.99	0.72	0.78	0.57	0.35	0.23
Rv1626	Rv1626	-	-	0.8	0.9	0.611	1.00	0.70	0.92	0.87	0.75	0.57	0.52	0.38	0.33	0.03	0.03	1.01	0.79	0.72	0.36	0.25	0.14
Rv2884	Rv2884	-	-	1.3	1.1	1.032	0.65	1.01	0.69	0.70	0.68	0.69	0.63	0.64	0.53	0.30	0.09	1.05	0.79	0.78	0.53	0.47	0.46
Rv3143	Rv3143	-	-	0.9	0.9	0.708	0.44	0.46	0.94	0.91	0.84	0.78	0.71	0.67	0.58	0.19	0.20	1.23	0.88	0.91	0.63	0.62	0.51

Table 4B. 2CR gene expression data (*in vivo*). 2CR gene expression data as measured *in vivo*. These data were collected from various papers and their supplementary data. When numbers are given, these represent fold-induction in microarray experiments. Light shading represents repression (fold change 0.5 or below), dark shading induction (fold change above 1.5), no shading indicates no clear change or absence of data. Two other types of data are considered, GFP-fusions, and SCOTS data. For these, only gene up-regulation can be measured, and level of expression is indicated by pluses (+: low expression to ++++: high expression). 'no sig.' means no signal detected, '-' means no data available. References are: (1) Rodriguez et al. (2002); (2) Manganelli et al. (2001); (3) Schnappinger et al. (2003); (4) Bacon et al. (2004); (5) Muttucumaru et al. (2004); (6) Voskuil et al. (2004); (7) Zahrt and Deretic (2001); (8) Haydel and Clark-Curtiss (2004).

	Condition:	BCG in J774 macrophages			H37Rv in J774 macrophages			H37Rv in human monocytes			H37Rv in human macrophages			*Mtb* clinical isolate 1254 in murine macrophages					
	Reference:	(7)			(7)			(7)			(8)			(3)					
	Data type:	GFP-fusions			GFP-fusions			GFP-fusions			SCOTS			Microarray					
														Naïve			**Activated**		
Rv number	**Gene name**	**2 h**	**3 dy**	**7 dy**	**2 h**	**3 dy**	**7 dy**	**2 h**	**3 dy**	**7 dy**	**18 h**	**2 dy**	**5 dy**	**4h**	**24h**	**48h**	**4h**	**24h**	**48h**
Rv0490	senX3	-	-	-	-	-	-	-	-	-	-	-	-	0.8	1.0	1.1	0.9	0.9	1.0
Rv0491	regX3	-	-	-	-	-	-	-	-	-	+	++++	no sig.	0.9	1.2	1.1	1.0	0.9	0.9
Rv0600	Rv0600c	-	-	-	-	-	-	-	-	-	-	-	-	1.3	1.6	2.1	1.3	2.7	1.9
Rv0601	Rv0601c	-	-	-	-	-	-	-	-	-	-	-	-	1.0	1.1	1.2	1.0	1.2	1.1
Rv0602	tcrA	-	-	-	-	-	-	-	-	-	no sig.	no sig.	no sig.	1.1	1.2	1.3	1.3	1.9	1.6
Rv0757	phoP	+++	+++	+++	+	+	+	+	+	+	++++	no sig.	++	0.8	0.9	0.8	0.7	0.6	0.7
Rv0758	phoR	-	-	-	-	-	-	-	-	-	-	-	-	-	-	-	-	-	-
Rv0844	narL	no sig.	no sig.	no sig.	no sig.	no sig.	no sig.	no sig.	no sig.	no sig.	no sig.	no sig.	no sig.	1.2	1.0	1.3	1.1	1.1	1.2
Rv0845	narS	-	-	-	-	-	-	-	-	-	-	-	-	1.4	1.2	1.3	1.6	1.8	1.5
Rv0902	prrB	-	-	-	-	-	-	-	-	-	-	-	-	0.9	1.0	1.0	1.1	1.1	1.1
Rv0903	prrA	no sig.	no sig.	no sig.	no sig.	no sig.	no sig.	no sig.	no sig.	no sig.	no sig.	++++	no sig.	1.0	1.3	1.1	1.0	1.0	1.0
Rv0981	mprA	no sig.	++	++	no sig.	no sig.	no sig.	no sig.	no sig.	no sig.	no sig.	++++	no sig.	1.0	1.1	1.0	1.1	1.0	1.1
Rv0982	mprB	-	-	-	-	-	-	-	-	-	-	-	-	1.0	1.3	1.4	1.0	1.4	1.9
Rv1027	kdpE	-	-	-	-	-	-	-	-	-	no sig.	+++	no sig.	1.1	1.2	1.3	1.3	1.6	1.5
Rv1028	kdpD	-	-	-	-	-	-	-	-	-	-	-	-	-	-	-	-	-	-
Rv1032	trcS	-	-	-	-	-	-	-	-	-	-	-	-	1.2	1.3	1.4	1.3	1.5	1.2
Rv1033	trcR	no sig.	no sig.	no sig.	no sig.	no sig.	no sig.	no sig.	no sig.	no sig.	+	no sig.	no sig.	1.5	1.5	1.8	1.6	2.0	1.6
Rv3132	dosS	-	-	-	-	-	-	-	-	-	-	-	-	0.9	1.1	1.3	1.1	4.0	2.2
Rv3133	dosR	no sig.	no sig.	no sig.	no sig.	no sig.	no sig.	no sig.	no sig.	++	no sig.	no sig.	+++	0.8	0.9	1.0	1.0	6.3	2.7
Rv3245	mtrB	-	-	-	-	-	-	-	-	-	-	-	-	1.0	1.1	1.1	1.1	1.1	1.2
Rv3246	mtrA	no sig.	+	+	no sig.	no sig.	no sig.	no sig.	no sig.	no sig.	++++	++	++++	0.6	0.7	0.7	0.6	0.6	0.4
Rv3764	tcrY	-	-	-	-	-	-	-	-	-	-	-	-	0.6	0.4	0.3	0.7	0.4	0.4
Rv3765	tcrX	no sig.	no sig.	no sig.	no sig.	no sig.	no sig.	no sig.	no sig.	no sig.	no sig.	no sig.	++	1.2	1.0	1.0	1.3	1.1	1.1
Unpaired sensors																			
Rv2027	Rv2027c	-	-	-	-	-	-	-	-	-	+++	+	+	1.3	1.1	1.3	1.4	1.5	1.5
Rv3220	Rv3220c	-	-	-	-	-	-	-	-	-	++++	no sig.	++++	1.4	1.8	2.0	1.5	2.1	2.1
Unpaired regulators																			
Rv0195	Rv0195	-	-	-	-	-	-	-	-	-	no sig.	no sig.	no sig.	1.6	1.6	1.7	1.6	1.9	1.8
Rv0260	Rv0260c	-	-	-	-	-	-	-	-	-	no sig.	no sig.	no sig.	1.5	1.9	1.9	2.2	2.6	2.8
Rv0818	Rv0818	+++	+++	+++	+	+	+	+	+	+	no sig.	no sig.	no sig.	1.2	1.1	1.1	1.3	1.1	1.1
Rv1626	Rv1626	no sig.	no sig.	no sig.	no sig.	no sig.	no sig.	no sig.	no sig.	no sig.	++++	++++	++	0.8	0.6	0.5	0.7	0.4	0.3
Rv2884	Rv2884	no sig.	no sig.	no sig.	no sig.	no sig.	no sig.	no sig.	no sig.	no sig.	no sig.	no sig.	no sig.	1.1	1.1	1.3	1.3	1.6	1.8
Rv3143	Rv3143	no sig.	+	+	no sig.	no sig.	no sig.	no sig.	no sig.	no sig.	no sig.	no sig.	no sig.	0.8	0.9	0.9	0.8	0.8	0.7

regX3-senX3 (*Rv0491*/*Rv0490*)

The *regX3-senX3* 2CR was isolated from *M. tuberculosis* and *M. bovis* BCG by the use of degenerate PCR before the sequence of the *M. tuberculosis* genome was available (Wren et al., 1992; Supply et al., 1997). Orthologues are found in all genomes analysed in Table 2. The structure of this operon is interesting in that the two genes are separated by a 230 bp region which consists of three repeats of a 77 bp MIRU (mycobacterial interspersed repeat unit). MIRUs are short repetitive sequences found within the *M. tuberculosis* complex and can be used for strain differentiation (Magdalena et al., 1998a). The number of repeats of the MIRU varies between different strains (Magdalena et al., 1998b), but it is not known whether this variation has an effect on virulence. A shorter, though related, MIRU is present in *M. avium* (Bull et al., 2003), but the genes from other sequenced genomes do not contain this region.

Computational analysis suggests that SenX3 contains an atypical PAS domain, similar to that of Mak2p in *Schizosaccharomyces pombe* (a peroxide stress sensor), and it has been suggested that it is the orthologue of *E. coli* ArcB, the sensor in the *arcAB* 2CR (Rickman et al., 2004). Respiration control protein ArcA acts mainly as a negative regulator of aerobic gene expression, but can also function as a positive activator in a few instances (Gunsalus and Park, 1994). PAS domains are widely distributed in proteins from members of the archaea and bacteria and from fungi, plants, insects, and vertebrates and these proteins sense oxygen, redox potential, light and some other stimuli (Taylor and Zhulin, 1999). Clearly light is an improbable stimulus for SenX3, so a more likely scenario is that the *regX3-senX3* system is involved in the sensing of an oxygen and/or redox-status signal. The nature of the signal detected by sensors is, in part, determined by the cofactors which are associated with the PAS domain, and the identification of such cofactors may help elucidate the nature of the signal.

Both the full length RegX3, and the N-terminally truncated SenX3 have been expressed and purified in *E. coli* as His-tagged fusion proteins (Himpens et al., 2000). His_6-SenX3 auto-phosphorylates *in vitro* in the presence of Mg^{2+} ions. Phosphoryl-transfer from His_6-SenX3 to His_6-RegX3 was also demonstrated *in vitro* confirming that these two proteins form a cognate pair. Site-directed mutagenesis was used to alter specific residues within the sensor and response regulator and indicated that His-167 and Asp-62 are the sites of phosphorylation in SenX3 and RegX3, respectively.

Experiments on promoter activity show some aspects of the system to be unsurprising, and others to be rather atypical. The genes appear to be expressed as an operon, with promoter activity detected upstream of *senX3* but not of *regX3* (Supply et al., 1997). There is also evidence that this system is autoregulatory, as His_6-RegX3 binds to its own promoter, and a 42 bp region was been identified as being important in binding (Himpens et al., 2000):

5'-ATGTGAACGGTAACCGAACAGCTGTGGCGTAGTGTGTGACTT-3'

In contrast, His_6-RegX3 does not bind to the intergenic MIRU. Though the binding suggests autoregulation, the type of regulation is not clear. Expression of plasmid-encoded *lacZ* under the control of the *SenX3* promoter region was increased two-fold in the presence of the *senX3-regX3* operon in *M. smegmatis* suggesting positive autoregulation (Himpens et al., 2000). In contrast, another group reported activity from a reporter construct to be 2.8-fold higher in an *M. tuberculosis* Δ*regX3* mutant than in the wild-type strain, suggesting that it acts as a repressor (Parish et al., 2003b). Promoter activity was reported to be low by both these groups, and we have seen that the *senX3-regX3* operon is expressed at low levels in a variety of conditions using RTq-PCR (our unpublished data). It is possible that this is the full range of expression regulation, implying a rather subtle mechanism of action. Alternatively we have still not identified conditions when the genes are fully induced.

We have only limited information regarding the expression of this system in infection models. mRNA was detected in early stages of infection (18h, 48h) of human macrophages, but not in the later stages (110h) by the SCOTS (Selective Capture of Transcribed Sequences) technique (Haydel and Clark-Curtiss, 2004). It also appears that sigma factor SigC plays a role in SenX3 expression since an *M. tuberculosis* CDC1551 Δ*sigC* knock-out mutant significantly underexpressed SenX3 (4.5-fold repression in stationary phase, confirmed by RTq-PCR) (Sun et al., 2004).

Mutants of *regX3*, *senX3* and *regX3-senX3* have been made, and there have been some reports of *in vitro* growth defects (Table 3). A Δ*regX3* mutant grew to a slightly lower optical density in stationary phase than the wild-type strain (Rickman et al., 2004), while a *regX3-senX3* mutant was described as erratic as growth defects were seen but not reproducible (Parish et al., 2003b). Attenuation was seen in human and mouse macrophages, and in immunocompromised and immunocompetent mice (Parish et al., 2003b; Rickman et al., 2004). In contrast, a transposon mutant of *regX3*, where the transposon insertion was in the 5' part of the *regX3* coding region, showed no difference in virulence on infection of C57BL/6J mice (Ewann et al., 2002). In TraSH experiments, the *regX3-senX3* system was not required for optimal growth *in vitro* (Sassetti et al., 2003), but *senX3* (though interestingly not *regX3*) was found to be required for survival *in vivo* (Sassetti and Rubin, 2003).

tcrA(B) (Rv0602c/Rv0601c+Rv0600c)

The *tcrA* gene (Two Component Regulator) was cloned from an *M. tuberculosis* chromosomal library in 1996. The clone was detected using a probe generated by PCR of *M. tuberculosis* genomic DNA with degenerate primers for the virulence-controlling *Salmonella* sp. *phoP* response regulator (N. Buchmeier, personal communication). Sequencing of the clone confirmed the presence of an open-reading frame with deduced amino acid sequence similar to the regulatory proteins OmpR and PhoP, and allowed for the generation of a Δ*tcrA* strain of *M. tuberculosis* Erdman by allelic exchange (Buchmeier, 1998).

TcrA has no apparent cognate sensor; the two genes downstream of *Rv0602c* (*tcrA*), namely *Rv0601c* (*tcrB1* herein) and *Rv0600c* (*tcrB2* herein), appear to be non-functional genes which resulted from a frameshift mutation in an originally unique open reading frame. TcrB1 (which in a 'functional' sensor would be the N-terminus of the protein) contains the SMART-database domains HAMP (for Histidine kinases, Adenylyl cyclases, Methyl-binding proteins and Phosphatases, and found in bacterial sensor and chemotaxis proteins and in eukaryotic histidine kinases), and 'Histidine Kinase Acceptor' (that includes the conserved phosphorylated histidine residue, which in this case would be His-131) (see http://smart.embl-heidelberg.de/ and Letunic et al., 2004). TcrB2 (which in a functional sensor would be the C-terminus of the protein) has the PFAM 'Histidine kinase' domain. It is possible to reconstitute the probable pre-frameshift-mutation sensor gene (which we call *tcrB*); the encoded protein is 363 amino acids long, with a convincing transmitter domain, and a short (29 residue) input domain located N-terminally to a single transmembrane region. It is fair to say that 29 residues is very short for an input domain when, in *M. tuberculosis*, the average length of the input domains is 212 residues, and perhaps even the unframeshifted TcrB product would be an inactive sensor with a truncated input domain. It has been suggested that *M. tuberculosis* CDC1551, rather than having separate homologues of *Rv0601c* and *Rv0600c*, encodes a unique 377 residue protein with a distinct N-terminal sequence (Tyagi and Sharma, 2004), and such a protein (MT0630) is indeed listed in the genome annotation. Nevertheless, we believe, both because of the amino acid homologies, and because of 3rd position GC-preference (J. Parkhill, personal communication), that the same frameshift mutation exists in *M. tuberculosis* CDC1551 as in *M. tuberculosis* H37Rv, and that MT0630 is misannotated. The same frameshift is also seen in *M. bovis* (Garnier et al., 2003).

The *tcrA* gene is present in all members of the *M. tuberculosis* complex (*M. tuberculosis*, *M. bovis*, *M. africanum*, *M. microti*), but appears absent by hybridization in non-tuberculosis-complex mycobacteria including *M. avium*, *M. smegmatis*, *M. intracellulare*, *M. fortuitum*, *M. gordonae*, *M. marinum* and *M. chelonae* (Buchmeier, 1998). A *tcrA* homologue is annotated in the *M. avium* subsp. *paratuberculosis* genome sequence (EMBL reference AE016958) (and the same gene is present in *M. avium*), but there is no conservation of the context, and we were unable to identify convincing Rv0601c and Rv0600c homologues (see Table 2). Therefore, TcrA may be unique to the *M. tuberculosis* complex.

The *tcrA* knock-out mutant was tested for growth in human monocytes, growth in Middlebrook and Proskauer-Beck media, and sensitivity to hydrogen peroxide, paraquat, protamine sulfate, mercury and arsenate but no phenotype was identified in these conditions (N. Buchmeier, personal communication). Nevertheless, TcrA has a regulatory effect on *M. tuberculosis* protein expression as evidenced by several changes in the synthesis of *M. tuberculosis* proteins between the wild-type strain and the mutant detected using 2-D PAGE (Buchmeier, 1998).

In a sense, *M. tuberculosis* is a natural *tcrB* mutant, and either its cognate response regulator TcrA is inactive (which may explain the lack of phenotype in the Δ*tcrA* mutant), or it is constitutively expressed and active in a non-phosphorylated form, or it is activated by cross-talk from another sensor. It would be interesting to know what the levels of expression and phosphorylation of TcrA are *in vivo* and *in vitro*. It would also be interesting to assess the effect of reinserting the putative functional *tcrB* gene into *M. tuberculosis*.

phoPR (Rv0757/Rv0758)

The *M. tuberculosis phoPR* 2CR was named because of its homology to the *Bacillus subtilis phoPR* system (S. Gordon, personal communication, and Cole et al., 1998). In *B. subtilis*, the *phoPR* system detects phosphate-limiting conditions; confusingly, the equivalent system in *E. coli* is called *phoBR*. There is an unrelated 2CR, *phoPQ*, originally identified in *Salmonella enterica* serovar Typhimurium, which responds to the levels of Mg^{2+} and Ca^{2+}, and controls virulence in this species (Groisman, 2001). The finding that an *M. tuberculosis phoP* mutant strain is attenuated in a mouse infection model led Perez et al., (2001) to propose that *M. tuberculosis phoPR* are more similar to the virulence-controlling *Salmonella* sp. *phoPQ* genes than they are to the *phoPR* genes of *B. subtilis*. More experimental data are needed before deciding which (if either) of the two systems is the most related non-mycobacterial system. Sequence similarity suggests the original *B. subtilis* prediction is correct (*M. tuberculosis* PhoR identities: 40.0% over 230 residues with *B. subtilis* PhoR, but only 24% over 346 residues for *S.* Typhimurium PhoQ; *M. tuberculosis* PhoP identities: 41.0% over 234 residues for *B. subtilis* PhoP; but only 32.9% over 228 residues for *S.* Typhimurium PhoP), and that whilst *M. tuberculosis phoPR* plays an important role in virulence, it may nevertheless have little analogy with *S.* Typhimurium *phoPQ*.

Reported gene experiments showed the *M. tuberculosis phoP* to be expressed in human and mouse macrophages at 2 h, 3 days and 7 days following infection; the *M. bovis* BCG gene was also expressed, but more strongly (Zahrt and Deretic, 2001). Experiments using SCOTS in human macrophages showed it to be expressed at 18 h and 5 days, but not at 2 days (Haydel and Clark-Curtiss, 2004). In microarrays, some up-regulation compared to normal culture was seen following treatment with SDS or H_2O_2, and in the Wayne model NRP2 (Non-Replicating Persistence stage 2; see Muttucumaru et al., 2004 and Table 4A).

PhoP was inactivated by insertion of a kanamycin-resistance cassette in the clinical *M. tuberculosis* isolate MT103 (Perez et al., 2001). *In vitro*, Δ*phoP* cells formed smaller colonies than wild-type cells when grown on 7H10, and exhibited loss of cording in Lowenstein-Jensen medium. The mutant was severely attenuated in resting mouse macrophages, and in BALB/c mice, indicating an important role in infection.

narLS (*Rv0844c/0845*)

The *M. tuberculosis narLS* 2CR was identified by homology to the *narXL* nitrate-sensing 2CR of *E. coli*, and the acronym derives from NitrAte Reductase (Stewart and MacGregor, 1982). However, in *M. tuberculosis*, the sensor gene (*Rv0845*) was named *narS* (Parish et al., 2003a) to distinguish it from *Rv1736c*, a probable nitrate reductase enzyme, which is sometimes called called *narX* (Cole et al., 1998).

This is the only paired system in *M. tuberculosis* where the two genes are not transcribed in the same direction. They are divergently transcribed, but nevertheless share a promoter region (Fig. 2A). We have identified probable *narLS* homologues in *M. marinum*, *M. avium*, *M. paratuberculosis*, and *M. smegmatis*, although there was no conservation of context in the last of these (Table 2). Interestingly, *M. marinum* appears to have duplicated *narS*, with a paralogue on the other side of *narL* (Fig. 3C).

We predict NarS to be a 6-transmembrane domain (6-TM) integral membrane protein, a topology that has been seen in other 2CR sensors (Galperin et al., 2001; Smirnova and Ullrich, 2004). We have not however detected convincing homology between NarS and the MHYT 6-TM proteins thought to coordinate with copper and sense oxygen, CO or NO status (Galperin et al., 2001), or the *Pseudomonas syringae* temperature-sensing CorS sensor (Smirnova and Ullrich, 2004). NarL is one of three LuxR-like response regulators in *M. tuberculosis*.

Very little positive phenotype has been determined for this 2CR. No expression of *narL* was seen in macrophages (Haydel and Clark-Curtiss, 2004), although some suggestion of *narS* expression was detected in NRP2 and activated macrophages in microarrays (Table 4). A Δ*narL* deletion mutant showed no alteration in virulence in SCID mice relative to the wild-type H37Rv strain (Parish et al., 2003a).

prrAB (*Rv0903c/Rv0902c*)

The *prrAB* system was first described following sequencing of the *M. tuberculosis* genome (Cole et al., 1998). Interest was aroused when the system was found to be expressed in the early stages of macrophage infection with *M. tuberculosis* and *M. bovis* BCG (Graham and Clark-Curtiss, 1999; Ewann et al., 2002), and its name is therefore an acronym for Phagocytosis Response Regulator (Graham and Clark-Curtiss, 1999).

Both the full length response regulator PrrA, and an N-terminally truncated construct of sensor PrrB, have been expressed and purified in *E. coli* as His-tagged fusion proteins (Ewann et al., 2004). His_6-PrrB autophosphorylates *in vitro* in the presence of Mn^{2+} ions and phosphotransfer from His_6-PrrB to His_6-PrrA has been demonstrated *in vitro* showing that these two proteins form a functional cognate pair.

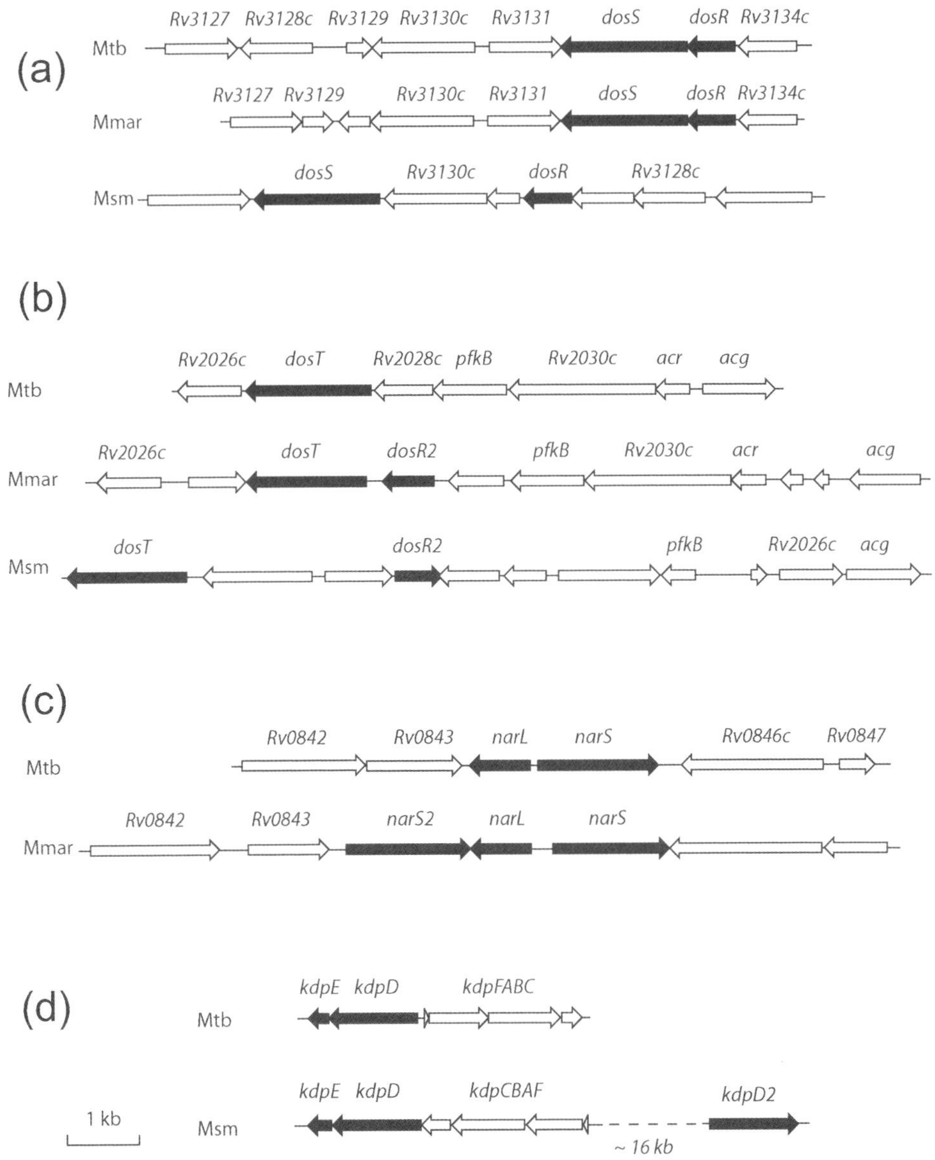

Fig. 3. Comparative genomics of selected *M. tuberculosis* 2CRs. (a) the *dosRS* region; (b) the *dosT* region; (c) the *narLS* region; (d) the *kdpDE* region. The 2CR genes are shown as filled arrows. Gene names or Rv numbers are indicated above the *M. tuberculosis* genes. When gene names or Rv numbers are indicated above non-*M. tuberculosis* genes, these indicate their most likely *M. tuberculosis* homologue. Mtb, *M. tuberculosis* H37Rv; Mmar, *M. marinum*; Msm, *M. smegmatis* mc^{2}155.

PrrA has been shown to specifically bind to a 317 bp fragment containing its own promoter region suggesting it is autoregulated. Both the phosphorylated and unphosphorylated forms are capable of binding DNA, but binding is more efficient with the phosphorylated regulator. Direct evidence for autoregulation was provided by the fact that a GFP reporter construct is transcribed in wild-type *M. tuberculosis* MT103, but not in a *prrA* mutant carring a transposon 5 bp upstream of the start codon (Ewann et al., 2004).

No effect was seen in the growth of this *prrA* mutant *in vitro*. However, in mouse macrophages, initial rates of growth were much lower than for the wild-type strain (Ewann et al., 2002). This is consistent with a lack of expression of this system in axenic culture (Ewann et al., 2004), and the induction of expression in early stages of mouse macrophage infection in both *M. bovis* BCG, as detected using GFP fusions (Ewann et al., 2002), and *M. tuberculosis*, as detected using SCOTS (Graham and Clark-Curtiss, 1999). In contrast, experiments using the GFP reporter showed no expression in macrophages (Zahrt and Deretic, 2001). The attenuation seen in macrophages, did not translate to the whole animal, as no phenotype was observed in BALB/c mice (Ewann et al., 2002).

mprAB (*Rv0981/Rv0982*)

Both the full length regulator MprA, and an N-terminally truncated construct of sensor MprB, have been expressed and purified in *E. coli* as GST- or His-tagged fusion proteins (Zahrt et al., 2003). Autophosphorylation of truncated MprB requires Mg^{2+} or Mn^{2+}, but not Ca^{2+}. Phosphorylated MprB transfers phosphate to MprA, its cognate regulator, but not to the non-cognate MtrA. In addition to kinase and phosphotransferase activities, many sensors also dephosphorylate their cognate response regulators; a truncated MprB was shown to effectively dephosphorylate the phosphorylated form of MprA (Zahrt et al., 2003).

In MprB, mutation of the His-249 to glutamine abolished phosphorylation showing that His-249 is the likely residue for phosphorylation for this system. Mutagenesis of His-249 also abolished the dephosphorylation activity of MprB. In MprA, conversion of the Asp-48 to alanine abolished transphosphorylation from MprB suggesting that this aspartate residue is the site for phosphorylation (Zahrt et al., 2003).

M. tuberculosis mprAB expression was detected after 48 h in human macrophages, though not earlier (18 h) or later (110 h) (Haydel and Clark-Curtiss, 2004). Expression of *M. tuberculosis mprA* was not detected in murine macrophages using GFP fusions, though low levels of expression were seen with *M. bovis* BCG (Zahrt and Deretic, 2001). Up-regulation of *mprAB* was seen in microarrays following various stresses, in particular SDS treatment (Manganelli et al. (2001) and Table 4A).

The name for these genes came from an experiment in which an *mprA* mutant was used in a low dose *in vivo* growth competition assay (Zahrt and Deretic, 2001).

An *mprA*::*Km*^{r} mutant was as able as H37Rv to establish an acute lung infection, but was less able to remain in the lungs of BALB/c mice during chronic infection following the activation of the cell-mediated immune system. The 2CR name therefore comes from the acronym for 'Mycobacterial Persistence Regulator'. Interestingly, an *M. tuberculosis mprA* mutant showed enhanced survival in resting (but not activated) murine macrophages (Zahrt and Deretic, 2001). In contrast, *M. bovis* BCG mutants with site-directed mutations that replaced His249 and Asp48 in MprB and MprA respectively led to attenuation in resting macrophages (Zahrt et al., 2003).

In TraSH experiments, the sensor gene *mprB*, but not the regulator *mprA*, was identified as being required for optimal growth *in vitro* (Sassetti et al., 2003). *MprA* was also not essential *in vivo*, but there were no data for *mprB* (Sassetti and Rubin, 2003).

kdpED (Rv1027c/Rv1028c)

The *kdpED* system was identified in the original *M. tuberculosis* genome annotation (Cole et al., 1998) and is homologous to the *E. coli* system, first identified by Epstein and Davies(1970), which controls the expression of the *kdpFABC* operon (31% and 49% identity to *E. coli* KdpD and KdpE respectively). In *E. coli*, the environmental signal for KdpD activation by autophosphorylation is a change in osmolarity or a drop in K^+ concentration, or indeed a combination of these (Sardesai and Gowrishankar, 2001). The phosphorylated KdpD transfers its phosphate group to KpdE, and the latter upregulates expression of the *kdpFABC* operon which encodes a high-affinity four-subunit K^+-transporting P-type ATPase (Gassel et al., 1999). This transport of K^+ maintains turgor pressure of the cell (Altendorf et al., 1994).

Although extensively studied in *E. coli*, the *kdpED* 2CR appears to be ubiquitous among bacteria (Hutchings et al., 2004). It is found in most sequenced mycobacteria, but is not present in *M. leprae* or in the corynebacteria (Table 2). *E. coli*, *M. marinum*, *M. avium* and *M. smegmatis* contain the *kdpFABCDE* genes in the same direction, whereas the *M. tuberculosis kdpED* genes, although contiguous to the *kdpFABC* genes, are divergently arranged with 192 bp separating the two clusters (Fig. 2A). In *M. smegmatis*, a second *kdpD* paralogue is seen approximately 16 kb upstream of the main gene cluster (Fig. 3D).

The *M. tuberculosis kpdED* system was the subject of extensive investigation by Steyn and colleagues (Steyn et al., 2003). Experimental and computational data suggest that KdpD has two cytoplasmic domains separated by four transmembrane regions. The N-terminal signal-sensing domain (N-KdpD) is 397 residues long, and the C-terminal histidine-kinase domain (C-KdpD) is 364 residues long. Interaction of the *M. tuberculosis* KdpD gene with other proteins was investigated using the yeast two-hybrid technique. N-KdpD was found to interact with two membrane lipoproteins: LprJ (Rv1690) and LprF (Rv1368). The yeast three-hybrid technique further extended these interactions and suggested that C-KdpD forms a ternary complex with N-KdpD/LprF and N-KdpD/LprJ. The work also demonstrated the dependence of *kdpFABC* expression on extracellular potassium concentration ($[K^+]_e$), with high levels of *kdpFABC* at low $[K^+]_e$, an effect amplified by overexpression of wild-type LprF and LprJ but reduced by overexpression of mutant LprF and LprJ. Thus whilst the work did not formally demonstrate that the up-regulation of *kdpFABC* at low $[K^+]_e$ is due to the activation of KdpD, and subsequent binding of phosphorylated KdpE to the *kdpFABC* promoter, it seems likely that the environmental signal for the *kdpED* 2CR is indeed the concentration of extracellular potassium with a response mediated by accessory lipoproteins LprF and LprJ.

In *M. tuberculosis*, *kdpE*, *lprJ* and *lprF* have been shown to be up-regulated in a nutrient starvation model of persistence (Betts et al., 2002). In macrophages, *kdpE* expression was detected after 48 h, but not earlier or later following infection (Haydel and Clark-Curtiss, 2004). In *M. avium*, *kdpD* was shown to be up-regulated in macrophages (Hou et al., 2002).

Deletion of the entire *kdpFABCDE* operon *in E.coli* has been shown to increase resistance to novobiocin and sensitivity to hygromycin (Zhou et al., 2003). TraSH experiments showed that *kdpE* is required for optimal growth *in vitro*, but *kdpD* is not (Sassetti et al., 2003). Following infection of SCID mice, a *kdpDE* mutant was shown to be hypervirulent (Parish et al., 2003a). In contrast, TraSH experiments indicated that the *kdpD* gene (but not the *kdpE* gene) is required for survival *in vivo* (Sassetti and Rubin, 2003) (curiously the converse of the *in vitro* data).

trcRS (*Rv1033c/Rv1032c*)

The *trcRS* 2CR was identified by the use of degenerate PCR before the sequence of the *M. tuberculosis* genome was available, and was named after the acronym for 'Tuberculosis Regulatory Component' (Haydel et al., 1999). The genes lie adjacent to the *kdp* cluster in *M. tuberculosis* (Fig. 2A). Though present in other mycobacteria (except for *M. leprae*) (Table 2), the location next to the *kdp* genes is not conserved.

The response regulator, TrcR, has been expressed and purified in *E. coli* as a 35 kDa His-tagged protein. Attempts to over-express the full-length sensor, TrcS, in *E. coli* were unsuccessful. This is possibly due to the two regions of hydrophobicity within the N-terminus that anchor the protein to the cytoplasmic membrane rendering it toxic to *E. coli*. Truncated forms of the sensor protein are not toxic, and two C-terminal fragments of TrcS, $TrcS^{C\text{-term1}}$ and $TrcS^{C\text{-term2}}$ (35kDa and 31kDa respectively) have been expressed and purified as His-tagged proteins (Haydel et al., 2002). Auto-phosphorylation of the purified C-terminal fragments of TrcS has been shown *in vitro* to be stimulated in the presence of Mn^{2+} or Ca^{2+}, whereas Mg^{2+} only minimally stimulates activity. This is unusual, as the vast majority of the histidine kinases that have been shown to auto-phosphorylate *in vitro* require Mg^{2+}. Phosphorylated $TrcS^{C\text{-term1}}$ and $TrcS^{C\text{-term2}}$ have been shown to effectively trans-phosphorylate a purified full length and truncated TrcR (TrcR $^{N\text{-term}}$). Western blotting has shown that TrcR runs as two forms in non-denaturing polyacrylamide gels, probably a monomeric and dimeric form. These two forms were seen in early- and mid-logarithmic growth in broth culture, however only the larger dimeric form was detected during late logarithmic phase (Haydel et al., 2002).

We have limited information regarding the expression of this two-component system. Intergenic RT-PCR, using primers that span the 9 bp intergenic region between *trcR* and *trcS,* has been used to show that *trcR* and *trcS* are co-transcribed as an operon in H37Rv and in virulent clinical isolate TB233 (Haydel et al., 2002). *TrcRS* was expressed in broth culture in both early and mid logarithmic phase though expression was higher in early log culture. These genes were also expressed in macrophages, but only at low levels after 18 h, and not at 48 h (Haydel et al., 2002; Haydel and Clark-Curtiss, 2004). These data contrast with GFP fusion experiments where no expression could be detected (Zahrt and Deretic, 2001); these inconsistencies could be due to differences in sensitivities of the methods used.

Purified recombinant TrcR has been shown to bind to its own promoter and induce its own expression (Haydel et al., 2002). A minimum 28 bp AT-rich region was required for binding:

5'-ATAAGCGACATTTGAAAAATTTATGAAT-3'

TrcR bound to its own promoter in both the phosphorylated and non-phosphorylated forms; however the phosphorylated TrcR had a ten-fold higher binding affinity over TrcR, and bound a larger sequence region.

A transposon mutant of *trcS*, where the transposon inserted into the 5' end of the *trcS* coding region, did not show any phenotype in mouse bone marrow-derived macrophages (Ewann et al., 2002). However, an *M. tuberculosis* strain carrying an 882bp deletion in *trcS* was hypervirulent in a SCID mouse model (Parish et al., 2003a). TraSH experiments detected no phenotype with either gene *in vitro* or *in vivo* (Table 3).

dosRS (*Rv3132c/Rv3133c*), *dosT* (*Rv2027c*)

This 2CR (composed of response regulator *Rv3133c* and sensor *Rv3132c*) was originally identified by subtractive hybridisation as being expressed at a higher level in the virulent *M. tuberculosis* H37Rv strain than in the avirulent H37Ra strain thus earning the name *devRS* (for 'Differentially Expressed in Virulent strain') (Kinger and Tyagi, 1993, Dasgupta et al., 2000). The gene name *dosR* came from the identification, in *M. bovis BCG*, of a homologous response regulator upregulated in hypoxia, and essential for hypoxic dormancy, hence *dosR* for 'DOrmancy Survival regulator' (Boon and Dick, 2002). By extension its cognate sensor gene *Rv3132c* can now also be found referred to as *dosS* (Kendall et al., 2004a; Roberts et al., 2004).

The existence of two different names is confusing and unnecessary. Historically, the gene names *devR* and *devS* should have precedence (and these are therefore the principal names used in Tuberculist), but in the literature both names are used. The recent discovery that Rv2027c, an orphan sensor with high similarity to DevS/DosS, can cross-talk with DosRS (Roberts et al., 2004; Saini et al., 2004), and the consequent renaming of *Rv2027c* as *dosT* (Roberts et al., 2004), add slight weight to the Dos nomenclature. Our group took to referring to *Rv3133c/Rv3132c* as the *dosRS* 2CR for historical reasons, and because of our interest in orphan sensor DosT. On reflection, we feel neither acronym is informative as to the direct biological role of the 2CR, and suggest (while acknowledging the historical precedence of *devRS*) that *dosRS* be used with the modified acronym of 'Dependent on Oxygen Status' (because of the more recent knowledge about the role of hypoxia in *dosR* activation, and with apologies to both groups originally coining the names).

The *dosRS* 2CR is the most investigated of mycobacterial 2CRs, and is the only 2CR in which conditions for induction have been identified and verified reproducibly. It also plays a key role in *M. tuberculosis* virulence and survival, hence this extended section.

THE *dosRS* REGULON

Of the 11 paired 2CRs in *M. tuberculosis*, only the *dosRS* regulon has been defined. The *dosRS* 2CR is critical in the hypoxic response of mycobacteria (Boon et al., 2001; Sherman et al., 2001; Boon and Dick, 2002; Bagchi et al., 2003; O'Toole et al., 2003; Park et al., 2003; Bacon et al., 2004), a condition associated circumstantially with mycobacterial persistence in macrophages. Genes under the control of the *dosRS* regulon are affected by hypoxia (Sherman et al., 2001; Bacon et al., 2004; Muttucumaru et al., 2004), and comparison of gene expression in *M. tuberculosis* H37Rv with a *dosR* mutant identified 50 genes induced by DosR (Park et al., 2003). It has since been shown that *M. tuberculosis dosR* is also induced following treatment with NO (Ohno et al., 2003; Voskuil et al., 2003), ethanol, or being moved from a rolling to a standing culture (Kendall et al., 2004a). Microarrays comparing gene expression in *M. tuberculosis* H37Rv with a *dosR* mutant after NO treatment (Voskuil et al., 2003) or leaving the cultures to stand (Kendall et al., 2004a), have been used to define the *dosR* regulon. There are a small number of differences in the lists published but, combining these experiments, it can be seen that there is a regulon of about 48 genes that are induced (directly or indirectly) by DosR.

In macrophage infection experiments, the *dosR* gene was shown to be induced in activated, but not in resting, macrophages (Schnappinger et al., 2003); as no induction was seen in $NOS2^-$ macrophages, it appears that NO produced in the respiratory burst (rather than hypoxia) was responsible for the *dosR* induction in this system. Other studies have shown *dosR* induction in resting macrophages, but only at late stages of infection (Zahrt and Deretic, 2001; Haydel and Clark-Curtiss, 2004), and it is unclear what the signal would be in these cases.

The gene most highly induced by DosR is *acr* (also known as *hspX*; *Rv2031*) which codes for a major antigen: alpha-crystallin. A mutation in *dosR* abolished the hypoxia induced induction of *acr* (Sherman et al., 2001; Park et al., 2003). A mutation in *dosS* had little or no effect on *acr* induction; this lack of phenotype already suggests that the DosS paralogue DosT (Rv2027c) might perform a complementary role in the hypoxic response (Sherman et al., 2001). This hypothesis was given further credence when studies using reporter constructs showed that while a mutation in *dosS* reduced the level of DosR-dependent induction of *acr*, it did not abolish it altogether (Roberts et al., 2004). In fact, it was shown that while *dosT* is not up-regulated by hypoxia, it still contributes to hypoxic induction of *acr* as a mutation in *dosT* reduced the level of DosR-dependent induction of *acr* from a reporter construct (Roberts et al., 2004).

Other genes under control of DosR include a number of genes showing homology to universal stress proteins, a fused nitrate reductase and ferredoxin. The precise role of these genes in adaptation to hypoxia remains to be elucidated.

Much of the work described so far has been done in *M. tuberculosis* but similar findings have been observed in the non-pathogenic *M. smegmatis* (Mayuri et al., 2002). Again, *acr* (*hspX*), *dosS* and *dosR* are all up-regulated in cultures grown under microaerobic conditions both at RNA-level (as assessed by RT-PCR) and at protein-level (as assessed by immunogold-staining). An *M. smegmatis dosR::hyg* mutant was constructed, and two-dimensional gel electrophoresis analysis was used to identify a number of proteins controlled by DosR under oxygen limitation: three universal stress proteins (USPs), a putative nitroreductase, and again Acr (O'Toole et al., 2003). These data clearly indicate a key role in hypoxia for the *dosR* homologue of *M. smegmatis*.

It has became apparent that a number of other studies on the induction of various protein and RNA species in different dormancy models were in fact looking at members of the *dosR* regulon (among others) (Hutter and Dick, 1999; Boon et al., 2001; Florczyk et al., 2001; Purkayastha et al., 2002), and lend support to the microarray studies although an added complexity to the hypoxic-response pattern is the observation that in a strain of *M. tuberculosis* with a *sigC* mutant, *acr* expression is down-regulated two-fold (Sun et al., 2004).

PROMOTER STUDIES

Park et al. (2003) identified a 20bp consensus motif found, with some variation, upstream of many of the genes rapidly induced by hypoxia, which they proposed was a binding site for the activated DosR regulator:

5'-TTSGGGACTWWAGTCCCSAA-3' (where S = C/G; W = A/T)

The same palindromic sequence was identified by Florczyk et al. (2003). Several genes contained two or three such motifs in their upstream region. This sequence has been shown to be bound by recombinant DosR (Park et al., 2003; Roberts et al., 2004).

Bagchi et al. (2003) cloned genomic regions upstream of *Rv3134c* ($P_{Rv3134c}$), and of *dosR* (including *Rv3134c* and its upstream region; $P_{Rv3134c\text{-}dosR}$) into plasmids with a promoterless *lacZ* reporter gene, and transformed these into *M. smegmatis*. The presence of the plasmids had no effect in normoxic conditions, but in hypoxic conditions, there was a detectable increase in beta-galactosidase activity in bacteria transformed with the $P_{Rv3134c}$-containing plasmid, and a greater increase still in activity with the $P_{Rv3134c\text{-}dosR}$ plasmid (Bagchi et al., 2003). These findings suggest that the *Rv3134c/dosR/dosS* operon is driven by at least two hypoxia-responsive promoters: one mapping upstream of *Rv3134c* and the other upstream of *dosR*. Interestingly, Park et al. (2003) identified two hypoxia-responsive motifs upstream of *Rv3134c*, but none upstream of *dosR*. Control of the *dosRS* system may take place both at the level of transcription, and of translation: Saini et al. (2004) found that in a *dosR* mutant strain of *M. tuberculosis* expressing only the receiver domain of DosR, induction was observed at the level of RNA expression, but no concomitant protein expression was detected.

MUTANT PHENOTYPES

Experiments looking at the phenotypes of different *dosRS* mutants are summarized in Table 3. In culture, Malhotra et al. (2004) reported reduced cell-to-cell adherence in a *dosR* mutant. Several studies reported no obvious difference in aerated culture, but the TraSH data (Sassetti et al., 2003) suggested a growth defect in *dosS* but not *dosR* mutants. Reduced survival in stationary phase or in hypoxic conditions was reported for both *dosS* and *dosR* mutants (Boon and Dick, 2002; O'Toole et al., 2003).

A possible role for *dosR* in the survival of mycobacteria in the hypoxia-induced stationary phase was suggested following the observation that, in *M. bovis* BCG Δ*Rv3134c* (with a polar-effect abolishing of *dosR* expression), recovery of colony forming units after 3 weeks *in vitro* growth was only 20% of the wild-type strain (Sherman et al., 2001). Furthermore, as discussed above, a kanamycin cassette *dosR* knock-out of *M. bovis* BCG lost the ability to survive hypoxia (Boon and Dick, 2002).

These results are in apparent contradiction with the demonstration that an *M. tuberculosis dosR* deletion mutant was hypervirulent in SCID mice, causing death more rapidly than the wild-type strain (Parish et al., 2003a). More rapid growth was also seen in activated macrophages and BALB/c mice. Bacterial counts were increased approximately 10-fold in early time points (15 days) in the lungs, liver and spleen. This suggested that the *dosR* response somehow slows down growth of wild-type bacteria under these conditions. However, a Δ*dosR::kan* mutant was attenuated in guinea pigs (Malhotra et al., 2004). The mutant strain was substantially less virulent to the guinea pigs with less pathology in lung, liver and spleen, and lower bacterial burden in the spleen. The TraSH experiments reported reduced survival for *dosR* mutants (Sassetti and Rubin, 2003).

PARALOGUES OF *dosRS*

One of the orphan sensor genes of *M. tuberculosis*, *Rv2027c*, is a *dosS* paralogue. It has been shown that this sensor (DosT) functions in the same way as DosS, in that it can phosphorylate DosR (Saini et al., 2004), and that both DosS and DosT must be inactivated to completely abrogate DosR expression (Roberts et al., 2004). Both DosS and DosT respond to hypoxia, activating DosR, but *dosT* transcription is not itself induced by hypoxia (as is the case with *dosS*), despite its location adjacent to the *acr* operon (see Fig. 2B). The precise signal to which DosS and DosT respond is unknown. Both are predicted to have two GAF domains in the N-terminal part of the proteins. GAF domains often bind nucleotides such as cGMP and cAMP (Aravind and Ponting, 1997); and it has been suggested that these may act as a signal, or that the proteins detect the redox state of a metabolite (O'Toole et al., 2003; Tyagi and Sharma, 2004). It is worth noting that GAF domains are evolutionarily and structurally related to PAS domains (Zhulin et al., 1997; Taylor and Zhulin, 1999); PAS domains can bind a variety of co-factors such as heme or flavins and an atypical PAS domain has been identified in SenX3 (Rickman et al., 2004). In sensors such as Aer, ArcB, FixL and NifL, PAS domains respond to changes in oxygen concentration (O'Toole et al., 2003).

Comparative genomics analysis shows that in *M. marinum,* which is closely related to the *M. tuberculosis* complex, the *dosRS* and *dosT* regions are highly similar to those of *M. tuberculosis* (see Fig. 3A and B). Remarkably however, a *dosR* homologue lies upstream of *dosT*, which we have called *dosR2*. This suggests that the *dosRS* system was duplicated at some point, and that *M. tuberculosis* lineage subsequently lost *dosR2*. In *M. smegmatis* the *dosRS* and *dosT* regions are recognizable, but there has been considerable rearrangement, which is made more difficult to interpret by the presence of multiple homologous universal stress proteins (USPs) in both regions. *DosR* and *dosS* are separated, but may lie within the same operon (Fig. 3A). There also appears to be a *dosR2* gene in *M. smegmatis* (Fig. 3B). The presence of the two *dosRS* paralogues, and of other paralogous genes within the clusters suggests that the whole region (rather than just *dosRS*) may at some point have duplicated.

The exact signals for DosS and/or DosT are unknown. The observed cross-talk between DosR and both DosS and DosT, and the observation that DosS and DosT are not induced in identical situations suggest that *M. tuberculosis* can fine-tune its response to the oxygen-limited environment, and therefore has important implications on its *in vivo* survival and virulence. Do DosS/DosT directly detect oxygen levels? Do they detect metabolic changes induced by the change in oxygenation, perhaps by binding a cytosolic compound (we have not detected a transmembrane domain in either DosS or DosT)? Alternatively, do they interact with other proteins? The fact that DosR, DosS, and the DosR-regulon can be reproducibly induced by a variety of signals, and the diversity in mutants available in *M. tuberculosis* and other mycobacteria means we have the tools to dissect the *dosRST* response and the possibility of identifying the exact signal detected: a single signal which has a commonality with all the conditions which activate the system and which will allow us to understand this complex response.

mtrAB (*Rv3246c/Rv3245c*)

The *mtrAB* 2CR was originally cloned in *M. tuberculosis* using response regulator *phoB* from *Pseudomonas aeruginosa* as a hybridization probe (hence *mtr* for '*Mycobacterium tuberculosis* Response regulator') (Via et al., 1996). *MtrAB* lies within a region that is highly conserved throughout the mycobacteria and corynebacteria. The *M. tuberculosis* genes are separated by 49 bp, and this intergenic region contains a short class-II MIRU. There is a space between the orthologues in *M. avium* and *M. paratuberculosis*, but not in other available sequenced genomes.

MtrAB is one of only four 2CRs still functional in *M. leprae* (see Table 2) suggesting a key role for the 2CR. Indeed, *mtrA* is an essential gene in *M. tuberculosis*. Zahrt and Deretic (2000) were unable to inactivate the *mtrA* gene in *M. tuberculosis* H37Rv in the absence of a plasmid-borne functional *mtrA*, and Parish et al. (2003a) failed to isolate *mtrB* or *mtrAB* mutants.

Low-level expression from the *mtrA* promoter was detected in mouse and human macrophages infected with *M. tuberculosis* (Via et al., 1996). In contrast, expression in *M. bovis* BCG, was inducible in mouse macrophages. Haydel and Clark-Curtiss (2004) identified *mtrA* as one of three constitutively expressed response regulators in macrophages (along with *dosT* and orphan regulator *Rv1626*). There is also some evidence that *mtrA* expression is up-regulated by iron in an IdeR-independent fashion (Rodriguez et al., 2002). As for *senX3*, sigma factor SigC plays a role, since *mtrA* expression was two-fold repressed in a Δ*sigC* mutant (Sun et al., 2004).

These data suggest a key role for the *mtrAB* 2CR, and its apparent essentiality makes it a prime therapeutic target, but the signal(s) recognised by MtrB remains unknown. It has been suggested that the *M. tuberculosis mtrAB* 2CR is homologous to a novel 2CR, *amrA-amkA*, involved in rifamycin SV production in *Amycolatopsis mediterranei* U32 (Wang et al., 2002), but we feel that the homology between these systems derives almost exclusively from similarity in the conserved transmitter, receiver and output domains, and not in the biologically determinant input domain of the sensors.

tcrXY (*Rv3765c/Rv3764c*)

Little work has been carried out with *tcrXY* (for 'Two-Component Regulator' see Parish et al., 2003a). The genes are conserved in all mycobacteria except for *M. leprae* (Table 2). Expression was seen in human macrophages late in macrophage infections (110 h), though no expression was detected using GFP fusions (Zahrt and Deretic, 2001).

A deletion of both genes was constructed in *M. tuberculosis*, and caused hypervirulence in a SCID mouse model (Parish et al., 2003a), suggesting this 2CR is anything but innocuous. However, in TraSH experiments, no effect was seen for either gene *in vitro* or *in vivo* (Table 3).

BIOLOGY OF ISOLATED GENES

M. tuberculosis has 2 orphan sensors, and 6 orphan response regulators. Clearly, if some of these proteins are capable of cross-talk with each other, or indeed with proteins of the paired 2CRs, this offers great scope for fine-tuning of environmental responses. Little is known of these proteins, other than for DosT (Rv2027c) which is discussed extensively above, and has been shown to cross-talk with the DosR response regulator (Roberts et al., 2004; Saini et al., 2004).

Rv0195

Response regulator *Rv0195* has some unusual features. It is the only 2CR gene, apart possibly from *tcrA(B)*, that appears to be specific to the *M. tuberculosis* complex (Table 2). It is also one of the three regulators with a LuxR-type DNA-binding domain. While its expression was not detected in macrophages by Haydel and Clark-Curtiss (2004), there is some evidence of up-regulation in microarray experiments, particularly in NRP2 in the Wayne model (Sherman et al., 2001; Bacon et al., 2004; Muttucumaru et al., 2004), and in macrophages (Schnappinger et al., 2003) (Table 4).

Rv0260C

Response regulator *Rv0260c* is found in all sequenced mycobacteria except for *M. leprae*. It lies in a small group of genes that is highly conserved even in *M. smegmatis* (Fig. 2B). Thus its location next to *narK3*, a predicted nitrite exclusion protein, may be significant. *NarK2*, another member of this family, is part of the *dosR* regulon induced in activated macrophages. In *M. tuberculosis*, *Rv0260c* overlaps with *Rv0259c*, and *Rv0260c*, *Rv0259c*, *Rv0258c* may form an operon. Unfortunately, the function of conserved hypothetical proteins Rv0259c and Rv0258c is unknown, but it is possible their function is related to the response controlled by Rv0260c. Some up-regulation of expression is seen in resting and activated macrophages (Schnappinger et al., 2003 and Table 4).

Rv0818

Response regulator *Rv0818* is found in all mycobacteria, except for *M. leprae* where it remains only as a pseudogene (Table 2). It overlaps with conserved hypothetical protein gene *Rv0819* and may form an operon with it. *Rv0818* expression is not activated in macrophages (Haydel and Clark-Curtiss, 2004), and microarray data show no variation in expression, with the exception of down-regulation in the late stages of the Wayne-model and of stationary phase culture which is observed in all 2CR genes (Table 4). Along with response regulator *phoP*, *Rv0181* was strongly expressed by *M. bovis* BCG ingested by murine macrophages, and up-regulated in *M. tuberculosis* H37Rv in both murine and human macrophages (Zahrt and Deretic, 2001).

Rv1626

Response regulator *Rv1626* is present in all mycobacteria including *M. leprae*. The C-terminal (output) domain of Rv1626 displays convincing homology to the ANTAR domain (for AmiR and NasR transcription antitermination regulators). ANTAR is an RNA-binding domain found in bacterial transcription antitermination regulatory proteins. ANTAR has been detected in various response regulators of two-component systems and in one-component sensory regulators from a variety of bacteria. Whilst most response regulators interact with DNA, ANTAR-containing regulators interact with RNA. The majority of the domain consists of a coiled-coil (O'Hara et al., 1999; Shu and Zhulin, 2002). As such, Rv1626 appears to be the only mycobacterial response regulator which targets RNA as opposed to DNA to effect its activity.

Rv1626 expression is shown to be strongly activated in macrophages by Haydel et al. (2004), yet microarray data show down-regulation of expression in activated macrophages (Schnappinger et al. (2003) and Table 4). TraSH data suggest *Rv1626* is required for *in vitro* growth (Sassetti et al., 2003; Sassetti and Rubin, 2003).

Rv2027C (dosT)

This orphan sensor is discussed extensively in the *dosRST* section above.

Rv2884

Response regulator *Rv2884* is present in all mycobacteria except in *M. leprae* where it is found as a pseudogene, and *M. smegmatis* in which we could not identify a convincing homologue. *Rv2884* was up-regulated in activated macrophages, but not in naïve macrophages (Schnappinger et al., 2003).

Rv3143

Rv3143 is the shortest response regulator in *M. tuberculosis* at only 133 amino-acids in length. Whilst the protein has an obvious receiver domain (which essentially accounts for the whole protein), there is no obvious output domain (such as a DNA binding domain). This may account for the lack of observed phenotypes related to *Rv3143* with the exception of down-regulation in NRP2 (Muttucumaru et al., 2004) and the late stages of the Wayne-model (Voskuil et al., 2004) (Table 4A).

Rv3220c

Rv3220c is the second of the orphan sensors in *M. tuberculosis* and is present in all mycobacteria including *M. leprae* (Table 2). The gene is strongly up-regulated early (18h) and late (110h) in growth in macrophages, although intriguingly it is not detected at 48h (Haydel and Clark-Curtiss (2004) and Table 4B). No such activation in murine or human macrophages was detected by Zahrt et al. (2001), although *M. tuberculosis* clinical isolate 1254 upregulated *Rv3220* in both naïve and activated murine macrophages (Schnappinger et al. (2003) and Table 4B). A Δ*Rv3220c* mutant showed no alteration in virulence in SCID mice relative to the wild-type H37Rv strain (Parish et al., 2003a).

CONCLUSIONS

In this chapter we have surveyed the literature available on the 30 *M. tuberculosis* proteins that compose this organism's 11 paired 2CRs, 2 orphan sensors and 6 orphan response regulators. With the exception of the *dosRST* system, the only system that we can at present reliably and reproducibly induce, there is relatively little information available on these proteins. Certainly, we have some suggestion of signal and function for a number of 2CRs, but these are mostly circumstantial, based on the results of quite dramatic environmental changes surveyed by microarray analysis (see Table 4), or hinted at by the phenotypes of mutants (see Table 3). Indeed, the limitations of the latter are illustrated by the conflicting phenotypes sometimes observed; for example the *dosRS* system has been found to be hypervirulent in SCID mice (Δ*dosR*, Parish et al., 2003b), attenuated in guinea pigs (Δ*dosR*, Malhotra et al., 2004), and have no phenotype in C57BL/6J mice (TraSH for *dosR*, Sassetti et al., 2003; Sassetti and Rubin, 2003).

Should we assume that the knock-out of a sensor should exhibit the same phenotype as the knock-out of its cognate response regulator? This hypothesis, assuming the 2CR components form a monogamous and exclusive couple, is tempting. However, this is not always the case, as for example with *kdpD* (predicted required for optimal growth) and *kdpE* (no phenotype) and others in the comprehensive TraSH experiments (Sassetti et al., 2003; Sassetti and Rubin, 2003) (Table 3). Such findings could be related to the mechanism of action of the 2CR. For example, if the response regulator binds to a regulon-gene promoter in a non-phosphorylated state and prevents transcription, but activates transcription when phosphorylated, the response regulator deletion will lead to derepression of the regulon-gene, but not the sensor deletion (because unphosphorylated response regulator will still bind).

Furthermore, more complex system may be operating, with 2CR activity controlled both upstream (*i.e.* at the level of activation of the 2CR sensor, for example by multiple signals, or by variations in signal level), and downstream (*i.e.* at the level of the response-regulator driven activation of the 2CRs' regulon, for example by cross-talk between 2CRs, or with genes under the control of more than one 2CR). We must consider scenarios in which transcription of each gene may be modulated separately, have separate controls, and where unphosphorylated response regulators are capable of activating (or repressing) some or all of the members of their regulon, albeit less effectively than when activated. Certainly it has been observed that unphosphorylated RegX3 (Himpens et al., 2000), PrrA (Ewann et al., 2004), TrcR (Haydel et al., 2002) and DosR (Roberts et al., 2004) can bind to their promoter regions. Conversely, auto-regulation may simply not operate, as is the case for example with the HilA response regulator in *Salmonella* Typhimurium (Bajaj et al., 1996).

It is important to know not only the relative level of expression of the 2CR genes (such as those determined between two different conditions in microarray experiments), but also the absolute levels. It is only in this context that the significance of up-regulation or the lack of it can be assessed; genes for which we have found no changes in expression to-date may actually be fully on, and no condition yet identified in which they are off.

What do the isolated 2CR genes do? We cannot imagine that sensor/response regulator adjacency is a *sine qua non* condition for 2CR activity, as it seems futile for *M. tuberculosis* to maintain 8 orphan components (including two conserved in the ruthlessly reductive *M. leprae*; Table 2). Clearly cross-talk can occur between orphan components and paired components, as demonstrated by the *dosRS* and *dosT* system. However, this represents a special case, in which the DosT component is a paralogue of the DosS sensor. The question is whether cross-talk can be identified between non-paralogous components as has been suggested to occur in eukaryotic filamentous ascomycetes (Catlett et al., 2003), or even in higher plants (Grefen and Harter, 2004).

Defining the regulons of the 2CRs is a key factor in our understanding what these systems do. This has been done only for *dosR* (see above), and circumstantially for *kdpED* (Steyn et al., 2003). Even these are likely to be more complex, with individual genes transcribed from more than one promoter, and with alternative sigma factors (for example *acr*, *senX3* and *mtrA* are all partially *sigC*-dependent).

The microarray data are certainly informative, and their use has revolutionised the analysis of 2CRs by permitting the large-scale analysis of up- and down-regulated genes (the regulon). However, the data do not so much identify the biological signal detected, as define the importance and influence of the 2CRs, and suggest an environment in which they may operate. Stress is clearly a factor in the activation of some 2CRs. As shown in Table 4, each stress condition described upregulates at least two 2CRs: low-iron (*mprA*, *tcrY*), SDS (*phoP*, *mprAB*, *tcrX*) or H_2O_2 (*phoP, mprA, Rv0260c*). Macrophage ingestion has a dramatic effect on the environment of the bacilli, and it comes as no surprise to see a number of the 2CRs up-regulated upon ingestion. Interestingly, a number of 2CRs appear to be induced only in activated macrophages, and not in naïve macrophages, namely *tcrA, narS, kdpE, tcrR, dosRS, Rv2884* (Table 4 and Schnappinger et al., 2003). Of course, down-regulation of gene expression is also a very important mode of control, but it is observed less often, with the exception of within bacilli kept for a relatively long time in a fixed state such as the Wayne low-oxygen persistence model, or in stationary phase (Voskuil et al., 2004).

With only one 2CR-activating environmental factor clearly identified –and even then the exact signal for *dosRS* induction still unknown– it seems clear that we are far from understanding the relatively limited set of *M. tuberculosis* 2CRs. Their dramatic influence on virulence and *in vivo* and *in vitro* survival makes them not only key players in tuberculosis pathogenesis, but by extension possible drug targets or candidates for mutations to generate novel vaccine strains. We look forward to the day when the full complement of detected signals will be known.

ACKNOWLEDGEMENTS

We thank Nancy Buchmeier for information on *tcrA*. Yukari Manabe for clarification on IdeR-controlled genes. Stephen Gordon for clarification on gene nomenclature. Julian Parkhill for his detailed analysis of the *tcrA(B)* system. We are grateful to the Sanger Centre for access to the unpublished *M. marinum* sequences, and for making their invaluable Artemis and ACT genome viewers available to all, and to TIGR for access to the unpublished *M. smegmatis* and *M. avium* sequences. SLK is funded by BBSRC grant 48/P18545, SCGR and FM are funded by Wellcome Trust grants 062508 and 073237 respectively.

REFERENCES

Altendorf, K., Voelkner, P., and Puppe, W. (1994). The sensor kinase KdpD and the response regulator KdpE control expression of the *kdpFABC* operon in *Escherichia coli*. Res. Microbiol. *145*, 374-381.

Aravind, L., and Ponting, C.P. (1997). The GAF domain: an evolutionary link between diverse phototransducing proteins. Trends Biochem. Sci. *22*, 458-459.

Bacon, J., James, B.W., Wernisch, L., Williams, A., Morley, K.A., Hatch, G.J., Mangan, J.A., Hinds, J., Stoker, N.G., Butcher, P.D., and Marsh, P.D. (2004). The influence of reduced oxygen availability on pathogenicity and gene expression in *Mycobacterium tuberculosis*. Tuberculosis (Edinb) *84*, 205-217.

Bagchi, G., Mayuri, and Tyagi, J.S. (2003). Hypoxia-responsive expression of *Mycobacterium tuberculosis Rv3134c* and *devR* promoters in *Mycobacterium smegmatis*. Microbiology *149*, 2303-2305.

Bajaj, V., Lucas, R.L., Hwang, C., and Lee, C.A. (1996). Co-ordinate regulation of *Salmonella typhimurium* invasion genes by environmental and regulatory factors is mediated by control of *hilA* expression. Mol. Microbiol. *22*, 703-714.

Bentley, S.D., Chater, K.F., Cerdeno-Tarraga, A.M., Challis, G.L., Thomson, N.R., James, K.D., Harris, D.E., Quail, M.A., Kieser, H., Harper, D., et al. (2002). Complete genome sequence of the model actinomycete *Streptomyces coelicolor* A3(2). Nature *417*, 141-147.

Betts, J.C., Lukey, P.T., Robb, L.C., McAdam, R.A., and Duncan, K. (2002). Evaluation of a nutrient starvation model of *Mycobacterium tuberculosis* persistence by gene and protein expression profiling. Mol. Microbiol. *43*, 717-731.

Boon, C., and Dick, T. (2002). *Mycobacterium bovis* BCG response regulator essential for hypoxic dormancy. J. Bacteriol. *184*, 6760-6767.

Boon, C., Li, R., Qi, R., and Dick, T. (2001). Proteins of *Mycobacterium bovis* BCG induced in the Wayne dormancy model. J. Bacteriol. *183*, 2672-2676.

Buchmeier, N.A. (1998). Unique association of a *Mycobacterium tuberculosis* regulatory gene (*tcrA*) with all members of the tuberculosis complex. InterSci. Conf. Antimicrob. Agents Chemother. *38*, 59 (abstract no. B-53).

Bull, T.J., Sidi-Boumedine, K., McMinn, E.J., Stevenson, K., Pickup, R., and Hermon-Taylor, J. (2003). Mycobacterial interspersed repetitive units (MIRU) differentiate *Mycobacterium avium* subspecies *paratuberculosis* from other species of the *Mycobacterium avium* complex. Mol. Cell Probes *17*, 157-164.

Catlett, N.L., Yoder, O.C., and Turgeon, B.G. (2003). Whole-genome analysis of two-component signal transduction genes in fungal pathogens. Eukaryot. Cell *2*, 1151-1161.

Cole, S.T., Brosch, R., Parkhill, J., Garnier, T., Churcher, C., Harris, D., Gordon, S.V., Eiglmeier, K., Gas, S., Barry, C.E., 3rd, et al. (1998). Deciphering the biology of *Mycobacterium tuberculosis* from the complete genome sequence. Nature *393*, 537-544.

Dasgupta, N., Kapur, V., Singh, K.K., Das, T.K., Sachdeva, S., Jyothisri, K., and Tyagi, J.S. (2000). Characterization of a two-component system, *devR-devS*, of *Mycobacterium tuberculosis*. Tuber. Lung Dis. *80*, 141-159.

Epstein, W., and Davies, M. (1970). Potassium-dependant mutants of *Escherichia coli* K-12. J. Bacteriol. *101*, 836-843.

Ewann, F., Jackson, M., Pethe, K., Cooper, A., Mielcarek, N., Ensergueix, D., Gicquel, B., Locht, C., and Supply, P. (2002). Transient requirement of the PrrA-PrrB two-component system for early intracellular multiplication of *Mycobacterium tuberculosis*. Infect. Immun. *70*, 2256-2263.

Ewann, F., Locht, C., and Supply, P. (2004). Intracellular autoregulation of the *Mycobacterium tuberculosis* PrrA response regulator. Microbiology *150*, 241-246.

Florczyk, M.A., McCue, L.A., Purkayastha, A., Currenti, E., Wolin, M.J., and McDonough, K.A. (2003). A family of *acr*-coregulated *Mycobacterium tuberculosis* genes shares a common DNA motif and requires *Rv3133c* (*dosR* or *devR*) for expression. Infect. Immun. *71*, 5332-5343.

Florczyk, M.A., McCue, L.A., Stack, R.F., Hauer, C.R., and McDonough, K.A. (2001). Identification and characterization of mycobacterial proteins differentially expressed under standing and shaking culture conditions, including Rv2623 from a novel class of putative ATP-binding proteins. Infect. Immun. *69*, 5777-5785.

Galperin, M.Y., Gaidenko, T.A., Mulkidjanian, A.Y., Nakano, M., and Price, C.W. (2001). MHYT, a new integral membrane sensor domain. FEMS Microbiol. Lett. *205*, 17-23.

Garnier, T., Eiglmeier, K., Camus, J.C., Medina, N., Mansoor, H., Pryor, M., Duthoy, S., Grondin, S., Lacroix, C., Monsempe, C., et al. (2003). The complete genome sequence of *Mycobacterium bovis*. Proc. Natl. Acad. Sci. USA *100*, 7877-7882.

Gassel, M., Mollenkamp, T., Puppe, W., and Altendorf, K. (1999). The KdpF subunit is part of the K(+)-translocating Kdp complex of *Escherichia coli* and is responsible for stabilization of the complex *in vitro*. J. Biol. Chem. *274*, 37901-37907.

Graham, J.E., and Clark-Curtiss, J.E. (1999). Identification of *Mycobacterium tuberculosis* RNAs synthesized in response to phagocytosis by human macrophages by selective capture of transcribed sequences (SCOTS). Proc. Natl. Acad. Sci. USA *96*, 11554-11559.

Grefen, C., and Harter, K. (2004). Plant two-component systems: principles, functions, complexity and cross talk. Planta *219*, 733-742.

Groisman, E.A. (2001). The pleiotropic two-component regulatory system PhoP-PhoQ. J. Bacteriol. *183*, 1835-1842.

Gunsalus, R.P., and Park, S.J. (1994). Aerobic-anaerobic gene regulation in *Escherichia coli*: control by the ArcAB and Fnr regulons. Res. Microbiol. *145*, 437-450.

Haydel, S.E., Benjamin, W.H., Jr., Dunlap, N.E., and Clark-Curtiss, J.E. (2002). Expression, autoregulation, and DNA binding properties of the *Mycobacterium tuberculosis* TrcR response regulator

In vitro evidence of two-component system phosphorylation between the *Mycobacterium tuberculosis* TrcR/TrcS proteins. J. Bacteriol. *184*, 2192-2203.

Haydel, S.E., and Clark-Curtiss, J.E. (2004). Global expression analysis of two-component system regulator genes during *Mycobacterium tuberculosis* growth in human macrophages. FEMS Microbiol. Lett. *236*, 341-347.

Haydel, S.E., Dunlap, N.E., and Benjamin, W.H., Jr. (1999). In vitro evidence of two-component system phosphorylation between the *Mycobacterium tuberculosis* TrcR/TrcS proteins. Microb. Pathog. *26*, 195-206.

Himpens, S., Locht, C., and Supply, P. (2000). Molecular characterization of the mycobacterial SenX3-RegX3 two-component system: evidence for autoregulation. Microbiology *146 Pt 12*, 3091-3098.

Hoch, J.A., and Silhavy, T.J., eds. (1995). Two-Component Signal Transduction (Washington, D.C.: ASM Press).

Hou, J.Y., Graham, J.E., and Clark-Curtiss, J.E. (2002). *Mycobacterium avium* genes expressed during growth in human macrophages detected by selective capture of transcribed sequences (SCOTS). Infect. Immun. *70*, 3714-3726.

Hutchings, M.I., Hoskisson, P.A., Chandra, G., and Buttner, M.J. (2004). Sensing and responding to diverse extracellular signals? Analysis of the sensor kinases and response regulators of *Streptomyces coelicolor* A3(2). Microbiology *In press*.

Hutter, B., and Dick, T. (1999). Up-regulation of *narX*, encoding a putative 'fused nitrate reductase' in anaerobic dormant *Mycobacterium bovis* BCG. FEMS Microbiol. Lett. *178*, 63-69.

Kendall, S.L., Movahedzadeh, F., Rison, S.C., Wernisch, L., Parish, T., Duncan, K., Betts, J.C., and Stoker, N.G. (2004a). The *Mycobacterium tuberculosis dosRS* two-component system is induced by multiple stresses. Tuberculosis (Edinb) *84*, 247-255.

Kendall, S.L., Rison, S.C.G., Movahedzadeh, F., Frita, R., and Stoker, N.G. (2004b). What do microarrays really tell us about *M. tuberculosis*? Trends Microb. *12*, 537-544

Kinger, A.K., and Tyagi, J.S. (1993). Identification and cloning of genes differentially expressed in the virulent strain of *Mycobacterium tuberculosis*. Gene *131*, 113-117.

Letunic, I., Copley, R.R., Schmidt, S., Ciccarelli, F.D., Doerks, T., Schultz, J., Ponting, C.P., and Bork, P. (2004). SMART 4.0: towards genomic data integration. Nucl. Acids Res. *32 Database issue*, D142-144.

Magdalena, J., Supply, P., and Locht, C. (1998a). Specific differentiation between *Mycobacterium bovis* BCG and virulent strains of the *Mycobacterium tuberculosis* complex. J. Clin. Microbiol. *36*, 2471-2476.

Magdalena, J., Vachee, A., Supply, P., and Locht, C. (1998b). Identification of a new DNA region specific for members of *Mycobacterium tuberculosis* complex. J. Clin. Microbiol. *36*, 937-943.

Malhotra, V., Sharma, D., Ramanathan, V.D., Shakila, H., Saini, D.K., Chakravorty, S., Das, T.K., Li, Q., Silver, R.F., Narayanan, P.R., and Tyagi, J.S. (2004). Disruption of response regulator gene, devR, leads to attenuation in virulence of *Mycobacterium tuberculosis*. FEMS Microbiol. Lett. *231*, 237-245.

Manganelli, R., Voskuil, M.I., Schoolnik, G.K., and Smith, I. (2001). The *Mycobacterium tuberculosis* ECF sigma factor sigmaE: role in global gene expression and survival in macrophages. Mol. Microbiol. *41*, 423-437.

Mayuri, Bagchi, G., Das, T.K., and Tyagi, J.S. (2002). Molecular analysis of the dormancy response in *Mycobacterium smegmatis*: expression analysis of genes encoding the DevR-DevS two-component system, Rv3134c and chaperone alpha-crystallin homologues. FEMS Microbiol. Lett. *211*, 231-237.

Mei, J.M., Nourbakhsh, F., Ford, C.W., and Holden, D.W. (1997). Identification of *Staphylococcus aureus* virulence genes in a murine model of bacteraemia using signature-tagged mutagenesis. Mol. Microbiol. *26*, 399-407.

Mizuno, T., Kaneko, T., and Tabata, S. (1996). Compilation of all genes encoding bacterial two-component signal transducers in the genome of the cyanobacterium, *Synechocystis* sp. strain PCC 6803. DNA Res. *3*, 407-414.

Movahedzadeh, F., Rison, S.C., Wheeler, P.R., Kendall, S.L., Larson, T.J., and Stoker, N.G. (2004). The *Mycobacterium tuberculosis Rv1099c* gene encodes a GlpX-like class II fructose 1,6-bisphosphatase. Microbiology *150*, 3499-3505.

Muttucumaru, D.G., Roberts, G., Hinds, J., Stabler, R.A., and Parish, T. (2004). Gene expression profile of *Mycobacterium tuberculosis* in a non-replicating state. Tuberculosis (Edinb) *84*, 239-246.

O'Hara, B.P., Norman, R.A., Wan, P.T., Roe, S.M., Barrett, T.E., Drew, R.E., and Pearl, L.H. (1999). Crystal structure and induction mechanism of AmiC-AmiR: a ligand-regulated transcription antitermination complex. EMBO J. *18*, 5175-5186.

Ohno, H., Zhu, G., Mohan, V.P., Chu, D., Kohno, S., Jacobs, W.R., Jr., and Chan, J. (2003). The effects of reactive nitrogen intermediates on gene expression in *Mycobacterium tuberculosis*. Cell Microbiol. *5*, 637-648.

O'Toole, R., Smeulders, M.J., Blokpoel, M.C., Kay, E.J., Lougheed, K., and Williams, H.D. (2003). A two-component regulator of universal stress protein expression and adaptation to oxygen starvation in *Mycobacterium smegmatis*. J. Bacteriol. *185*, 1543-1554.

Pao, G.M., and Saier, M.H., Jr. (1995). Response regulators of bacterial signal transduction systems: selective domain shuffling during evolution. J. Mol. Evol. *40*, 136-154.

Parish, T., Smith, D.A., Kendall, S., Casali, N., Bancroft, G.J., and Stoker, N.G. (2003a). Deletion of two-component regulatory systems increases the virulence of *Mycobacterium tuberculosis*. Infect. Immun. *71*, 1134-1140.

Parish, T., Smith, D.A., Roberts, G., Betts, J., and Stoker, N.G. (2003b). The *senX3-regX3* two-component regulatory system of *Mycobacterium tuberculosis* is required for virulence. Microbiology *149*, 1423-1435.

Park, H.D., Guinn, K.M., Harrell, M.I., Liao, R., Voskuil, M.I., Tompa, M., Schoolnik, G.K., and Sherman, D.R. (2003). *Rv3133c/dosR* is a transcription factor that mediates the hypoxic response of *Mycobacterium tuberculosis*. Mol. Microbiol. *48*, 833-843.

Parkinson, J.S. (1993). Signal transduction schemes of bacteria. Cell *73*, 857-871.

Parkinson, J.S., and Kofoid, E.C. (1992). Communication modules in bacterial signaling proteins. Annu. Rev. Genet. *26*, 71-112.

Perez, E., Samper, S., Bordas, Y., Guilhot, C., Gicquel, B., and Martin, C. (2001). An essential role for *phoP* in *Mycobacterium tuberculosis* virulence. Mol. Microbiol. *41*, 179-187.

Purkayastha, A., McCue, L.A., and McDonough, K.A. (2002). Identification of a *Mycobacterium tuberculosis* putative classical nitroreductase gene whose expression is coregulated with that of the *acr* aene within macrophages, in standing versus shaking cultures, and under low oxygen conditions. Infect. Immun. *70*, 1518-1529.

Rickman, L., Saldanha, J.W., Hunt, D.M., Hoar, D.N., Colston, M.J., Millar, J.B., and Buxton, R.S. (2004). A two-component signal transduction system with a PAS domain-containing sensor is required for virulence of *Mycobacterium tuberculosis* in mice. Biochem. Biophys. Res. Commun. *314*, 259-267.

Roberts, D.M., Liao, R.P., Wisedchaisri, G., Hol, W.G., and Sherman, D.R. (2004). Two sensor kinases contribute to the hypoxic response of *Mycobacterium tuberculosis*. J. Biol. Chem. *279*, 23082-23087.

Rodriguez, G.M., Voskuil, M.I., Gold, B., Schoolnik, G.K., and Smith, I. (2002). *ideR*, An essential gene in mycobacterium tuberculosis: role of IdeR in iron-dependent gene expression, iron metabolism, and oxidative stress response. Infect. Immun. *70*, 3371-3381.

Saini, D.K., Malhotra, V., and Tyagi, J.S. (2004). Cross talk between DevS sensor kinase homologue, Rv2027c, and DevR response regulator of *Mycobacterium tuberculosis*. FEBS Lett. *565*, 75-80.

Sardesai, A.A., and Gowrishankar, J. (2001). trans-acting mutations in loci other than *kdpDE* that affect *kdp* operon regulation in *Escherichia coli*: effects of cytoplasmic thiol oxidation status and nucleoid protein H-NS on *kdp* expression. J. Bacteriol. *183*, 86-93.
Sassetti, C.M., Boyd, D.H., and Rubin, E.J. (2001). Comprehensive identification of conditionally essential genes in mycobacteria. Proc. Natl. Acad. Sci. USA *98*, 12712-12717.
Sassetti, C.M., Boyd, D.H., and Rubin, E.J. (2003). Genes required for mycobacterial growth defined by high density mutagenesis. Mol. Microbiol. *48*, 77-84.
Sassetti, C.M., and Rubin, E.J. (2003). Genetic requirements for mycobacterial survival during infection. Proc. Natl. Acad. Sci. USA *100*, 12989-12994.
Schnappinger, D., Ehrt, S., Voskuil, M.I., Liu, Y., Mangan, J.A., Monahan, I.M., Dolganov, G., Efron, B., Butcher, P.D., Nathan, C., and Schoolnik, G.K. (2003). Transcriptional adaptation of *Mycobacterium tuberculosis* within macrophages: Insights into the phagosomal environment. J. Exp. Med. *198*, 693-704.
Sherman, D.R., Voskuil, M., Schnappinger, D., Liao, R., Harrell, M.I., and Schoolnik, G.K. (2001). Regulation of the *Mycobacterium tuberculosis* hypoxic response gene encoding alpha -crystallin. Proc. Natl. Acad. Sci. USA *98*, 7534-7539.
Shu, C.J., and Zhulin, I.B. (2002). ANTAR: an RNA-binding domain in transcription antitermination regulatory proteins. Trends Biochem. Sci. *27*, 3-5.
Smirnova, A.V., and Ullrich, M.S. (2004). Topological and deletion analysis of CorS, a *Pseudomonas syringae* sensor kinase. Microbiology *150*, 2715-2726.
Stephens, R.S., Kalman, S., Lammel, C., Fan, J., Marathe, R., Aravind, L., Mitchell, W., Olinger, L., Tatusov, R.L., Zhao, Q., et al. (1998). Genome sequence of an obligate intracellular pathogen of humans: *Chlamydia trachomatis*. Science *282*, 754-759.
Stewart, V., and MacGregor, C.H. (1982). Nitrate reductase in *Escherichia coli* K-12: involvement of chlC, chlE, and chlG loci. J. Bacteriol. *151*, 788-799.
Steyn, A.J., Joseph, J., and Bloom, B.R. (2003). Interaction of the sensor module of *Mycobacterium tuberculosis* H37Rv KdpD with members of the Lpr family. Mol. Microbiol. *47*, 1075-1089.
Sun, R., Converse, P.J., Ko, C., Tyagi, S., Morrison, N.E., and Bishai, W.R. (2004). *Mycobacterium tuberculosis* ECF sigma factor *sigC* is required for lethality in mice and for the conditional expression of a defined gene set. Mol. Microbiol. *52*, 25-38.
Supply, P., Magdalena, J., Himpens, S., and Locht, C. (1997). Identification of novel intergenic repetitive units in a mycobacterial two-component system operon. Mol. Microbiol. *26*, 991-1003.
Taylor, B.L., and Zhulin, I.B. (1999). PAS domains: internal sensors of oxygen, redox potential, and light. Microbiol. Mol. Biol. Rev. *63*, 479-506.
Tyagi, J.S., and Sharma, D. (2004). Signal transduction systems of mycobacteria with special reference to *M. tuberculosis*. Curr. Sci. *86*, 93-102.

Via, L.E., Curcic, R., Mudd, M.H., Dhandayuthapani, S., Ulmer, R.J., and Deretic, V. (1996). Elements of signal transduction in *Mycobacterium tuberculosis*: *in vitro* phosphorylation and *in vivo* expression of the response regulator MtrA. J. Bacteriol. *178*, 3314-3321.

Voskuil, M.I., Schnappinger, D., Visconti, K.C., Harrell, M.I., Dolganov, G.M., Sherman, D.R., and Schoolnik, G.K. (2003). Inhibition of respiration by nitric oxide induces a *Mycobacterium tuberculosis* dormancy program. J. Exp. Med. *198*, 705-713.

Voskuil, M.I., Visconti, K.C., and Schoolnik, G.K. (2004). *Mycobacterium tuberculosis* gene expression during adaptation to stationary phase and low-oxygen dormancy. Tuberculosis (Edinb) *84*, 218-227.

Wang, W., Zhang, W., Chen, H., Chiao, J., Zhao, G., and Jiang, W. (2002). Molecular and biochemical characterization of a novel two-component signal transduction system, amrA- amkA, involved in rifamycin SV production in *Amycolatopsis mediterranei* U32. Arch. Microbiol. *178*, 376-386.

Wren, B.W., Colby, S.M., Cubberley, R.R., and Pallen, M.J. (1992). Degenerate PCR primers for the amplification of fragments from genes encoding response regulators from a range of pathogenic bacteria. FEMS Microbiol. Lett. *78*, 287-291.

Zahrt, T.C., and Deretic, V. (2000). An essential two-component signal transduction system in *Mycobacterium tuberculosis*. J. Bacteriol. *182*, 3832-3838.

Zahrt, T.C., and Deretic, V. (2001). *Mycobacterium tuberculosis* signal transduction system required for persistent infections. Proc. Natl. Acad. Sci. USA *98*, 12706-12711.

Zahrt, T.C., Wozniak, C., Jones, D., and Trevett, A. (2003). Functional analysis of the *Mycobacterium tuberculosis* MprAB two-component signal transduction system. Infect. Immun. *71*, 6962-6970.

Zhou, L., Lei, X.H., Bochner, B.R., and Wanner, B.L. (2003). Phenotype microarray analysis of *Escherichia coli* K-12 mutants with deletions of all two-component systems. J. Bacteriol. *185*, 4956-4972.

Zhulin, I.B., Taylor, B.L., and Dixon, R. (1997). PAS domain S-boxes in Archaea bacteria and sensors for oxygen and redox. Trends Biochem. Sci. *22*, 331-333.

Chapter 3

Protein Secretion and Export in *Mycobacterium tuberculosis*

Sherry Kurtz and Miriam Braunstein

ABSTRACT

The secreted and exported proteins of *Mycobacterium tuberculosis* are protein subsets exposed to the host cell environment, making them attractive candidates for virulence factors and immunogenic antigens. In this Chapter, we review the current understanding of the translocation systems responsible for localizing these extracytoplasmic proteins and the available methods for identifying them. Also discussed are examples of secreted and exported *M. tuberculosis* proteins shown to contribute to virulence and/or the immune response in the host.

INTRODUCTION

Secreted and exported proteins have important roles in the physiology and virulence of *Mycobacterium tuberculosis* and in the development of a host immune response during *M. tuberculosis* infection. These subsets of proteins have great potential for being targets of new drugs, vaccines, and diagnostics for tuberculosis and they are of special interest to the mycobacterial research community.

Secreted proteins, which we will define as proteins released from the tubercle bacillus, were the subject of some of the earliest *M. tuberculosis* studies. This is largely due to the fact that secreted proteins of *M. tuberculosis* are relatively easy to purify in a safe manner. Although the earliest preparations were crude, refinements to media and growth conditions have since enabled collection of culture filtrates from broth cultures of *M. tuberculosis* containing the proteins secreted into media after short growth periods and with limited contamination by proteins released via cell lysis.

Exported proteins, as we define them, are the proteins located beyond the cytoplasm (Braunstein and Belisle, 2000). These proteins are primarily found on the cell surface, but they are occasionally released into culture medium over time. Thus, the exported/surface proteins are either wholly or partially cell envelope associated and include proteins localized to the cytoplasmic membrane (integral membrane proteins), the cell wall core, and the outer lipid layer of the mycobacterial cell wall. In comparison to secreted proteins, exported proteins are not easy to enrich. However, improved differential centrifugation protocols for subcellular fractionation and new solubilization protocols for extraction of proteins from membranes and cell walls are now being used for detailed analyses of *M. tuberculosis* surface proteins (Gu et al., 2003; Rosenkrands et al., 2000b).

The distinction between secreted and exported proteins of *M. tuberculosis* can be difficult to make. Many proteins have been identified in both locations. This could reflect a number of possibilities: 1) the existence of multiple mechanisms to position a protein in two locations, 2) the observation of a cell envelope species en route to its final extracellular destination, or 3) the artificial release of some surface proteins as a result of *in vitro* culture conditions. A detailed understanding of the processes of protein localization in mycobacteria will help to distinguish between these possibilities. In the mean time, it is useful to keep in mind the similarities between secreted and exported proteins; they both cross the cytoplasmic membrane and end up exposed to the environment.

Our current understanding of how secreted and exported *M. tuberculosis* proteins are transported across the cytoplasmic membrane to the cell wall and beyond is limited. Mycobacteria possess a cell wall with many unique elements (Fig. 1) (reviewed in Brennan and Nikaido, 1995; Daffe and Draper, 1998), and it is likely that novel translocation systems exist to transport proteins into and across this unusual and complex cell wall. Beyond the cytoplasmic membrane is the cell wall core of mycobacteria which is composed of a network of crosslinked peptidoglycan and arabinogalactan to which long chain mycolic acids are covalently attached. Associated with the cell wall core is an outer layer of free lipids. Finally, under certain growth conditions, these outer lipids may be part of an exopolysaccharide capsule-like structure (Daffe and Etienne, 1999; Draper, 1998; Ortalo-Magne et al., 1995).

The secreted and exported proteins of *M. tuberculosis* represent a substantial percentage of the *M. tuberculosis* proteome; yet, the majority of these proteins remain uncharacterized. At least some of these proteins are important to mycobacterial physiology, as exemplified by the Antigen 85 mycolyltransferase complex which is involved in the synthesis of the unique mycobacterial cell wall (Belisle et al., 1997). Other proteins function in virulence. Studies of diverse bacterial pathogens demonstrate that virulence factors tend to be secreted or surface-associated proteins (Finlay and Falkow, 1997). As an intracellular pathogen, *M. tuberculosis* survives and grows within macrophages. The secreted and exported proteins are ideally positioned to promote this intracellular survival either by modifying the antimicrobial host environment or protecting the bacillus from host defenses and macrophage attacks. By constructing *M. tuberculosis* mutants lacking individual secreted and exported proteins and evaluating the virulence of such mutants in cultured macrophages and animal models, it is now possible to directly test the role of these proteins in *M. tuberculosis* pathogenesis. Such studies are underway and they are generating a growing list of secreted and surface proteins shown to function in the virulence of *M. tuberculosis*. Secreted and exported proteins also act as antigens presented to the immune system. Secreted proteins of *M. tuberculosis* can function in the development of protective immunity as shown by experiments in which mice and guinea pigs vaccinated with these proteins were protected against challenge with *M. tuberculosis* (Andersen, 1994; Hubbard et al., 1992; Pal and Horwitz, 1992; Roberts et al., 1995). Furthermore, some secreted and exported proteins are specifically recognized by serum from tuberculosis patients and could be exploited in the future as diagnostic antigens to report on infection (Garg et al., 2003; Samanich et al., 1998; Silva et al., 2003).

Our objectives in this Chapter are to discuss the current understanding of the translocation systems responsible for transporting *M. tuberculosis* secreted and exported proteins to their extracytoplasmic location, to provide an overview of the methods used to identify these subsets of proteins, and to review those proteins already shown to function in virulence and/or the immune response.

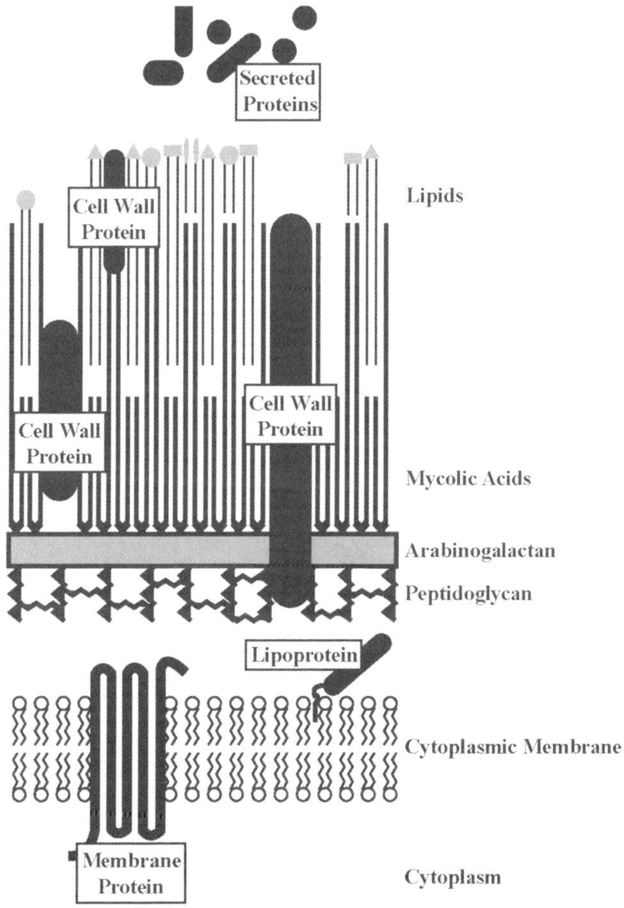

Fig. 1. Model of the mycobacterial cell envelope with secreted and exported proteins. Secreted proteins are defined as those proteins released from the cell. Exported proteins are defined as those proteins localized beyond the cytoplasm and include integral membrane proteins, lipoproteins, and proteins inserted throughout the cell wall.

PROTEIN SECRETION AND EXPORT SYSTEMS OF *M. TUBERCULOSIS*

Sec PATHWAY

The Sec pathway is the primary and essential pathway for transporting proteins across the cytoplasmic membrane of bacteria. The ultimate destination of proteins exported by this pathway is either the cell envelope or complete release (secretion) from the cell. Research conducted primarily in *Escherichia coli* has produced an advanced understanding of this pathway (Reviewed in de Keyzer et al., 2003; Pugsley, 1993).

Proteins transported by the Sec pathway are synthesized as precursors and are recognized by the presence of a tripartite Sec signal sequence at their amino terminus. On average a signal sequence is 18-30 amino acids in length. During translocation, a signal peptidase cleaves the signal sequence from the precursor to generate the mature protein. There are two types of Sec signal sequences: standard signal sequences and lipoprotein signal sequences. Both possess a charged amino-terminus, a hydrophobic core, and a polar region containing the cleavage site (Pugsley, 1993). They differ in the cleavage site sequence. The standard signal sequence cleavage site follows short chain amino acids (most often alanine or glycine) at positions -3 and -1 and is cleaved by the signal (leader) peptidase LepB. The lipoprotein signal sequence cleavage site contains a large hydrophobic amino acid (generally a leucine) at position -3 and a universally conserved cysteine at position +1 (the +1 residue is the first amino acid of the mature protein). Prior to cleavage by the lipoprotein signal peptidase (LspA), the cysteine residue must be modified to diacylglycerylcysteine by a prolipoprotein diacylglyceryl transferase (Lgt) (Wu, 1996). Following lipoprotein signal sequence cleavage the N-terminal cysteine may be further modified by additional fatty acids. The presence of a lipoprotein signal sequence has important implications for the subsequent localization of the mature protein since the lipid modification can tether the protein to the membrane. In Gram-negative bacteria lipoproteins can be localized to the inner or outer membrane based on the amino acid in the second position of the mature protein and the Lol sorting system (involving the LolA,B,C,D,E proteins) (reviewed in Tokuda and Matsuyama, 2004). In Gram-positive bacteria, which lack an outer membrane, all lipoproteins are believed to be anchored to the cytoplasmic membrane (Sutcliffe and Russell, 1995).

There are a number of Sec proteins involved in the recognition and translocation of Sec precursors across the cytoplasmic membrane. SecA is a multifunctional protein with many roles in the export pathway. SecA binds to precursor proteins, to chaperones such as SecB that deliver precursors, to acidic phospholipids in the membrane, and to components of the membrane embedded translocase channel (SecY and SecE) (Economou, 1998). Furthermore, SecA is an ATPase that provides energy for protein translocation. Through cycles of ATP binding and hydrolysis,

SecA delivers precursor proteins to the integral membrane components of the translocase and undergoes conformational changes that drive stepwise export of the precursor across the membrane (Economou and Wickner, 1994; Hartl et al., 1990; Pugsley, 1993). The precursor protein is in an unfolded state during export by the Sec pathway. The energy from ATP hydrolysis is required to initiate translocation and it alone is sufficient to carry out the process *in vitro*; however, the proton motive force (PMF) can help drive protein export (Schiebel et al., 1991). SecG and the complex of SecD, SecF, and YajC are proteins that increase the efficiency of Sec-dependent export (Duong and Wickner, 1997).

The Sec pathway also participates in the delivery and insertion of integral cytoplasmic membrane proteins. Hydrophobic transmembrane domains of nascent membrane proteins are recognized at an early stage in translation, when the protein is still associated with the ribosome, by the signal recognition particle (SRP). Bacterial SRP is comprised of a protein Ffh (fifty-four homologue) and a 4.5S RNA (encoded by the *ffs* gene). The complex of ribosome-nascent chain-SRP then interacts with the SRP receptor FtsY to deliver integral membrane proteins to the Sec translocase (de Gier and Luirink, 2003; Valent et al., 1998). During translocation, when a transmembrane domain of the nascent protein reaches the Sec translocon it halts export and transits into the lipid bilayer. The YidC protein functions in the stable integration of membrane proteins in the lipid bilayer (Samuelson et al., 2000). YidC can also function independent of the Sec translocase to promote integration of certain Sec-independent membrane proteins (Serek et al., 2004; van der Laan et al., 2004).

Sec PATHWAY IN *M. TUBERCULOSIS*

M. tuberculosis appears to possess a functional Sec pathway. The genome sequence includes homologues of all the Sec pathway components as well as SRP and YidC (Braunstein and Belisle, 2000; Cole et al., 1998). The only Sec factor absent is the SecB chaperone, but SecB is not found in all Sec systems (Cole et al., 1998; Fekkes and Driessen, 1999). Consistent with the absence of SecB, the mycobacterial SecAs (discussed below) lack the C-terminal SecB binding site (Fekkes et al., 1997). Presumably in mycobacteria other chaperones fulfill the function of SecB in delivering unfolded precursors to SecA and the other translocase components.

Additional support for the existence of a functional Sec pathway in mycobacteria comes from examples of experimentally proven secreted and surface proteins of *M. tuberculosis* that are synthesized with standard or lipoprotein Sec signal sequences (Tables 1 and 2). Upon analyzing a subset of known and predicted secreted *M. tuberculosis* proteins with standard Sec signal sequences, Wiker et al.(2000) observed that a high proportion of these proteins possess an aspartic acid (D) in the +1 position and a proline (P) in the +2 position of the mature protein (Wiker et al., 2000). The significance of this DP motif remains to be tested.

LIPOPROTEIN PROCESSING AND EXPORT IN *M. TUBERCULOSIS*

As predicted by homology, the lipoprotein signal peptidase (LspA) homologue in *M. tuberculosis* cleaves lipoprotein signal sequences. A *lspA* mutant of *M. tuberculosis* is defective in lipoprotein processing as shown by the accumulation of lipid modified but uncleaved lipoprotein precursors (Sander et al., 2004). Interestingly, this mutant is defective for growth in macrophages and in the mouse model of tuberculosis indicating that at least some lipoproteins contribute to virulence and that cleavage of the lipoprotein signal sequence is required for their proper function. The precise role of lipoprotein processing in virulence is not clear. It could be that lipoprotein signal sequence cleavage is important to lipoprotein stability, further lipid modification, proper localization of lipoproteins in the mycobacterial cell envelope, or that it is required for the activity of specific lipoprotein(s). Because lipoproteins are associated with an array of immunomodulatory activities (discussed later in the Chapter), additional analysis of the *lspA* mutant will help to understand the role of *M. tuberculosis* lipoproteins in host interactions.

To date, there has been no systematic study of lipoprotein location in the mycobacterial cell envelope. Processed lipoproteins could end up anchored to the cytoplasmic membrane or sorted to more external locations, such as the outer lipid layer. In support of the latter possibility are reports of lipoproteins observed in cell wall fractions and culture filtrates, although the possibility of contamination with cell membranes or cell lysis products in these preparations can not be ruled out. If lipoproteins are localized to the outer lipid layer of the mycobacterial cell wall the responsible sorting mechanism is not clear, since there are no obvious homologues of the Lol lipoprotein sorting system in the *M. tuberculosis* genome.

SecA PROTEINS IN *M. TUBERCULOSIS*

The SecA protein has been studied in numerous bacteria and repeatedly found to be encoded by a single copy essential gene (Schmidt and Kiser, 1999). It was a surprise to discover that the *M. tuberculosis* genome contains two *secA* homologues (*secA1* and *secA2*) (Cole et al., 1998). This property appears to be shared by all *Mycobacterium* species, including the nonpathogenic *M. smegmatis*. More recently, additional examples of bacteria with multiple *secA* genes have been identified. All are Gram-positive bacteria and many are pathogenic species (*Staphylococcus aureus, Streptococcus pneumoniae*, *Streptococcus gordonii*, *Streptococcus parasanguis, Listeria monocytogenes*, *Corynebacterium diphtheriae,* and *Bacillus anthracis*) (Bensing and Sullam, 2002; Chen et al., 2004; Lenz and Portnoy, 2002; Mazmanian et al., 2001; Sanger Institute, 2003; The Institute for Genomic Research, 2003).

Both SecA1 and SecA2 of *M. tuberculosis* share significant amino acid similarity with other SecA homologues and both contain the hallmark ATP-binding motifs (Braunstein et al., 2001). However, of the two proteins SecA1 is more similar to SecA in other bacteria. The available data indicates that SecA1 is the primary "housekeeping" SecA that functions in a manner similar to the well characterized SecA proteins of other bacteria. For example, *secA1* is essential in mycobacteria like *secA* in *E. coli*. This was directly shown in *M. smegmatis* by demonstrating that the chromosomal *secA1* gene cannot be deleted in a wild-type strain unless a plasmid expressing *secA1* is present (Braunstein et al., 2001). In *M. tuberculosis secA1* also appears to be a gene that cannot be disrupted as indicated by transposon site hybridization (TraSH) (Sassetti et al., 2003). Furthermore, it was shown that the N-terminal region of SecA1 can substitute for the corresponding region of *E. coli* SecA to complement an *E. coli secA51*(Ts) mutant (Owens et al., 2002) indicating a conservation of function.

SecA2 IN MYCOBACTERIA

Unlike the essential *secA1*, *secA2* of mycobacteria is non-essential as demonstrated by the successful construction of *secA2* deletion mutants in *M. smegmatis* and *M. tuberculosis* (Braunstein et al., 2001; Braunstein et al., 2003). Two different approaches demonstrate that SecA2 functions in protein export in mycobacteria. First, it was shown that the *secA2* mutant of *M. smegmatis* has a partial defect in exporting certain PhoA fusion proteins (Braunstein et al., 2001). Because the PhoA fusion proteins tested possess recognizable Sec signal sequences, this data suggests that SecA2 can assist SecA1 and the general Sec pathway to export some proteins with consensus Sec signal sequences. However, SecA2 cannot replace the function of SecA1 in the Sec pathway as shown by the inability to delete *secA1* even in a strain constitutively expressing SecA2 (Braunstein et al., 2001). Second, *M. tuberculosis* proteins secreted into the media by wild-type and *secA2* mutant strains were compared by two-dimensional polyacrylamide gel electrophoresis (2D-PAGE) and a small number of proteins whose release depends upon *secA2* were identified (Braunstein et al., 2003). These proteins lack Sec signal sequences which suggests a second role for SecA2 in nonconventional export of a specific subset of proteins. 2D-PAGE analysis also identified three proteins upregulated in the *M. tuberculosis secA2* mutant, the basis of which remains to be established. So far, this comparative analysis has been limited to culture filtrates (secreted proteins). Since the conserved Sec pathway exports proteins to the cell surface as well as serving as the first step in secretion, additional unidentified substrates that depend upon SecA2 for export may exist in cell envelope fractions.

One of the proteins identified as being released into media in a SecA2 dependent fashion is superoxide dismutase (SodA) (Braunstein et al., 2003). Multiple laboratories had previously identified SodA in the culture media of growing *M. tuberculosis*, with 20-92% of SodA being reported as extracellular (Abou-Zeid et al., 1988b; Andersen et al., 1991a; Harth and Horwitz, 1999a; Raynaud et al., 1998; Sonnenberg and Belisle, 1997; Tullius et al., 2001; Zhang et al., 1991). Because SodA lacks a recognizable Sec signal sequence it was long debated whether the release was attributable to a specific secretion pathway or autolysis. Tullius et al. (2001) concluded that SodA release by *M. tuberculosis* is not due to a specific export mechanism (Tullius et al., 2001). This conclusion was based on the demonstration that high level expression of stable proteins (for example, expression from multi-copy vectors) could lead to detection in culture medium in the absence of secretion. However, our results indicate that release of endogenous SodA, expressed from its chromosomal locus, is dependent upon the accessory secretion factor SecA2 (Braunstein et al., 2003). Thus, it appears that *M. tuberculosis* SodA is secreted by an unconventional secretion pathway involving SecA2.

The mechanistic basis of SecA2-dependent secretion remains to be defined. Proteins that work with SecA2 in export are undefined as is the mechanism for substrate recognition and targeting to SecA2. The virulence of the *M. tuberculosis secA2* mutant is decreased in the mouse model of infection (Braunstein et al., 2003) indicating that some SecA2-dependent proteins are exported virulence factors. The demonstration that SecA2 in *L. monocytogenes* contributes to virulence and that SecA2 exports an adhesin in *S. gordonii* and *S. parasanguis* (Bensing and Sullam, 2002; Chen et al., 2004; Lenz et al., 2003; Lenz and Portnoy, 2002) suggests that an accessory SecA is a newly appreciated strategy used by some Gram-positive bacterial pathogens and mycobacteria to localize specific subsets of virulence factors.

TWIN-ARGININE TRANSLOCATION (Tat) PATHWAY

The twin-arginine translocation (Tat) pathway is a recent discovery in bacteria, and our understanding of this pathway is relatively limited. Like the Sec pathway, the Tat system transports proteins across the cytoplasmic membrane and the final destination is either localization to the cell envelope or secretion from the cell (Berks et al., 2000; Hatzixanthis et al., 2003; Voulhoux et al., 2001).

Tat substrates are synthesized as precursors with an amino-terminal signal sequence that resembles a Sec signal sequence in tripartite structure and in possessing a cleavage site (Palmer and Berks, 2003). However, Tat signal sequences can be distinguished from Sec signal sequences by the presence of a 'twin arginine' motif (S/T)-R-R-X-#-# (# = hydrophobic residue) in the charged amino-terminus which directs the precursor to the Tat pathway (Palmer and Berks, 2003). Additional differences exist between Sec and Tat signal sequences although these other features are still being defined. Site-directed mutagenesis experiments indicate that the twin arginines play an important role in export as simultaneous replacement of both with conservative lysines eliminates export of numerous Tat substrates (Alami et al., 2002; DeLisa et al., 2002; Ize et al., 2002; Stanley et al., 2000; Yahr and Wickner, 2001). However, the requirement for both arginines may not be as indispensable to Tat export as originally thought. Replacement of one of the twin arginines is tolerated with some Tat signal sequences and there are two reports of native Tat substrates lacking a twin arginine pair (Buchanan et al., 2001; DeLisa et al., 2002; Hinsley et al., 2001; Ignatova et al., 2002; Ize et al., 2002; Stanley et al., 2000).

Another distinctive feature of the Tat pathway is that the precursor protein is exported in a fully folded state. Some substrates are even exported folded with bound cofactors and others are exported as assembled multimeric complexes (Rodrigue et al., 1999; Santini et al., 1998; Sargent et al., 1998; Weiner et al., 1998). Tat export also differs from the Sec pathway in being exclusively driven by the energy of the proton motive force (PMF).

The Tat pathway operates independently of the Sec pathway (Santini et al., 1998), and Tat-dependent export requires a minimum of two membrane proteins, TatA and TatC (Dilks et al., 2003; Yen et al., 2002). Some bacteria additionally possess TatB and TatE proteins, both of which share regions of homology and similar structural organization to TatA. The function of each Tat factor in translocation is not yet clear. Recently, TatC was shown to cross link to the Tat signal sequence at an early step in translocation leading to the proposal that TatC functions as a substrate recognition and specificity factor (Alami et al., 2003). TatA, which is present in excess compared to the other Tat factors and self associates to form large homooligomeric structures when overexpressed, (Porcelli et al., 2002; Sargent et al., 2001) is proposed to be the major component of the Tat translocase channel (Palmer and Berks, 2003).

Tat PATHWAY IN *M. TUBERCULOSIS*

Based on genomic analysis, the Tat pathway appears to be present in mycobacteria. The *M. tuberculosis* genome sequence reveals two genes in a predicted operon which encode homologues of TatA and TatC. A less convincing TatB homologue is located elsewhere in the genome (Cole et al., 1998; Dilks et al., 2003; Yen et al., 2002). Furthermore, the TATFIND program, a Tat substrate recognition program based on sequence information from published Tat signal sequences, identified potential Tat substrates in *M. tuberculosis* (31 in H37Rv and 29 in CDC1551) (Dilks et al., 2003). Candidate Tat substrates include phospholipase C which is localized to the cell wall and contributes to the virulence of *M. tuberculosis* (Raynaud et al., 2002). Thus, it is likely that the Tat pathway functions in localizing virulence factors in *M. tuberculosis*. A role for the Tat pathway in virulence is already demonstrated in *Pseudomonas aeruginosa,* enterohemorrhagic *E. coli,* and *Agrobacterium tumefaciens* (Ding and Christie, 2003; Ochsner et al., 2002; Pradel et al., 2003).

ESAT-6 SECRETION

In 1995, Andersen et al. identified a small highly immunogenic protein secreted by *M. tuberculosis* which was named the Early Secreted Antigenic Target 6kDa (ESAT-6) (Andersen et al., 1995; Sorensen et al., 1995). Since then, ESAT-6 has received much attention for its roles in virulence and immunity. Interestingly, the *esat6* gene (also called *Rv3875* and *esxA*) is located within RD1, a region of the *M. tuberculosis* and *M. bovis* chromosome (spanning genes *Rv3871-Rv3879c*) that is deleted in *M. bovis* BCG (Fig. 2A) (Behr et al., 1999; Mahairas et al., 1996). Thus, ESAT-6 is absent in attenuated BCG vaccine strains (Harboe et al., 1996).

Mycobacterial genomes encode for a family of ESAT-6-like proteins (22 members in *M. tuberculosis*) and genomic analysis recently defined a superfamily

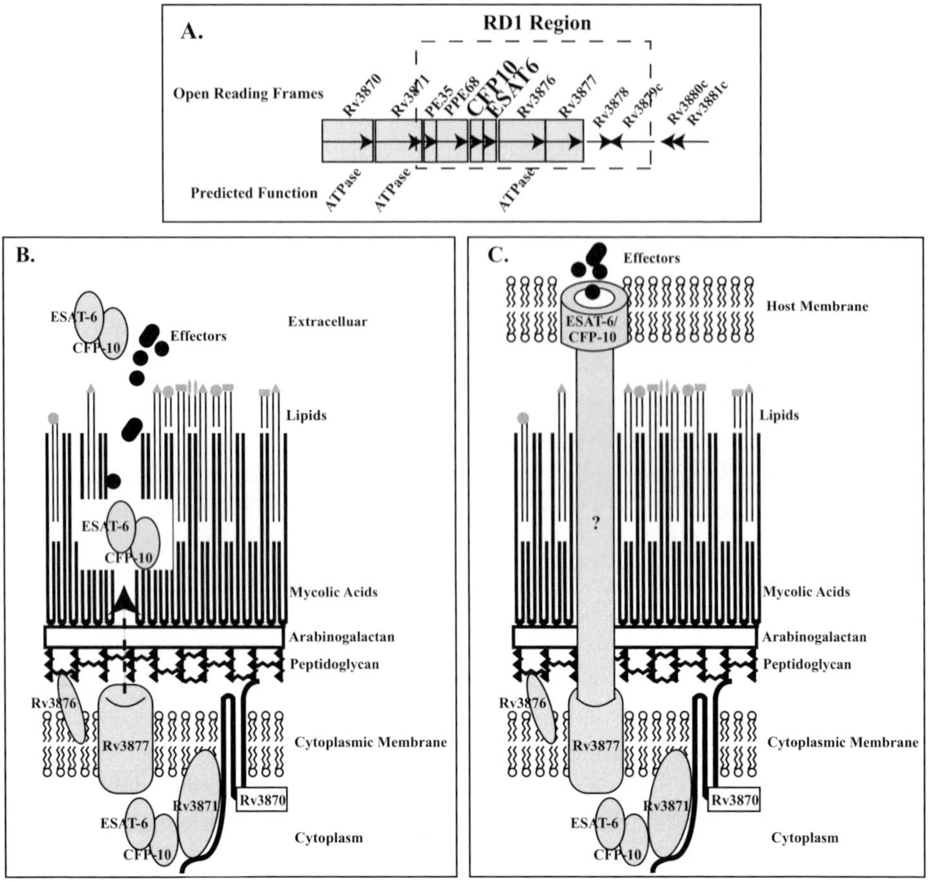

Fig. 2. ESAT-6 secretion system of *M. tuberculosis*. A) Schematic of the RD1 region of *M. tuberculosis* H37Rv. ORFs are depicted as black arrows showing the direction of transcription. ORF names are listed above the diagram. Shaded boxes indicate genes that are conserved members of the ESAT-6 cluster regions. B) Cartoon of the cell wall diffusion model of *M. tuberculosis* ESAT-6 secretion. In this model demonstrated secretion factors (Rv3870, Rv3871, Rv3876, Rv3877) function to translocate the protein across the cytoplasmic membrane and ESAT-6 and CFP-10 subsequently diffuse across the cell wall network. Hypothetical effector molecules transported by the ESAT-6 secretion system are shown. C) Cartoon of the cell wall channel model for *M. tuberculosis* ESAT-6 secretion. In this model ESAT-6 and CFP-10 traverse the cell wall in a hypothetical channel and localize to the tip of the channel where they act as regulators or pore forming molecules to promote secretion of effectors, possibly across host membranes. Hypothetical effector molecules transported by the ESAT-6 secretion system are shown. See text for additional details of these models.

of ESAT-6/WXG100 proteins in mycobacteria and some Gram-positive bacteria (Gey Van Pittius et al., 2001; Pallen, 2002; Tekaia et al., 1999). Although the overall sequence similarity among the ESAT-6/WXG100 proteins can be low, the superfamily members exhibit strong conservation around a WXG motif, small size (approximately 100 amino acids), and clustering with homologous genes. In *M. tuberculosis*, genes encoding ESAT-6 like proteins are generally found as pairs, as exemplified by the *esat6* gene in RD1 being in an operon with another family member *cfp10* (also referred to as *Rv3874*, *esxB* and *lhp*) (Berthet et al., 1998b).

ESAT-6 is released into culture filtrates by *M. tuberculosis* after short periods of growth and in the absence of obvious autolysis leading to its designation as a secreted protein; however, ESAT-6 also appears to be surface localized as demonstrated by its presence in cell wall fractions (Andersen et al., 1995; Andersen et al., 1991a; Pym et al., 2003; Rosenkrands et al., 2000b; Sorensen et al., 1995). ESAT-6 lacks a recognizable signal sequence; thus, the responsible translocation system was not immediately obvious. It is likely that most ESAT-6-like proteins are secreted since there are reports of other family members (CFP-10, Rv0287, and Rv0288) being secreted by *M. tuberculosis* (Berthet et al., 1998b, Skjot, 2000; Rosenkrands et al., 2000b). Again these other examples of secreted ESAT-6 family members lack recognizable signal sequences.

Early analysis of the *M. tuberculosis* genome sequence led to the proposal that the genes adjacent to *esat6* participate in the secretion process (Cole et al., 1998). Often genes encoding components of a specialized secretion system are clustered in the genome with the genes encoding the secreted substrates. The ORFs flanking *esat6* encode predicted membrane proteins, nucleotide binding proteins, and ATPases, all of which are compelling candidates for components of an ESAT-6 translocation channel and transducers of energy to drive the process (Gey Van Pittius et al., 2001; Tekaia et al., 1999). Experiments involving deletions and transposon insertions in individual RD1 genes in *M. tuberculosis* and "knock in" experiments where subsets of genes were added to BCG support this hypothesis. *M. tuberculosis* mutants with deletions or insertions within *Rv3870* (*snm1*), *Rv3871* (*snm2*), *Rv3876*, and *Rv3877* (*snm4*) are defective for ESAT-6 secretion (Guinn et al., 2004; Hsu et al., 2003; Stanley et al., 2003). "Knock-in" constructs added to BCG generated consistent results revealing that at least one gene upstream of *Rv3872* and at least one gene downstream of *esat6* (*Rv3875*) is required for the secretion process (Pym et al., 2003). Interestingly, these same RD1 genes are required for CFP-10 secretion as well (Guinn et al., 2004; Pym et al., 2003; Stanley et al., 2003). It is important to note that not all genes in RD1 appear to function in ESAT-6/CFP-10 secretion. The secretion of ESAT-6 was unaffected in deletion mutants lacking *Rv3872* and *Rv3873*, which encode for the PE35 and PPE68 protein members of the PE/PPE multi-gene family of *M. tuberculosis* (Hsu et al., 2003).

Additional genes are also likely to contribute to the ESAT-6 secretion system. Many ESAT-6 family members in mycobacteria are located in large clusters of conserved genes that extend beyond the borders of RD1 (referred to as ESAT-6 cluster regions 1 to 5), and these additional genes may function in the secretion system (Gey Van Pittius et al., 2001; Tekaia et al., 1999). In *M. marinum* transposon insertions in RD1 and the neighboring region identified mutants in the homologous

Rv3876, *Rv3878*, *Rv3879c*, and *Rv3881c* genes that are defective in ESAT-6 secretion (Gao et al., 2004). *Rv3878*, *Rv3879c*, and *Rv3881c* are not conserved members of the ESAT-6 gene clusters (Gey Van Pittius et al., 2001; Tekaia et al., 1999). These results await confirmation in *M. tuberculosis*. An unresolved difference between the *M. marinum* and *M. tuberculosis* studies is that in *M. marinum* mutants with defects in ESAT-6 secretion did not exhibit a corresponding CFP-10 secretion defect (Gao et al., 2004).

Although genes involved in this secretion pathway are now known, we are only starting to understand the mechanistic basis of the secretion machinery and the function of individual components. Rv3870 and Rv3871 possess homology to YukA, a predicted membrane bound ATPase of *Bacillus subtilis*. Homologues of *yukA* are often found near *esat6* family members in bacterial genomes (Pallen, 2002). Stanley et al. (2003) used a yeast 2-hybrid system to show that Rv3870 interacts with Rv3871 (Stanley et al., 2003). This is consistent with each protein representing a functional domain of YukA. Rv3876 is also a predicted ATPase. It is tempting to speculate that these putative ATPases provide energy to the secretion process. Rv3877 is a predicted polytopic membrane protein (10-12 transmembrane domains). In this regard, Rv3877 is reminiscent of the SecY protein which is part of the Sec translocase that acts as a channel for protein export across the cytoplasmic membrane.

ESAT-6 and CFP-10 physically interact according to yeast two-hybrid studies, far-Western blotting, and *in vitro* experiments in which they form a 1:1 complex (Okkels and Andersen, 2004; Renshaw et al., 2002; Stanley et al., 2003). Rv0287 and Rv0288, another tandem pair of secreted ESAT-6 family members in *M. tuberculosis*, also interact (Okkels and Andersen, 2004). Whether these complexes are important for secretion, stability and/or function remains to be resolved. An interesting possibility is that one of the proteins is acting as a chaperone to promote assembly of a translocation competent structure or targeting to the secretion machinery. Using the yeast two-hybrid system CFP-10 was shown to interact with the C-terminus of Rv3871 (Stanley et al., 2003). This result is significant in that it links a secreted substrate with the secretion machinery. It may also reflect a role for CFP-10 in delivering ESAT-6 to the secretion apparatus.

There are unanswered questions regarding the workings of the ESAT-6 secretion system. Is the pathway simply responsible for exporting the proteins across the cytoplasmic membrane followed by diffusion through the cell wall or does it transport the proteins across the cytoplasmic membrane and cell wall (Fig. 2B, 2C)? Are ESAT-6 and CFP-10 the sole proteins secreted by the system? What role does ESAT-6 play in distantly related bacteria?

An attractive hypothesis for a common role of ESAT-6 in different bacteria is that ESAT-6 is a conserved component of the secretion system, as opposed to being a secreted effector molecule with a conserved function. This idea is consistent with specialized secretion systems in other bacteria in which secreted components of the machinery and the components of the secretion apparatus tend t conserved than the substrates themselves (Plano et al., 2001). ESAT-6 a component of the secretion apparatus that is secreted during assembly or for which an extracellular location is required for function, perhaps as a regulator that triggers secretion in the appropriate environment. A function as a regulator is consistent with the recent report that the ESAT-6 secretion system in nonpathogenic *M. smegmatis* negatively regulates conjugal DNA transfer (Flint et al., 2004). Alternatively, ESAT-6-like proteins could function as a plug for the secretion system or as a pore forming molecule responsible for delivering the true secreted effector proteins across host membranes, such as the phagosome membrane in the case of intracellular *M. tuberculosis* (Fig. 2C). In these last two proposals ESAT-6 would be located at the tip of the ESAT-6 secretion apparatus and later released. In the *Yersinia* Type III secretion system the YopB, YopD and LcrV proteins are proposed to form a channel in the host cell membrane for delivery of Yop effector proteins from *Yersinia* into host cells (Buttner and Bonas, 2002). A similar role for ESAT-6-like proteins would be consistent with the report that purified ESAT-6 can disrupt artificial lipid bilayers and that RD1 genes, *esat6* and *cfp10* in particular, are required for cytolytic activity of *M. tuberculosis* and *M. marinum* (Gao et al., 2004; Hsu et al., 2003; Lewis et al., 2003).

METHODS FOR LARGE-SCALE IDENTIFICATION OF SECRETED AND EXPORTED PROTEINS OF *M. TUBERCULOSIS*

Methods to identify secreted and exported *M. tuberculosis* proteins have greatly improved over the past 10 years. Various proteomic methods can be applied to identify the proteins in subcellular fractions (cytoplasmic membrane, cell wall, and secreted) and new technologies are continually being developed. Genetic methods involving the use of gene fusions and reporters for protein export represent an alternative approach. Finally, *in silico* biology can be used to predict protein location by analyzing an amino acid sequence for the presence of signal sequences, transmembrane domains, and posttranslational modification sites associated with surface proteins. None of these methods is perfect, but, if used in concert with each other, they will help achieve a comprehensive understanding of all the secreted and exported proteins of *M. tuberculosis*.

PROTEOMICS

One standard method to study secreted and exported proteins of bacteria is to analyze and identify the protein constituents of the culture filtrate and cell envelope. Early on, this approach was applied to *M. tuberculosis* but the available technology for resolving proteins in a given fraction and for the identification of individual proteins was limited. The majority of early studies involved culture filtrates of *M. tuberculosis* and *M. bovis* BCG and they can be placed into one of two categories: 1) evaluating culture filtrate proteins for reactivity to a panel of *M. tuberculosis* specific antibodies as an early means of classification/identification and 2) identifying antigens among the culture filtrate proteins with assays that included lymphocyte and delayed-type hypersensitivity responses (Abou-Zeid et al., 1987; Andersen et al., 1991b; Closs et al., 1980; Collins et al., 1988; Nagai et al., 1991; Wiker et al., 1988; Yoneda and Fukui, 1965; Young et al., 1992).

In 1991 Nagai et al. applied a combination of 2D-PAGE, protein purification by chromatography, and identification by amino-terminal amino acid sequencing to *M. tuberculosis* culture filtrates (Nagai et al., 1991). Although only a small number of culture filtrate proteins were identified, some of which were previously known, this study is highly significant in being the first to generate a 2D map of the *M. tuberculosis* culture filtrate and to initiate characterization on a proteomic level. More comprehensive analyses of the *M. tuberculosis* culture filtrate proteome have since been undertaken (Jungblut et al., 1999; Mattow et al., 2003; Rosenkrands et al., 2000a; Rosenkrands et al., 2000b; Sonnenberg and Belisle, 1997; Weldingh et al., 1998). The level of resolution of the 2D-PAGE maps has steadily improved and the number of protein spots observed in these studies ranges from 100 – 1250. It is important to point out that multiple spots corresponding to the same protein, presumably a reflection of post-translational modification, processing, or sample preparation, were identified in nearly all of these studies making the actual number of distinct proteins in the culture filtrate lower than the observed number of spots. For example, of 381 proteins spots identified by mass spectrometry or amino-terminal sequencing only 137 different proteins were represented (Mattow et al., 2003). Nonetheless, the question arises of whether this surprisingly large number of proteins in *M. tuberculosis* culture filtrates represents true secreted proteins. In all likelihood not all the culture filtrate proteins identified by the sensitive proteomic methods are truly secreted. Some of the proteins detected may have been released from the cell envelope or from dead cells during growth of the culture. As alluded to earlier, the definition of true secreted proteins is currently a challenge and it represents an issue that will need to be addressed in the future. It is our hope that a better appreciation of the protein secretion systems of *M. tuberculosis* will help to resolve this issue. Two mycobacterial proteome databases that include 2D maps of culture filtrates are available on the Internet: 1) the proteome database for microbial research at the Max Planck Institute for Infection Biology (Mollenkopf et al., 1999) and 2) the *M. tuberculosis* proteome database at the Department of Infectious Disease Immunology Statens Serum Institute (Rosenkrands et al., 2000a; Rosenkrands et al., 2000b) (See Web Resources).

So far, the application of a proteomic approach to identify *M. tuberculosis* cell envelope proteins has been limited. This is due to intrinsic difficulties in protein extraction from membranes and cell walls, as well as solubility issues for these classes of proteins. However, application of new methodologies is aiding proteomic based experiments to identify surface proteins. Rosenkrands et al. (2000) generated a 2D map of cell wall proteins and they have begun to identify the proteins in this fraction (Rosenkrands et al., 2000b). To overcome the difficulty of resolving membrane proteins by 2D-PAGE, Gu et al. (2003) studied the membrane fraction of *M. tuberculosis* by means of 1D SDS-PAGE coupled to liquid chromatography-nanospray-tandem mass spectrometry (Gu et al., 2003). This latter study identified 450 different proteins which include integral membrane proteins with transmembrane domains and membrane-associated proteins. Only six of the 85 transmembrane proteins identified were previously reported emphasizing the need for such studies.

LIMITATIONS OF A PROTEOMIC APPROACH

When analyzing a large number of proteins by 2D-PAGE, overlapping protein spots and proteins with properties that prevent them from being visualized can pose difficulty. Furthermore, 2D-PAGE of membrane proteins has consistently proved to be a formidable obstacle. More advanced technology in protein sample and gel preparation along with improvements in analysis are gradually overcoming these problems. Difficulty in preparing pure subcellular fractions (*i.e.* culture filtrates devoid of contaminating intracellular or cell wall products and cell wall fractions free of membrane proteins) poses a further challenge.

However, in our opinion, a major limitation of the current proteomic approaches is the dependence on *in vitro* grown cultures as the source of culture filtrates and subcellular fractions. This is because an important and interesting class of proteins - those only expressed and/or exported in the context of the host during an infection - will be missed. The identification of *M. tuberculosis* proteins expressed while the bacilli reside in macrophages is a difficult proteomic problem of its own due to contaminating host proteins. The goal of identifying proteins secreted and exported by *M. tuberculosis* from within macrophages is even more challenging and one which will require the development of new technologies.

To our knowledge, there is only one report of a method developed and used to identify mycobacterial proteins secreted while the bacilli reside in macrophages. This method of Beatty and Russell (2000) involved labeling mycobacterial surface proteins of *M. bovis* BCG with a fluorescein-tagged succinimidyl ester and infecting macrophages with the labeled bacilli (Beatty and Russell, 2000). The infected macrophages were then fractionated and subcellular compartments other than mycobacterial phagosomes were analyzed. Seven fluorescein labeled proteins were detected in these non-mycobacteria containing compartments and two were identified: fibronectin attachment protein (FAP, also known as the 45/47-kDa antigen) and Antigen 85 (Beatty and Russell, 2000). Identification of a *M. tuberculosis* protein outside of the phagosome indicates that the protein was secreted and trafficked out of the phagosome. This method represents a creative way to begin approaching the problem of identifying *M. tuberculosis* proteins secreted in the host. However, the requirement of pre-labeling *in vitro* grown bacilli prior to infection means that this method will still miss secreted proteins exclusively expressed in the host.

GENETICS

With the advances in mycobacterial genetic methods it has become possible to study protein export in mycobacteria using gene fusions and reporters of protein location. These reporters are enzymes that are only active when exported beyond the cytoplasm. By removing the endogenous export signal of a reporter enzyme, the reporter will only be active if fused to a protein that provides the requisite signals for export. Such reporters provide a powerful way to identify exported proteins on a whole genome scale or to test a prediction that a given protein is exported. Genetic screens with the export reporters discussed below have identified close to 100 secreted and exported *M. tuberculosis* proteins.

ALKALINE PHOSPHATASE (PhoA) REPORTER

E. coli produces an alkaline phosphatase (PhoA) which is normally localized to the periplasm, and the enzyme is only active when it is exported beyond the cytoplasm. Manoil and Beckwith were the first to exploit this property and use PhoA as a reporter for protein export in *E. coli* (Manoil and Beckwith, 1985). They did so by creating a Tn5 transposon that contains a truncated *phoA* gene (*'phoA*) lacking the amino-terminal signal sequence required for protein export. If the transposon inserts in the correct reading frame within the coding sequence of an exported protein, a PhoA fusion protein will be synthesized and exported. Using a colorimetric substrate for alkaline phosphatase activity, export of the fusion protein can be assayed on agar plates.

The PhoA reporter system functions in *M. smegmatis* (Timm et al., 1994). By screening a *M. tuberculosis* genomic library cloned upstream of a truncated *'phoA* on a plasmid, *M. tuberculosis* proteins capable of promoting export of 'PhoA were identified (Lim et al., 1995). This early screen was conducted in *M. smegmatis* and succeeded in identifying four exported fusions. Notable among them were the 19kDa lipoprotein (LpqH) and the Exported repetitive protein (Erp), both of which are discussed later in the Chapter. Other laboratories have since performed similar genetic screens involving genomic libraries of *M. tuberculosis* and *M. avium* cloned upstream of *'phoA* (Carroll et al., 2000; Wiker et al., 2000). Individual *phoA* fusion constructs have also been tested for export using *E. coli* as a host (Gomez et al., 2000). Furthermore, an *in vitro 'phoA* transposition system was developed for mycobacteria (Braunstein et al., 2000). This system involves targeting a *M. tuberculosis* cosmid library *in vitro* with a Tn*552'phoA* transposon to create a transposon mutagenized library which is then screened in *M. smegmatis*. Using this method 32 different exported *M. tuberculosis* proteins were initially identified and the list has since been extended to 52 such proteins (Braunstein et al., 2000; Braunstein, M. Jalapathy, K., and Jacobs, W.R. unpublished). These studies establish the utility of a PhoA reporter for studying mycobacterial export. However, due to the presence of endogenous activities in *M. tuberculosis* that act on the color substrate of alkaline phosphatase, the PhoA reporter can not be directly employed in *M. tuberculosis* (Braunstein, M. unpublished). For this reason *M. smegmatis* or

E. coli must be used as the surrogate host, and any exported proteins only expressed and/or exported in *M. tuberculosis* will be overlooked.

β-LACTAMASE (Bla) REPORTER

The TEM-1 β-Lactamase (Bla) from *E. coli* catalyzes the hydrolysis of β-lactam antibiotics, which target cell wall synthesis. In order to protect the bacterium from attack by β-lactams, β-lactamases must be extracytoplasmic. When a truncated β-lactamase (lacking its endogenous export signal) is fused to a secreted/exported protein, β-lactamase activity can be used as a reporter of protein location. Enzymatically active fusions will enable growth in the presence of β-lactam antibiotics. A truncated Bla, lacking a signal sequence for export was first developed as a reporter for protein export in *E. coli* (Broome-Smith et al., 1990). The Bla reporter was used to screen a *M. tuberculosis* genomic library in *E. coli* (Chubb et al., 1998). A drawback of the Bla reporter is that it cannot be employed directly in mycobacteria due to endogenous β-lactamase activities (Kasik, 1979).

NUCLEASE (Nuc) REPORTER

Because the PhoA and Bla reporters fail to work directly in *M. tuberculosis*, other export reporters have been sought. Initially the *Staphylococcus aureus* nuclease (Nuc) reporter appeared to be an attractive solution. The nuclease activity of Nuc, which can be detected by halo production on DNAse agar, is dependent on export from the cytoplasm. A truncated Nuc reporter, lacking a signal sequence for export, was successfully used in Gram-positive bacteria (Poquet et al., 1998). Nuc was first developed as a reporter for mycobacteria by Downing et al. (1999) and was reported to work directly in *M. smegmatis* and *M. tuberculosis* (Downing et al., 1999). The Nuc reporter was used to screen a *M. tuberculosis* genomic library in *M. smegmatis* leading to the identification of three proteins that promote export of Nuc (Downing et al., 1999). However, in a subsequent study the truncated Nuc reporter was observed to be secreted by *M. smegmatis* even in the absence of an export signal (Recchi et al., 2002). Thus, the usefulness of the Nuc reporter is now unclear.

LIMITATIONS OF A GENETIC APPROACH

Genetic tools for studying exported proteins can be extremely powerful, but the reporter systems have limitations. The non-native reporter molecules used in the fusions could be problematic and prevent recognition or export by the appropriate translocation system. Thus, the inability to demonstrate export with an exogenous reporter system should not serve as a conclusive finding. Another problem is that the reporter systems, with the possible exception of Nuc, are limited to surrogate hosts for screening. Finally, as with the proteomic approaches, genetic methods suffer in their use of *in vitro* growth conditions. The genetic reporter systems are conducted in mycobacteria growing on agar plates. Again this means that proteins only expressed and/or exported in the context of the host during an infection will be missed.

BIOINFORMATICS

In silico biology is leading researchers into a new era of protein discovery. Computer programs, many of which are available on the Internet, allow predictions of protein location, post-translational modification, structure, and function (See Web Resources). These bioinformatics tools can help identify exported proteins. Programs such as SignalP, SPScan, PSORT, and TATFIND can identify signal sequences for export (Bendtsen et al., 2004; Dilks et al., 2003; GeneticsComputerGroup, 1997; Nakai and Kanehisa, 1991). TMPred will identify transmembrane domains and predict the membrane topology of a given protein (Hofmann and Stoffel, 1993). PrositeScan will predict sites of posttranslational modification including acylation sites for lipoprotein modification (Bucher and Bairoch, 1994).

Some prediction programs of protein location have been applied to *M. tuberculosis* in a genome-wide analysis. *LocateProtein* and *ExProt* predicted 23% (863 proteins) and 22% of *M. tuberculosis* proteins as being exported (located beyond the cytoplasm), respectively (Saleh et al., 2001; Schneider, 1999). This prediction is on the order of that proposed for *E. coli* and most other bacteria (Pugsley, 1993; Schneider, 1999).

In another study conducted by Gomez et al. (2000), the first 70 amino acids from each of the 3924 predicted protein coding sequences from *M. tuberculosis* was subjected to successive rounds of evaluation with a set of bioinformatics programs (SignalP, SPScan, TMpred, and PrositeScan) to generate a list of 52 proteins "most likely secreted" (Gomez et al., 2000). The exported nature of ten of these proteins was tested by fusing them to the PhoA reporter and assaying export in *E. coli*. Nine of the ten proteins were exported as indicated by active PhoA fusions. This demonstrates the useful nature of the prediction programs in directing subsequent research studies.

As discussed later in the Chapter, secreted and exported proteins are often immunoantigens that possess T cell epitopes. These peptides can be experimentally identified using *M. tuberculosis* specific T cells. However, researchers have also sought *in silico* methods to identify T cell epitopes. Programs such as Optimer and Epimer, TEPITOPE, ProPred Epimatrix, Patent-Blast, and BlastiMer (See Web resources) all predict T cell epitopes based on known class I and class II MHC binding motifs and/or libraries of peptides recognized by T cells (Davenport et al., 1995; De Groot et al., 2001; Meister et al., 1995; Schafer et al., 1998; Singh and Raghava, 2001; Sturniolo et al., 1999). Several groups have used these programs to identify new T cell epitopes within known *M. tuberculosis* antigens (Al-Attiyah and Mustafa, 2004; Meister et al., 1995; Panigada et al., 2002; Vordermeier et al., 2003). Using a subset of these tools, a whole genome evaluation was undertaken by DeGroot et al. (2001), who narrowed down the *M. tuberculosis* predicted proteome to 3000 potential T cell epitopes (De Groot et al., 2001).

LIMITATIONS OF A BIOINFORMATIC APPROACH

Prediction programs are developed from databases of experimental data derived from diverse bacteria. However, each protein is a unique entity and each organism brings its own challenges to the technology. This is particularly relevant to mycobacteria which possess a unique cell envelope structure that may demand novel protein export mechanisms. It is important to recognize that the prediction programs mentioned above, none of which were developed from studies of mycobacterial export, may not accurately predict all secreted and exported proteins of mycobacteria. This may explain some of the secreted and exported *M. tuberculosis* proteins identified by proteomic or genetic methods that are not identified by the available *in silico* prediction programs, which mostly rely on predictions of classical signal sequences or transmembrane domains. For now it is important that bioinformatic predictions of exported proteins continue to be validated by other means.

SECRETED AND EXPORTED *M. TUBERCULOSIS* PROTEINS INVOLVED IN VIRULENCE

Secreted and exported proteins of *M. tuberculosis* are ideally positioned to sense the environment, interact with the host, and/or protect the bacillus from attack during infection. Thus, these subsets of proteins are good candidates as virulence factors of *M. tuberculosis*. Table 1 provides a list of secreted and exported proteins reported to function in virulence as shown by experiments in which a *M. tuberculosis* mutant lacking the protein of interest was evaluated in macrophage, mice, or guinea pig infections. It should be mentioned that not all the mutants evaluated were subject to complementation analysis to definitively demonstrate that the absence of the secreted/exported protein in question is responsible for the observed attenuation in virulence. In an attempt to present as comprehensive a list as possible we have tried to include any virulence protein reported to be secreted or exported by proteomic, genetic, or bioinformatic methods. Thus, the secreted or exported nature for some of the proteins listed awaits independent confirmation. A subset of these virulence factors are highlighted below.

Erp

The Exported repetitive protein (Erp) was the first exported virulence determinant described for *M. tuberculosis*. Erp was identified as an exported protein in an initial '*phoA* screen conducted in mycobacteria (Lim et al., 1995). Later, Berthet et al. (1998) confirmed the exported nature of Erp by immunoblotting culture filtrates with anti-Erp antibodies (Berthet et al., 1998a). Additionally, immunoelectron microscopy showed Erp on the surface of *in vitro* grown bacilli as well as intracellular *M. tuberculosis* in macrophage phagosomes (Berthet et al., 1998a). Furthermore, Erp was observed both released into the phagosome lumen and located on the surface of small vesicles outside of the phagosome. These results suggest that Erp is secreted and traffics out of the phagosome during infection. *M. tuberculosis*

Table 1. Secreted and exported proteins of *M. tuberculosis* with a demonstrated role in virulence*.

Protein[a]	Export[b] Signal Prediction	Function	Reference(s) for Role in Virulence
Rv0072 (2)	TM	Hypothetical ABC transporter	(Hisert et al., 2004)
Rv0153c/PtbB/MPtbB (1)	None	Tyrosine phosphates	(Singh et al., 2003b)
Rv0169/Mce1A (1)	TM	Unknown	(Arruda et al., 1993; Flesselles et al., 1999; Shimono et al., 2003)
Rv0204c (2)	TM	Unknown	(Camacho et al., 1999)
Rv0405/Pks6 (2)	TM	Hypothetical polyketide synthase	(Camacho et al., 1999; Hisert et al., 2004)
Rv0410c/PknG (1)	TM	Protein kinase	(Cowley et al., 2004; Walburger et al., 2004)
Rv0432/SodC (1)	SSP	Superoxide dismutase	(Piddington et al., 2001)
Rv0450c/MmpL4 (1)	TM	Hypothetical transporter	(Camacho et al., 1999)
Rv0475/HbhA (1)	TM	Heparin binding	(Menozzi et al., 1998; Menozzi et al., 1996; Pethe et al., 2001)
Rv0507c/MmpL2 (2)	TM	Hypothetical transporter	(Camacho et al., 1999)
Rv0642c/MmaA4 (1)	None	Oxygenated mycolic acid biosynthesis	(Dubnau et al., 2000)
Rv0899/OmpA (1)	TM	Porin	(Raynaud et al., 2002)
Rv1395 (2)	TM	Hypothetical transcriptional activator	(Camacho et al., 1999)
Rv1411c/LprG (1)	LSP	Unknown	(Bigi et al., 2004)
Rv1811/MgtC (2)	TM	Magnesium transporter	(Buchmeier et al., 2000)
Rv1857/ModA	TM	Hypothetical molybdate binding protein	(Camacho et al., 1999)
Rv1908c/KatG (1)	TM	Catalase-peroxidase	(Ng et al., 2004)
Rv2031c/Acr/HspX/ 16kDa Ag (1)	None	Chaperone	(Yuan et al., 1998)
Rv2220/GlnA1 (1)	None	Glutamine synthetase	(Harth and Horwitz, 1999b; Tullius et al., 2003)
Rv2351c/PlcA/MpcA (1)	TM	Phospholipase C	(Raynaud et al., 2002)
Rv2350c/PlcB/MpcB (1)	SSP, TM	Phospholipase C	(Raynaud et al., 2002)
Rv2349c/PlcC (1)	TM	Phospholipase C	(Raynaud et al., 2002)
Rv2383c/MbtB (2)	TM	Mycobactin biosynthesis	(De Voss et al., 2000)
Rv2452c (2)	TM	Unknown	(Camacho et al., 1999)
Rv2930/FadD26 (1)	TM	PDIM biosynthesis	(Camacho et al., 1999; Rousseau et al., 2004)
Rv2938/DrrC (2)	TM	PDIM transport	(Camacho et al., 1999)
Rv2941/FadD28 (1)	TM	PDIM biosynthesis	(Camacho et al., 1999; Cox et al., 1999)
Rv2942/MmpL7 (2)	TM	PDIM transport	(Camacho et al., 1999; Cox et al., 1999)
Rv2958c (2)	TM	Hypothetical glycosyl transferase	(Hisert et al., 2004)
Rv3018c/PPE46 (2)	TM	Unknown	(Camacho et al., 1999)
Rv3487c/LipF (2)	TM	Hypothetical lipase	(Camacho et al., 1999)

Rv3804c/Ag8A/FbpA, MPT44 (1)	SSP, TM	Mycolyl transferase, fibronectin binding protein	(Armitige et al., 2000)
Rv3810/Erp/PirG (1)	SSP	Unknown	(Berthet et al., 1998a)
Rv3823c/MmpL8 (2)	TM	Sulfolipid biosynthesis/export	(Converse et al., 2003; Domenech et al., 2004)
Rv3846/SodA	TM	Superoxide dismutase	(Edwards et al., 2001)
Rv3870/Snm1 (2)	TM	ESAT-6 secretion	(Guinn et al., 2004; Sassetti et al., 2003; Stanley et al., 2003)
Rv3871/Snm2 (2)	TM	ESAT-6 secretion	(Guinn et al., 2004; Hsu et al., 2003; Sassetti et al., 2003; Stanley et al., 2003)
Rv3874/CFP10/EsxB/ Lph (1)	None	Unknown	(Guinn et al., 2004; Hsu et al., 2003; Sassetti et al., 2003; Stanley et al., 2003)
Rv3875/ESAT-6/EsxA (1)	None	Unknown	(Guinn et al., 2004; Hsu et al., 2003; Sassetti et al., 2003; Stanley et al., 2003)
Rv3876 (2)	TM	ESAT-6 secretion	(Guinn et al., 2004; Hsu et al., 2003; Sassetti et al., 2003)
Rv3877/Snm4 (2)	TM	ESAT-6 secretion	(Hsu et al., 2003; Sassetti et al., 2003; Stanley et al., 2003)
Sensor Kinase Members of Two-Component Regulatory Systems:[c]			
Rv0490/SenX3 (2)	SSP, TM	Sensor kinase	(Parish et al., 2003)
Rv0757/PhoR (1)	SSP, TM	Hypothetical sensor kinase	(Perez et al., 2001)
Rv0902c/PrrB (2)	TM	Hypothetical sensor kinase	(Ewann et al., 2002)
Rv0982/MprB (1)	TM	Hypothetical sensor kinase	(Zahrt and Deretic, 2001)
Rv1028c/KdpD (2)	TM	Hypothetical sensor kinase	(Parish et al., 2003)
Rv1032c/TrcS (2)	TM	Sensor kinase	(Parish et al., 2003)
Rv3132c/DevS (1)	TM	Sensor kinase	(Malhotra et al., 2004; Parish et al., 2003)
Rv3764c/TcrY (2)	TM	Hypothetical sensor kinase	(Parish et al., 2003)

*A role in virulence was determined by testing the corresponding *M. tuberculosis* mutant lacking the protein of interest in macrophage, mice, or guinea pig models of tuberculosis infection.

[a] The secreted or exported nature of a given protein was determined by either (1) Genetic reporter fusions and/or Proteomic methods involving protein localization in subcellular fractions (culture filtrates, cell wall, or cell membrane) or (2) Bioinformatic prediction only.

[b] Export signal prediction was made using PSORT and TMPred prediction. Standard signal peptide (SSP), Lipoprotein signal peptide (LSP), Transmembrane domain (TM).

[c] For some of the sensor kinases a role in virulence is implicated from studies of the cognate response regulator.

and *M. bovis* BCG *erp* mutants are defective for growth in macrophage and mouse models of infection (Berthet et al., 1998a; de Mendonca-Lima et al., 2003). There is no doubt that Erp participates in the pathogenesis of *M. tuberculosis*; however, Erp has no informative homology and its role in virulence remains unknown.

PLC ENZYMES

There is a precedent for bacterial pathogens to possess exported phospholipases that function in virulence, for example *P. aeruginosa* (Ostroff et al., 1990; Terada et al., 1999) and *L. monocytogenes* (Schwarzer et al., 1998; Vazquez-Boland et al., 1992). The first two *M. tuberculosis* phospholipase C (PLC) genes cloned produce enzymatic activities characteristic of PLCs (Johansen et al., 1996; Leao et al., 1995). There are a total of four phospholipase C homologues in the *M. tuberculosis* genome designated *plcA* (*mpcA*), *plcB* (*mpcB*), *plcC* and *plcD*.

Subcellular localization studies with His-tagged PlcA, PlcB and PlcC showed each protein to be associated with the cell wall, with no protein detected in the culture filtrate (Raynaud et al., 2002). This is consistent with early reports of PLC activity being cell associated and absent in the culture filtrate of *M. tuberculosis* (Johansen et al., 1996). Individual *M. tuberculosis* mutants with transposon insertions in each of the *plc* genes have been isolated and mutants with multiple *plc* genes simultaneously deleted (a triple *plcABC* deletion mutant and a quadruple *plcABCD* mutant) have been constructed (Parish and Stoker, 2000; Raynaud et al., 2002). Using these mutants, all four genes were shown to contribute to the total phospholipase C activity of the bacillus (Raynaud et al., 2002). Although all of the *plc* mutants grew like wild-type *M. tuberculosis* in THP-1 macrophages, a role for PLC in virulence was revealed by the demonstration that both the triple and quadruple *plc* mutants exhibit a defect in persistence in the lungs of mice infected by the aerosol route (Raynaud et al., 2002). It is interesting to note that in the early molecular analysis of *plcA* and *plcB* the putative signal sequences for export were described as having unusual amino acid composition (Johansen et al., 1996). These unusual signal sequences now make sense in the context of these phospholipases being likely substrates of the Twin arginine translocation (Tat) system.

ANTIGEN 85

Antigen 85 (Ag85), also known as fibronectin binding protein (Fbp), is one of the most abundant proteins in the culture filtrate of *M. tuberculosis*, comprising as much as 25% of the total protein in the culture filtrate (De Bruyn et al., 1989; Horwitz et al., 1995; Wiker et al., 1986). Antigen 85 has also been observed in the cell wall, although to lesser extent (Florio et al., 1997). Three closely related proteins comprise the Ag85 complex (also referred to as the 30-32-33 kDa antigen complex): Ag85A (FbpA), Ag85B (FbpB) and Ag85C (FbpC).

Multiple functions have been attributed to the Ag85 complex. Members of the Ag85 complex bind fibronectin (Abou-Zeid et al., 1991; Abou-Zeid et al., 1988a), which could play a role in tissue attachment and cellular adherence during infection, and they also possess mycolyltransferase activity. Mycolyltransferases are believed to be involved in the transfer of mycolyl residues onto the cell wall arabinogalactan (Jackson et al., 1999). *In vitro* assays demonstrate that these mycolyltransferases catalyze an integral step in the biosynthesis of α,α'- trehalose dimycolate (TDM), which are abundant molecules in the cell wall of *M. tuberculosis* (Belisle et al., 1997). A competitive inhibitor of trehalose, shown to inhibit Ag85 activity, is bacteriostatic in *M. aurum* (Belisle et al., 1997). Hence, Ag85 is an attractive target for anti-tuberculosis drug development.

Immunoelectron microscopy with anti-Ag85 antibodies demonstrated that Ag85 is associated with the cell surface of intracellular *M. tuberculosis* (Beatty and Russell, 2000; Harth et al., 1996). Additionally, Ag85 is found secreted into the phagosomal space and associated with cytoplasmic organelles distinct from the mycobacterial phagosome (Beatty and Russell, 2000; Harth et al., 1996). The observed increase in the amount of Ag85 observed in the phagosomal space of macrophages infected with live versus killed bacilli indicates that secretion of Ag85 is an active process occurring in the phagosome (Harth et al., 1996).

M. tuberculosis strains with single mutations in *fbpA*, *fpbB*, or *fbpC* have been constructed. Consistent with a role in cell wall biosynthesis the *fbpC* mutant has a 40% decrease in cell wall linked mycolates, and an increased permeability to small hydrophobic and hydrophilic molecules (Jackson et al., 1999). Interestingly, all three mutants did not behave similarly when tested in macrophages. Both *fbpB* and *fbpC* mutants multiply normally in macrophages (Armitige et al., 2000; Jackson et al., 1999). In contrast, the *fbpA* mutant exhibits a dramatic growth defect in both murine and human macrophage cell lines (Armitige et al., 2000). The *fbpA* mutant also exhibited a marked growth defect *in vitro* in minimal Sauton media, but not in fully supplemented 7H9 media. These results suggest that FbpA, at least, contributes to the ability of *M. tuberculosis* to grow in macrophages. It remains possible that the other Ag85 proteins play similar roles that will only be revealed when multiple *fbp* genes are simultaneously mutated.

ESAT-6/CFP-10

The immunodominant ESAT-6 and CFP-10 antigens found in the culture filtrate of *M. tuberculosis* also play a role in virulence. The location of the corresponding genes (*esat6* and *cfp10*) in the RD1 region led to the hypothesis that deletion of these genes and other RD1 genes in *M. bovis* BCG vaccine strains contributes to the attenuated virulence phenotype of BCG. Recent studies by numerous investigators have, in fact, proven that genes in RD1 contribute to virulence (Gao et al., 2004; Guinn et al., 2004; Hsu et al., 2003; Lewis et al., 2003; Pym et al., 2002; Sassetti and Rubin, 2003; Stanley et al., 2003; Wards et al., 2000). Two research groups set out to recreate the entire RD1 deletion in the H37Rv strain of *M. tuberculosis* (Hsu et al., 2003; Lewis et al., 2003). In comparison to wild-type H37Rv, the resulting RD1 deletion mutants exhibit reduced growth in macrophages, and mice infected with the RD1 mutant survive for a longer period of time and exhibit reduced lung pathology. Interestingly, although the RD1 mutant of H37Rv is greatly attenuated in comparison to H37Rv it remains more virulent than BCG indicating that the attenuated phenotype of BCG is not solely a result of the RD1 deletion (Sherman et al., 2004). Using the opposite approach, Pym et al. (2002) evaluated the significance of RD1 to virulence through "knock in" experiments involving the introduction of a wild type copy of the RD1 locus from *M. tuberculosis* into BCG. The resulting strain exhibits increased virulence in comparison to BCG in mice (Pym et al., 2002).

M. tuberculosis, M. bovis and *M. marinum* mutants with mutations in *esat6* and/or *cfp10* have also been constructed and exhibit reduced virulence. (Gao et al., 2004; Guinn et al., 2004; Hsu et al., 2003; Sassetti and Rubin, 2003; Stanley et al., 2003; Wards et al., 2000). The results with individual *esat6* and *cfp10* mutants more specifically identify a role for ESAT-6 and CFP-10 in *M. tuberculosis* virulence. However, the function of the corresponding proteins in pathogenesis is not clear. Mutations in *esat6* and *cfp10* in *M. tuberculosis* and *M. marinum* were reported to be associated with reduced lysis and cytotoxicity of host cells (Gao et al., 2004; Guinn et al., 2004; Hsu et al., 2003). Consistent with these results Hsu et al.(2003) demonstrated that purified ESAT-6 will disrupt an artificial lipid bilayer (Hsu et al., 2003). Thus, one proposal is that these proteins function in virulence through a membranolytic activity that promotes bacterial spreading, tissue invasion and dissemination. An alternative function is suggested by the observation that *esat6* mutants elicit greater levels of IL-12, TNFα, and nitric oxide (NO) by macrophages which could indicate a role for ESAT-6 in the inhibition of early macrophage responses by *M. tuberculosis* during infection (Stanley et al., 2003). Finally, as we suggested earlier, the function of ESAT-6 and CFP-10 could be as components of the specialized ESAT-6 secretion system with their role in pathogenesis being to promote the secretion of as yet unidentified effector molecule(s) with direct role(s) in virulence.

Individual mutations in other genes of the *M. tuberculosis* RD1 region have also been shown to be attenuated for virulence in cultured macrophages and in the murine model of infection (*Rv3870, Rv3871, Rv3876, Rv3877*) (Guinn et al., 2004; Hsu et al., 2003; Sassetti and Rubin, 2003; Stanley et al., 2003). These mutants are also defective in secretion of ESAT-6 and CFP-10, as discussed earlier in the chapter. The four corresponding RD1 proteins all possess predicted transmembrane domains making it likely that these ESAT-6 secretion factors are cell envelope localized.

PknG

Protein Kinase G (PknG) is a eukaryotic-like serine threonine kinase of *M. tuberculosis* with demonstrated kinase activity (Av-Gay et al., 1999; Koul et al., 2001; Walburger et al., 2004). This type of kinase was initially thought to be unique to eukaryotes, but several bacterial homologues have been found including examples in *Yersinia pseudotuberculosis* (YopO) (Hakansson et al., 1996) and *P. aeruginosa* (PSTK) (Wang et al., 1998), both of which participate in virulence. The *M. tuberculosis* genome sequence contains eleven potential serine threonine kinases (Av-Gay et al., 1999; Cole et al., 1998).

Two research groups constructed and evaluated *pknG* mutants (Cowley et al., 2004; Walburger et al., 2004). In one study a *pknG* mutant of *M. tuberculosis* exhibited an *in vitro* growth defect in liquid media (Cowley et al., 2004). When tested in immunodeficient and immunocompetent mice, the *pknG* mutant was less virulent on the basis of bacterial burden and increased survival time of infected animals (Cowley et al., 2004). However, differing degrees of attenuation was seen in different mouse strains. The observation that *pknG* appears to be in an operon with *glnH*, a glutamine binding protein, led the authors to test whether PknG functions in the regulation of glutamine levels (Cowley et al., 2004; Harth and Horwitz, 1999b; Harth et al., 2000). In fact, the *pknG* mutant was reported to exhibit decreased *de novo* glutamine synthesis in comparison to wild-type *M. tuberculosis* (Cowley et al., 2004).

A second group constructed and tested a *pknG* mutant of *M. bovis* BCG (Walburger et al., 2004). No *in vitro* growth defect for this mutant was reported. This latter study demonstrated a role for PknG in bacterial replication within macrophages. Further, unlike wild-type *M. tuberculosis* and *M. bovis* BCG, the *pknG* mutant was deficient in inhibiting phagosome-lysosome fusion. The *pknG* mutant was observed in LAMP positive phagosomes (indicative of phagosome-lysosome fusion) and in cell fractionation experiments the *pknG* mutant partitioned with the lysosomal fraction. The role of PknG in controlling lysosomal trafficking is attributed to the kinase activity of the protein, as a *M. smegmatis* strain expressing wild-type PknG inhibits lysosomal trafficking while a *M. smegmatis* strain expressing PknG with a mutation in the kinase domain does not (Walburger et al., 2004). This study provides an attractive explanation for PknG function in *M. tuberculosis* pathogenesis (i.e that PknG regulates phagosome maturation by way of host protein phosphorylation).

PknG possesses a predicted transmembrane domain but lacks an amino-terminal signal sequence. Localization experiments of BCG-infected macrophages showed that PknG is located both inside and outside of phagosomes in the host cell (Walburger et al., 2004). This was shown by immunoblotting fractions from infected cells with anti-PknG antibodies and immunoelectron microscopy experiments (Walburger et al., 2004). When killed bacilli were used PknG was only observed in the mycobacterial phagosome suggesting that PknG is actively secreted and trafficked out of the phagosome by intracellular *M. tuberculosis* (Walburger et al., 2004). The responsible PknG secretion system remains unknown.

An exciting implication of these results is that PknG may be useful in the development of new drugs for tuberculosis. Along these lines, Walburger et al. (2004) screened a compound library and found a specific inhibitor of PknG activity. When added to *M. tuberculosis* and *M. bovis* BCG infected macrophages this inhibitor increased phagosome-lysosome fusion and led to more efficient clearing of the intracellular bacilli (Walburger et al., 2004).

HbhA

The mycobacterial heparin-binding hemagluttanin (HbhA) was discovered during a search for *M. tuberculosis* and *M. bovis BCG* molecules involved in binding to sulfated glycoconjugates on the surface of mammalian epithelial cells (Menozzi et al., 1996). Such adherence molecules could be important in *M. tuberculosis* infection either in the initial seeding of the lung or in the process of dissemination to other organs. HbhA was initially purified from culture filtrates, but the protein is also observed in cell wall fractions and, more specifically, is present on the bacterial cell surface as shown by immunoelectron microscopy (Menozzi et al., 1996). There is one predicted transmembrane domain in the amino acid sequence of HbhA which helps explain the observed surface localization. However, HbhA lacks a recognizable amino-terminal signal sequence and it is not clear how it reaches the culture filtrate (Menozzi et al., 1998; Menozzi et al., 1996). HbhA is glycosylated, and this glycosylation is essential to HbhA-mediated hemagluttination (Menozzi et al., 1998).

HbhA is an epithelial cell adhesin, and a *M. tuberculosis hbhA* mutant exhibits decreased binding to A549 epithelial cells, while still retaining full infectivity for macrophage cell lines (Menozzi et al., 1996; Pethe et al., 2001). The *hbhA* mutant is further defective for extrapulmonary spread in an intranasal murine model of infection, as shown by wild-type levels of growth in lungs of infected animals but decreased colonization of spleens. These data suggest that HbhA is essential for attachment to epithelial cells and for a full disseminated infection (Pethe et al., 2001).

In a separate study, an independently constructed *hbhA* mutant of *M. tuberculosis* was evaluated (Mueller-Ortiz et al., 2002). Mice infected with this *hbhA* mutant by the aerosol route revealed reduced growth in lung, liver, and spleen. Although the two studies of *hbhA* mutants gave somewhat different results in the mouse model they agree on a role of HbhA in virulence.

MmpL7 AND DrrC

The unusual mycobacterial cell wall has long been thought to contribute to the ability of *M. tuberculosis* to cause disease (reviewed in Barry, 2001; Daffe and Etienne, 1999; Draper, 1998). Associated with the cell wall core is a diverse collection of highly unusual lipids which include phthiocerol dimycocerosate (PDIM). PDIM is unique to pathogenic mycobacteria (Brennan and Nikaido, 1995; Kolattukudy et al., 1997). Two independent signature tagged mutagenesis screens identified transposon insertion mutants of *M. tuberculosis* with defects in their ability to grow in lungs of mice and defects in the synthesis or export of PDIM to the cell wall (Camacho et al., 1999; Cox et al., 1999). Sequence analysis revealed the transposon insertion sites to be in a region of the chromosome with genes previously implicated in PDIM synthesis (Azad et al., 1997). One of the attenuated transposon insertions mapped to *mmpl7*, which encodes a putative integral membrane protein with 12 predicted transmembrane domains. Biochemical characterization demonstrated that the *mmpL7* mutant can properly synthesize PDIM but it remains cytosolic and is not exported to the cell wall, suggesting that MmpL7 is required for PDIM export to the cell wall (Camacho et al., 1999; Cox et al., 1999). Another transposon mutant mapped to *drrC*, which encodes for a predicted integral membrane protein with homology to ABC transporters. Like the *mmpl7* mutant, a *drrC* mutant is defective in PDIM export (Camacho et al., 2001). These results implicate an important role for cell wall localized PDIM and membrane localized PDIM exporters in *M. tuberculosis* pathogenesis. Using mutants defective in PDIM synthesis, PDIM was shown to participate in protecting *M. tuberculosis* from extracellular stresses, such as detergents and reactive nitrogen species, and to contribute to the low permeability of the mycobacterial cell envelope. These PDIM activities could protect the bacillus from intracellular stresses during infection and may help explain the role of cell wall PDIM in *M. tuberculosis* pathogenesis (Camacho et al., 2001; Rousseau et al., 2004). Furthermore, properly localized PDIM appears important to down regulation of the host immune response which may also explain its role in virulence (Rousseau et al., 2004).

M. tuberculosis is a complex pathogen that subverts the host in many different ways. During infection the bacillus must attach and enter the cell, mold the phagosome into a suitable environment and protect itself from attack of the host immune response. At this point, only a few of the molecules involved in these processes have been identified, but the importance of secreted and exported proteins to *M. tuberculosis* pathogenesis is clear. Additional study of *M. tuberculosis* secreted and exported proteins will not only uncover more examples of virulence factors but serve to expand our understanding of the mechanisms employed by *M. tuberculosis* to cause disease.

SECRETED AND EXPORTED *M. TUBERCULOSIS* PROTEINS THAT ELICIT AN IMMUNE RESPONSE

As compared to cytosolic proteins, secreted and exported proteins of *M. tuberculosis* are relatively accessible and they are among the earliest molecules presented to the immune system during infection. A growing list of these *M. tuberculosis* proteins are reported to elicit a host immune response (Table 2). These antigens are of interest to mycobacterial researchers for two important reasons. They will help us understand the immune response to *M. tuberculosis* and facilitate the development of improved vaccines and diagnostics for tuberculosis. As with Table 1, we attempted to present as comprehensive a list as possible. All of the proteins in Table 2 were reported as being secreted or cell-envelope associated by proteomic methods. Due to the limitations of proteomic methods some of these proteins may not be true secreted or exported proteins. Specifically, we did not evaluate the purity of the culture filtrates or cell envelope fractions used to localize each of the listed proteins.

Both innate and adaptive immune responses are elicited during *M. tuberculosis* infection. Key components of innate immunity are the mammalian pattern recognition receptors, such as Toll-like receptors (TLR), that specifically recognize microbial

Table 2. Secreted and exported proteins of *M. tuberculosis* that elicit an immune response*.			
Protein[a]	Export[b] Signal Prediction	Function	Reference(s) for Role in Immunity
Rv0009/ PpiA/ CFP22	None	Hypothetical Peptidyl-prolyl *cis-trans* isomerase	(Weldingh et al., 1998)
Rv0040c/MTC28	SSP	Unknown	(Manca et al., 1997a)
Rv0125/PepA/ MTB32A	SSP, TM	Hypothetical Serine protease	(Skeiky et al., 1999)
Rv0129c/Ag85C/ FbpC/ MPT45	SSP, TM	Mycolyl transferase, fibronectin binding	(Nagai et al., 1991; Rumschlag et al., 1988; Wiker et al., 1986; Worsaae et al., 1987)
Rv0164/MTSP17/TB18.5	None	Unknown	(Lim et al., 2004)
Rv0169/Mce1A	TM	Unknown	(Ahmad et al., 1999; Panigada et al., 2002)
Rv0288/EsxH/TB10.4/CFP7 (1)	None	Unknown	(Skjot et al., 2002; Skjot et al., 2000)
Rv0350/DnaK	None	Hypothetical chaperone	(Garsia et al., 1989; Pope et al., 1991)
Rv0358/PTRP	None	Unknown	(Singh et al., 2001)
Rv0440/Hsp65/65kDaAg/ GroEL2/Hsp60/CIE Ag82	TM	Hypothetical chaperone	(Havlir et al., 1991; Shinnick et al., 1987)
Rv0475/HbhA	TM	Heparin-binding hemagluttinin	(Temmerman et al., 2004)
Rv0577/CFP32/TB27.3	None	Unknown	(Covert et al., 2001; Huard et al., 2003)
Rv0798c/CFP29	None	Unknown	(Rosenkrands et al., 1998; Weldingh et al., 1998)
Rv0824c/DesA1	None	Hypothetical stearoyl-acp desaturase	(Jackson et al., 1997)
Rv0867c/RpfA	SSP	Resuscitation promoting factor	(Yeremeev et al., 2003)
Rv0934/PhoS1/ PstS1/ 38kDa/CIE Ag78/Antigen 5	LSP, TM	Phosphate transport	(Andersen and Hansen, 1989; Andersen et al., 1988; Covert et al., 2001; Verbon et al., 1992b; Young et al., 1991)
Rv0983/PepD/MTB32B	TM	Hypothetical serine protease	(Skeiky et al., 1999)
Rv1009/RpfB	LSP	Resuscitation promoting factor	(Yeremeev et al., 2003)
Rv1352	SSP	Unknown	(Covert et al., 2001)
Rv1411c/LprG	LSP	Unknown	(Gehring et al., 2004; Hovav et al., 2003)
Rv1759c/Wag22	TM	Fibronectin binding	(Espitia et al., 1999)
Rv1810	SSP, TM	Unknown	(Covert et al., 2001)
Rv1818c/PE_PGRS33	TM	Unknown	(Delogu and Brennan, 2001)
Rv1827/CFP17	None	Unknown	(Covert et al., 2001; Weldingh et al., 1998)
Rv1837c/GlcB/88kDa	TM	Hypothetical malate synthase	(Hendrickson et al., 2000; Samanich et al., 1998)
Rv1860/Apa/Mpt32/45/ 47kDa/FAP/ ModD	SSP, TM	Fibronectin binding	(Coates et al., 1989)

Rv1886c/Ag85B/FbpB/ HYT27/MPT59	SSP, TM	Mycolyl transferase, fibronectin binding	(Abou-Zeid et al., 1988a; Borremans et al., 1989; Godfrey et al., 1992; Huygen et al., 1990a; Huygen et al., 1990b; Nagai et al., 1991)
Rv1926c/MPT63	SSP	Unknown	(Manca et al., 1997b; Nagai et al., 1991)
Rv1932/Tpx/CFP20	TM	Hypothetical thiol peroxidase	(Covert et al., 2001; Weldingh et al., 1998)
Rv1979c	SSP, TM	Hypothetical permease	(Cockle et al., 2002)
Rv1980c/MPT64	SSP	Unknown	(Nagai et al., 1991)
Rv1983/PE_PGRS35	TM	Unknown	(Cockle et al., 2002)
Rv1984c/CFP21	SSP, TM	Unknown	(Weldingh et al., 1998)
Rv1986	SSP, TM	Hypothetical lysine transporter	(Cockle et al., 2002)
Rv2031c/Acr/ HspX/16kDa Ag	None	Chaperone	(Lee et al., 1992; Verbon et al., 1992a; Vordermeier et al., 1993)
Rv2220/GlnA1	None	Glutamine synthetase	(Samanich et al., 1998)
Rv2301/Cut2/CFP25	SSP, TM	Hypothetical cutinase	(Weldingh et al., 1998)
Rv2351c/PlcA/ MpcA	SSP,TM	Phospholipase C	(Falla et al., 1991; Leao et al., 1995; Parra et al., 1991)
Rv2376c/Cfp2/MTB12	SSP	Unknown	(Webb et al., 1998)
Rv2389c/RpfD	TM	Resuscitation promoting factor	(Yeremeev et al., 2003)
Rv2416c/Eis	TM	Unknown	(Dahl et al., 2001)
Rv2450c/RpfE	SSP	Resuscitation promoting factor	(Yeremeev et al., 2003)
Rv2608/PPE42	TM	Unknown	(Chakhaiyar et al., 2004)
Rv2654c	TM	Unknown	(Aagaard et al., 2004)
Rv2780/AldH/44kDa	None	Alanine dehydrogenase	(Ljungqvist et al., 1988; Worsaae et al., 1988)
Rv2873/MPT83	LSP	Unknown	(Hewinson et al., 1996)
Rv2875/MPT70	SSP, TM	Unknown	(Terasaka et al., 1989)
Rv2878c/MPT53/DsbE	SSP	Unknown	(Nagai et al., 1991)
Rv2945c/LppX/24kDa	LSP	Unknown	(Huygen et al., 1993; Lefevre et al., 2000)
Rv3044/FecB	LSP	Hypothetical Fe-III dicitrate binding protein	(Covert et al., 2001)
Rv3148c/GroES/MPT57/ 12kDa	None	Hypothetical chaperone	(Baird et al., 1988; Covert et al., 2001)
Rv3204/MTSP11	None	Unknown	(Lim et al., 2004)
Rv3367/PE_PGRS51	TM	Unknown	(Singh et al., 2001)
Rv3417c/ GroEL1/ cpn60-1	TM	Hypothetical chaperone	(Lewthwaite et al., 2001; Monahan et al., 2001)
Rv3763/19kDa/LpqH	LSP, TM	Unknown	(Ashbridge et al., 1989; Sonnenberg and Belisle, 1997)
Rv3803c/FbpD/FbpC1/ MPB51/ MPT51	SSP, TM	Hypothetical fibronectin binding protein	(Nagai et al., 1991; Wiker et al., 1990)

Rv3804c/Ag85A/FbpA/ MPT44	SSP, TM	Mycolyl transferase, fibronectin binding	(Abou-Zeid et al., 1988a; Borremans et al., 1989; Godfrey et al., 1992; Huygen et al., 1990a; Huygen et al., 1990b; Nagai et al., 1991)
Rv3841/BfrB	None	Bactoferritin	(Covert et al., 2001)
Rv3846/SodA/23kDa/ MPT58/CIE Ag62	TM	Superoxide dismutase	(Zhang et al., 1991)
Rv3873/PPE68	TM	Unknown	(Demangel et al., 2004)
Rv3874/CFP10/EsxB/Lhp/ MTSA-10	None	Unknown	(Berthet et al., 1998b; Demangel et al., 2004; Ulrichs et al., 1998)
Rv3875/ESAT-6/EsxA	None	Unknown	(Andersen et al., 1995; Sorensen et al., 1995)
Rv3881c/MTB48	TM	Unknown	(Lodes et al., 2001)
Rv3914/MPT46/TrxC	None	Thioredoxin	(Nagai et al., 1991)

*An immune response was determined by either T cell or antibody reactivity to the protein.
[a]The secreted or exported nature of proteins in this Table was determined by reports of identification in culture filtrates, cell wall, or cell membrane fractions. The purity of the fractions used in the individual studies was not taken into account in compiling this list.
[b]Export signal prediction was made using PSORT and TMPred prediction. Standard signal sequence peptide (SSP), Lipoprotein signal sequence peptide (LSP), Transmembrane domain (TM).

ligands. Upon binding, TLRs initiate a signaling cascade that leads to expression of cytokines and regulatory molecules, such as TNFα and IL-12. This can in turn activate macrophage antimicrobial effector mechanisms, such as inducible nitric oxide synthase (iNOS), and modulate the developing adaptive immune response by influencing dendritic cell (DC) maturation and T-cell activation (Quesniaux et al., 2004). Furthermore, TLR signaling appears to play a role in the regulation of phagosome maturation (Blander and Medzhitov, 2004). *M. tuberculosis* is able to activate cells through interactions with TLR2 and TLR4, and progress has been made in identifying *M. tuberculosis* surface molecules that are TLR ligands (reviewed in Heldwein and Fenton, 2002; Quesniaux et al., 2004; Stenger and Modlin, 2002). TLR pathways play an important role in the host response to several pathogens, and studies of the specific role(s) of this system in host protection against *M. tuberculosis* are underway (Abel et al., 2002; Branger et al., 2004; Drennan et al., 2004; Fremond et al., 2003; Heldwein et al., 2003; Kamath et al., 2003; Reiling et al., 2002; Shim et al., 2003; Sugawara et al., 2003).

The protective immune response to *M. tuberculosis* is an adaptive cell-mediated Th1 response involving both CD4+ and CD8+ T cells (Caruso et al., 1999; Flynn et al., 1992; Mogues et al., 2001; Sousa et al., 2000; Tascon et al., 1998). There are many examples of exported and secreted *M. tuberculosis* antigens that are presented by class I and/or class II MHC molecules and recognized by CD8+ and/or CD4+ T cells. Such proteins represent good candidates for being protective antigens Furthermore, experiments in which mice and guinea pigs were vaccinated with *M. tuberculosis* secreted proteins demonstrated protection against challenge with virulent *M. tuberculosis* (Andersen, 1994; Hubbard et al., 1992; Pal and Horwitz, 1992; Roberts et al., 1995). These experiments plus others focusing on individual *M. tuberculosis* secreted proteins demonstrate that at least some of these proteins are capable of eliciting an immune response. Adding to their appeal as targets for new vaccines is the early presentation of secreted and exported antigens to the immune system that occurs during infection.

In addition to a T cell response, *M. tuberculosis* infection elicits a humoral response. Although the antibody response to *M. tuberculosis* is not likely to contribute to the control of infection, antigens recognized by host antibodies could be the basis of serodiagnostics for tuberculosis (Garg et al., 2003; Gennaro, 2000; Johnson et al., 1997).

To tackle the problem of identifying *M. tuberculosis* antigens, various immunologic assays have been applied to individual purified proteins or attempted on a proteomic level. Because of the higher probability that secreted and surface proteins are presented to the immune system, these protein fractions are often used as the starting material for these studies. Early studies relied upon immune sera from tuberculosis patients and infected animals or antibodies raised against *M. tuberculosis* protein subsets to identify antigens in culture filtrates, whole cell extracts, or recombinant expression libraries (reviewed in Young et al., 1992). More recently, immunoblotting *M. tuberculosis* culture filtrate proteins separated by 2D-PAGE with tuberculosis patient serum expanded the number of proteins identified by this approach (Samanich et al., 1998). *M. tuberculosis* antigens with the ability to elicit reactivity from *M. tuberculosis* specific T cells have also been identified from culture filtrates (reviewed in Belisle et al., 2004). Bioinformatic T cell epitope prediction programs, mentioned earlier, can facilitate these efforts. In addition to antibody and T cell antigens, researchers have also started to identify immunomodulatory antigens of *M. tuberculosis*. We highlight below some of the better characterized secreted and exported antigens.

19KDA LIPOPROTEIN (LpqH)

The 19kDa lipoprotein (LpqH) of *M. tuberculosis* is a glycosylated surface protein which is occasionally observed in culture filtrates (Rosenkrands et al., 2000b; Sonnenberg and Belisle, 1997). The immunogenic properties of this protein have been extensively studied. The 19kDa lipoprotein elicits both humoral and cell mediated immune responses in infected mice and humans, with both CD4+ and CD8+ T cell epitopes having been mapped in the protein (Faith et al., 1991; Harris et al., 1993; Harris et al., 1991; Jackett et al., 1988; Mohagheghpour et al., 1998; Vordermeier et al., 1997).

The 19kDa lipoprotein is also an immunomodulatory antigen, and the relationship between the 19kDa lipoprotein and the host cell is proving to be a complex one. The protein was identified from a cell wall fraction of *M. tuberculosis* as an inducer of IL-12 production by human macrophages and dendritic cells and as an inducer of iNOS and NO by the RAW macrophage cell line. This cell activation by 19kDa is mediated by TLR2 (Brightbill et al., 1999; Fortune et al., 2004; Thoma-Uszynski et al., 2000). A role for the 19kDa lipoprotein in promoting apoptosis in a TLR2 dependent manner has also been suggested by some, but not all, studies (Ciaramella et al., 2004; Gehring et al., 2003; Lopez et al., 2003). Another intriguing property of the 19kDa lipoprotein is that, like intact *M. tuberculosis*, it can inhibit macrophage expression of class II MHC and antigen processing. This effect is observed with both untreated and IFN-γ treated macrophages and it is also TLR2 dependent (Fortune et al., 2004; Fulton et al., 2004; Gehring et al., 2003; Noss et al., 2001; Pai et al., 2003). In addition, the 19kDa lipoprotein was reported to inhibit antigen processing via the alternate (TAP-independent) class I MHC pathway and inhibit phagosome maturation in a TLR2 dependent manner (Tobian et al., 2003). Finally, the 19kDa lipoprotein also blocks INFγ induced upregulation of FcγRI receptors by macrophages in a TLR2 dependent fashion (Gehring et al., 2003). The apparently opposing influences on the immune response (macrophage and dendritic cell activation versus suppression of antigen presentation) indicate a complex relationship between the 19kDa lipoprotein, TLR2, and modulation of the host response. It suggests that *M. tuberculosis* exploits the same 19kDa-TLR2 interaction to initially induce a pro-inflammatory response and to later down-regulate the host response.

The activities described above for the 19kDa lipoprotein were all demonstrated with purified protein applied to macrophages or dendritic cells. This is an important detail in light of conflicting data obtained when the 19kDa lipoprotein is expressed and exported in the context of intact mycobacteria. While purified 19kDa lipoprotein induces IL-12 production in non-activated macrophages, wild-type *M. tuberculosis* expressing endogenous 19kDa lipoprotein induces relatively little IL-12 and suppresses IL-12 upregulation (Brightbill et al., 1999; Nau et al., 2002). Furthermore, when the19kDa lipoprotein from *M. tuberculosis* is expressed in *M. smegmatis*, which does not have a 19kDa lipoprotein homologue, it inhibits induction of IL-12 (Post et al., 2001). Thus, it seems that the way in which the protein is presented to the host cell, either as a protein in solution or exported and anchored to the bacterial cell surface, affects the host cell response.

The influence of the 19kDa lipoprotein on establishment of protective immunity has also been tested. Expression of the *M. tuberculosis* 19kDa lipoprotein in *M. vaccae* and *M. smegmatis* has a detrimental effect on the efficacy of these strains as vaccines (Yeremeev et al., 2000). This is consistent with an immunosuppressive role of the protein when expressed in mycobacteria. However, deletion of the gene encoding the 19kDa lipoprotein in BCG did not create a BCG strain with improved protection capability (Yeremeev et al., 2000).

LprG

LprG (also called the 27kDa lipoprotein or P27) was initially identified as an *M. bovis* antigen recognized by immune sera from infected cattle (Bigi et al., 1997).

LprG is found in the membrane fraction of strains from the *M. tuberculosis* complex (Bigi et al., 1997). Studies of LprG indicate that it is another example of a lipoprotein with apparently opposing interactions (suppressive and stimulatory) with the immune system. In a manner similar to that undertaken with the 19kDa lipoprotein, LprG was identified in a cell wall fraction of *M. tuberculosis* as a protein that inhibits class II MHC antigen processing in a TLR-2 dependent manner (Gehring et al., 2004). However, purified LprG also stimulates TNFα production by macrophages through TLR-2, and mice immunized with purified LprG elicit a Th1 immune response to the antigen (Gehring et al., 2004; Hovav et al., 2003). Interestingly, LprG vaccination (by way of protein or DNA vaccine) is not protective and actually seems to decrease the level of protection achieved by BCG vaccination in mice (Hovav et al., 2003).

Ag85

The Antigen 85 complex (Ag85A, Ag85B, Ag85C) plays a role in the developing immune response during *M. tuberculosis* infection. As described, this antigen family represents a substantial portion of the total protein in the *M. tuberculosis* culture filtrate and the proteins are exported in the host environment, making them ideal candidates to interact with the immune system. Members of the Antigen 85 complex induce an antigen specific T cell response with reactive T cells being identified in tuberculosis patients, with both CD4+ and CD8+ T cell epitopes identified in the proteins (Geluk et al., 1998; Geluk et al., 2000; Klein and Fox, 2001; Klein et al., 2001; Lee and Horwitz, 1999; Valle et al., 2001). A humoral response is also generated to Ag85, as shown by the presence of Ag85-specific antibodies in tuberculosis and leprosy patients (Bentley-Hibbert et al., 1999; Espitia et al., 1992; Naito et al., 1998; Samanich et al., 1998). These antibodies may one day be exploited as a diagnostic tool for *M. tuberculosis* infection (Garg et al., 2003; Samanich et al., 1998).

The potent antigenic properties of the Antigen 85 complex make it a particularly attractive candidate for vaccine development, and it is one of the most thoroughly studied vaccine candidates (reviewed in Mustafa, 2002; Olsen and Andersen, 2003). Several groups have tested purified Ag85 protein alone or in concert with other exported proteins for the ability to elicit protection against *M. tuberculosis* challenge. These studies showed that these preparations give some protection, but it is no more efficacious than BCG nor is the protection as enduring as BCG (Andersen, 1994; Brooks et al., 2001; Horwitz et al., 1995; Roberts et al., 1995). DNA vaccine constructs with Ag85 have also been evaluated and showed protection at a level less than or equal to BCG (Baldwin et al., 1998; Huygen et al., 1996; Kamath et al., 1999a; Kamath et al., 1999b; Lozes et al., 1997). Another vaccination strategy utilized a modified Vaccinia virus as a delivery vehicle to express Ag85A in the host cytoplasm as a way to boost BCG protection (Goonetilleke et al., 2003; McShane et al., 2001). Finally, a recombinant BCG strain overexpressing and secreting *M. tuberculosis* Ag85B represents one of the most promising new tuberculosis vaccines. Vaccination with this recombinant BCG elicits a higher level of protection than traditional BCG. Following challenge with virulent *M. tuberculosis* the recombinant BCG vaccinated guinea pigs exhibit lower bacterial burden, fewer lesions in their organs and demonstrate increased length of survival when compared to animals vaccinated with traditional BCG (Horwitz and Harth, 2003; Horwitz et al., 2000).

ESAT-6/CFP-10

ESAT-6 and CFP-10 elicit immune responses during *M. tuberculosis* infection. Once again the location of *esat6* and *cfp10* genes in the RD1 region has interesting implications. Due to the presence of the corresponding proteins in *M. tuberculosis* but absence in BCG, they are candidates for being antigens that elicit *M. tuberculosis* specific responses. This trait could be useful for both serodiagnostic and vaccine development.

Both ESAT-6 and CFP-10 elicit T cell responses during infection as shown by the response of T cells derived from animals and humans infected with *M. tuberculosis* (Andersen et al., 1995; Arend et al., 2000; Boesen et al., 1995; Pollock and Andersen, 1997; Skjot et al., 2000). Both CD4+ and CD8+ T cell epitopes have been characterized in ESAT-6 (Brodin et al., 2004; Brookes et al., 2003; Majlessi et al., 2003; Mustafa et al., 2003). ESAT-6 and CFP-10 also elicit antibodies in *M. tuberculosis* infected animals and humans (Brusasca et al., 2001; Brusasca et al., 2003; Dillon et al., 2000). In addition, ESAT-6 and CFP-10 may also possess immunomodulatory activities. However, the details of the specific effects and mechanism are only starting to be worked out (Singh et al., 2003a; Stanley et al., 2003; Trajkovic et al., 2002).

Furthermore, the immune response to ESAT-6 subunit and DNA vaccines is protective as revealed by immunization and *M. tuberculosis* challenge experiments. However, the level of protection observed is not greater than that observed with the current BCG vaccine (Brandt et al., 2000; Kamath et al., 1999a). Efforts to develop ESAT-6 into a new vaccine continue (Brandt et al., 2000; Doherty and Andersen, 2002; Lowrie et al., 1997; Malin et al., 2000; McShane et al., 2001; Mollenkopf et al., 2001; Morris et al., 2000). One such effort involves a subunit vaccine based on an Ag85B-ESAT-6 fusion protein which was shown to elicit protection on the order of that observed with BCG and better than that observed by immunization with either Ag85B or ESAT-6 alone (Weinrich Olsen et al., 2001).

Another vaccination strategy being explored involves recombinant BCG strains constructed to express and secrete ESAT-6. Specifically, a "knock-in" BCG recombinant in which the entire RD1 region was introduced, described earlier, was tested for its potential utility as a vaccine. In comparison to vaccination with standard BCG, this recombinant BCG::RD1 strain exhibits better protection against tuberculosis challenge on the basis of reduced pathology and bacterial load in spleens of mice (Pym et al., 2003). Through the course of testing various recombinant BCG "knock-in" strains Pym et al. (2003) demonstrated that secretion is specifically required for these strains to induce an ESAT-6 specific T-cell response (Pym et al., 2003). This represents the first direct demonstration in *M. tuberculosis* that secretion of an antigen, as opposed to expression alone, is specifically required for proper interaction with the immune system. In a separate study a different recombinant BCG strain, rBCG-2, which was reported to express and secrete ESAT-6 demonstrated no increased protection over that observed with BCG vaccination (Bao et al., 2003). The reason for the discrepancy between the two studies is not clear, but a key difference is the way in which ESAT-6 was secreted by the recombinant BCG strains. In the first study BCG::RD1 secretes ESAT-6 by way of its native specialized secretion system involving RD1 (discussed earlier in the Chapter). Thus, functional ESAT-6 is correctly localized. In addition, CFP-10 and possibly other effectors are secreted by this recombinant. In the second study the recombinant BCG strain likely secretes ESAT-6 by way of the Sec pathway since the RD1 region is absent in this strain (only the ESAT-6 gene was provided) and because the ESAT-6 expression construct included a Sec signal sequence fused to the amino-terminus of ESAT-6 (Bao et al., 2003). It will be interesting to determine if and how the different secretion mechanisms explain the effects of these recombinant BCG vaccines on production of protective immunity.

Studies of secreted and exported *M. tuberculosis* antigens have interesting implications beyond increasing our understanding of the immune responses to tuberculosis. Secreted *M. tuberculosis* proteins that elicit an immune response in the host are likely to represent antigens that are expressed during infection. As mentioned earlier, proteins secreted in the host are particularly difficult to identify. Another outcome of studying *M. tuberculosis* immunoantigens relates to antigens that elicit CD8+ T cell responses. Antigens presented by class I MHC molecules are generally processed in the host cell cytosol in a TAP-dependent manner, and this pathway is important for control of *M. tuberculosis* infection (Behar et al., 1999; Sousa et al., 2000). Hence, it is likely that some of that class I secreted antigens of *M. tuberculosis* reach the host cytosol. Since there are intrinsic difficulties in identifying *M. tuberculosis* proteins secreted or exported during infection and little is known about how proteins traffic in the host cell, studies of *M. tuberculosis* immunoantigens may help understand *M. tuberculosis* protein secretion from within host cells.

FUTURE TRENDS

STUDIES OF TRANSPORT ACROSS THE UNIQUE MYCOBACTERIAL CELL WALL

A significant unanswered question relevant to secretion in *M. tuberculosis* is how proteins cross the mycobacterial cell wall. The unique network of covalently linked molecules and associated lipids which comprise the cell wall of mycobacteria create a protective barrier for the bacillus (Fig. 1). This barrier is highly impermeable to small molecules, even more so than the outer membrane of Gram-negative *E. coli* (reviewed in Brennan and Nikaido, 1995). This cell wall barrier is likely to pose difficulty for both the intracellular influx of molecules and for protein secretion. Some models to explain secretion across the *M. tuberculosis* cell wall and release into the extracellular space include the following: 1) active transport mechanisms traffic proteins across the cell wall, 2) proteins transit the network by passive diffusion, and 3) following localization to the cell wall, culture filtrate proteins are released in sloughed off portions of the cell wall or cell wall vesicles. Studies of the relationship between the mycobacterial cell wall and protein secretion should help clarify the processes at work.

SPECIALIZED SECRETION SYSTEMS

We are only at the beginning of understanding protein export and secretion by *M. tuberculosis*. There are many reports of *M. tuberculosis* proteins lacking predicted signal sequences being found in culture filtrates (Braunstein and Belisle, 2000). Although not all of these proteins are likely to be true secreted products, the identification of secretion systems that function in the release of proteins lacking signal sequences (the SecA2 secretion system and ESAT-6 secretion system) establishes the existence of unconventional secreted proteins in *M. tuberculosis*. We believe there are additional examples of secreted *M. tuberculosis* proteins whose localization system remains a mystery.

The identification of the SecA2 and ESAT-6 secretion systems in *M. tuberculosis* is exciting and warrants additional study. For both pathways, basic questions remain unanswered. For example, not all the proteins that comprise the respective secretion apparatuses and the manner in which secreted substrates are recognized are known. It is also unclear whether these systems are solely responsible for exporting proteins across the cytoplasmic membrane allowing them to diffuse across the cell wall or if these systems directly translocate proteins across the cytoplasmic membrane and cell wall into the extracellular environment. In regards to ESAT-6 secretion it is also worthwhile considering the possibility that the system not only secretes the proteins across the *M. tuberculosis* cell wall but that the system also functions to traffic proteins across additional host membranes.

IDENTIFICATION OF *M. TUBERCULOSIS* PROTEINS SECRETED DURING INFECTION

In our opinion, the secreted and exported proteins of greatest interest have yet to be thoroughly investigated. These are the proteins secreted only during the course of infection. There are a small number of proteins shown to be secreted during intracellular growth of *M. tuberculosis* in macrophages. For most of these examples the proteins were first identified as being secreted *in vitro* followed by immunoelectron microscopy to localize the protein in *M. tuberculosis* or *M. bovis*

BCG infected macrophages (Beatty and Russell, 2000; Berthet et al., 1998a; Harth et al., 1996; Neyrolles et al., 2001; Walburger et al., 2004). This two-step method is limited to secreted proteins first identified though another method, and will not identify proteins exclusively expressed or secreted in macrophages. As mentioned above, the method of Beatty and Russell (2000) does enable the specific identification of *M. tuberculosis* proteins secreted in the host (Beatty and Russell, 2000). However, this method specifically focuses on a class of secreted *M. tuberculosis* proteins which traffic to subcellular compartments other than mycobacterial phagosomes and it is additionally dependent on proteins exported to the cell surface during *in vitro* growth of the bacilli prior to use in infection assays.

The development of methods for the comprehensive identification of bacterial proteins secreted from within macrophages has proven a difficult problem even for intracellular pathogens other than *M. tuberculosis*. Below we describe a method, recently used in the study of *Yersinia pestis* proteins transported into eukaryotic cells via a Type III secretion system, that we believe can be adapted to identify *M. tuberculosis* proteins secreted while the bacillus resides in macrophages (Day et al., 2003). This method utilizes a newly developed reporter system involving a phosphorylatable peptide tag derived from the Elk-1 eukaryotic transcription factor. The Elk tag reporter is comprised of a nuclear localization sequence (NLS) fused to a 17 amino acid portion of the Elk protein that can be phosphorylated, but only by a host kinase in the eukaryotic nucleus. When fused to a bacterial protein, the Elk reporter will be phosphorylated only if the fusion construct is secreted by the bacterium into the host cytosol and can reach the nucleus. The NLS will direct the fusion protein to the nucleus. For this system to work the fusion protein must reach the cytosol. In the case of *M. tuberculosis* this means trafficking out of the phagosome. The fusion protein will also have to be compatible with nuclear import. Following successful secretion and nuclear import phosphorylated fusion proteins can be detected with phospho-specific anti-Elk antibodies. We believe there is potential for this reporter system to be developed into a large-scale screen for *M. tuberculosis* proteins secreted from within macrophages.

TRAFFICKING OF *M. TUBERCULOSIS* SECRETED PROTEINS IN MACROPHAGES

Our knowledge of the *M. tuberculosis* proteins secreted while the bacilli reside in macrophages is minimal. Even less is known about the final cellular destination of proteins secreted by *M. tuberculosis* within the host. Given that *M. tuberculosis* is an intracellular pathogen that resides in phagosomes, the most obvious location for these proteins is the phagosome lumen. However, other destinations are possible including the phagosome membrane, host cytosol, nucleus, other organelles, and even the potential for release from the host cell by an exocytic mechanism. There is accumulating data showing some secreted proteins of *M. tuberculosis* trafficking beyond the phagosome. Immunoelectron microscopy experiments identified the 19-kDa lipoprotein, Ag85, FAP (45/47-kDa antigen), PknG, and Erp proteins both inside and outside of mycobacterial phagosomes (Beatty and Russell, 2000; Berthet et al., 1998a; Harth et al., 1996; Neyrolles et al., 2001; Walburger et al., 2004). Furthermore, fractionation of infected macrophages has identified *M. tuberculosis* proteins in host cell compartments other than the mycobacterial phagosome and host cytosol (Beatty and Russell, 2000; Walburger et al., 2004). Finally, since antigens presented by class I MHC molecules are generally processed in the host cell cytosol, the existence of class I secreted antigens of *M. tuberculosis* that elicit $CD8^+$ T cell responses is yet another reason to consider protein trafficking out of the phagosome. Mycobacterial lipids traffic out of the phagosome and have even been seen to escape from the host cell (Beatty et al., 2000; Fischer et al., 2001). Thus, it is possible that protein trafficking occurs via lipid trafficking. Experiments comparing the trafficking patterns of individual secreted proteins and lipids should help determine if different cell wall components traffic together.

CONCLUSION

Although we remain far from fully appreciating protein secretion and export in *M. tuberculosis*, the importance of these extracytoplasmic subsets of proteins in *M. tuberculosis* virulence and immunity is already clear. Continuation of the studies underway to comprehensively identify and characterize the secreted and cell envelope proteins will surely add to the already impressive list of known virulence factors and immunoantigens and increase our understanding of *M. tuberculosis* pathogenesis in the process. While less is known about the protein translocation systems that operate in *M. tuberculosis*, in just five years time we have started to appreciate the processes involved. Assisted by the recent advances in mycobacterial proteomic and genetic methods and the availability of the *M. tuberculosis* genome sequence we are optimistic that this momentum will continue in the future. In light of the significance of secreted and exported proteins to drug, vaccine and diagnostic development, continued study of *M. tuberculosis* protein secretion and export should help to reach the goal of controlling tuberculosis in the future.

WEB RESOURCES RELEVANT TO THE TOPICS COVERED IN THIS REVIEW

SIGNAL SEQUENCE PREDICTION PROGRAMS:

PSORT WWW Server:
http://psort.nibb.ac.jp/
(Nakai and Kanehisa, 1991)

Signal P 3.0 Server:
http://www.cbs.dtu.dk/services/SignalP/
(Bendtsen et al., 2004)

Exprot Server:
http://laurentian.ca/biology/saleh/exprot
(Saleh et al., 2001)

PrositeScan Server:
http://www.expasy.ch/sprot/prosite.html
(Bucher and Bairoch, 1994)

SPScan:
http://molbio.unmc.edu/gcg-help/spscan.html
(Genetics Computer Group, 1997)

TRANSMEMBRANE PREDICTION PROGRAMS:
TMpred Server:
http://www.ch.embnet.org/software/TMPRED_form.html
(Hofmann and Stoffel, 1993)

PREDICTED TAT SUBSTRATES IN BACTERIAL GENOMES:
TATFIND Genome analysis:
http://www.sas.upenn.edu/~pohlschr/tatprok.html
(Dilks et al., 2003)

PREDICTED *M. TUBERCULOSIS* SECRETED PROTEINS:
http://tbsp.phri.org/
(Gomez et al., 2000)

SECRETED PROTEOME DATABASES FOR *M. TUBERCULOSIS*:
Proteome Database for Microbial Research at Max Planck Institute for Infection Biology:
http://www.mpiib-berlin.mpg.de/2D-PAGE/EBP-PAGE/index.html
(Mollenkopf et al., 1999)

Department of Infectious Disease Immunology Statens Serum Institute:
http://www.ssi.dk/graphics/html/tbimmun/Protein_database/protein_database.htm
(Rosenkrands et al., 2000a; Rosenkrands et al., 2000b)

T CELL EPITOPE PREDICTION PROGRAMS:
ProPred:
http://www.imtech.res.in/raghava/propred
(Singh and Raghava, 2001)

TEPITOPE:
http://www.vaccinome.com
(Sturniolo et al., 1999)

ACKNOWLEDGEMENTS

We thank the members of the Braunstein laboratory for helpful comments, and acknowledge the current support of the National Institute of Health (NIH RO1AI054540).

REFERENCES

Aagaard, C., Brock, I., Olsen, A., Ottenhoff, T.H., Weldingh, K., and Andersen, P. (2004). Mapping immune reactivity toward Rv2653 and Rv2654: two novel low-molecular-mass antigens found specifically in the *Mycobacterium tuberculosis* complex. J. Infect. Dis. *189*, 812-819.

Abel, B., Thieblemont, N., Quesniaux, V.J., Brown, N., Mpagi, J., Miyake, K., Bihl, F., and Ryffel, B. (2002). Toll-like receptor 4 expression is required to control chronic *Mycobacterium tuberculosis* infection in mice. J. Immunol. *169*, 3155-3162.

Abou-Zeid, C., Filley, E., Steele, J., and Rook, G.A. (1987). A simple new method for using antigens separated by polyacrylamide gel electrophoresis to stimulate lymphocytes *in vitro* after converting bands cut from Western blots into antigen-bearing particles. J. Immunol. Methods *98*, 5-10.

Abou-Zeid, C., Garbe, T., Lathigra, R., Wiker, H.G., Harboe, M., Rook, G.A., and Young, D.B. (1991). Genetic and immunological analysis of *Mycobacterium tuberculosis* fibronectin-binding proteins. Infect. Immun. *59*, 2712-2718.

Abou-Zeid, C., Ratliff, T.L., Wiker, H.G., Harboe, M., Bennedsen, J., and Rook, G.A. (1988a). Characterization of fibronectin-binding antigens released by *Mycobacterium tuberculosis* and *Mycobacterium bovis* BCG. Infect. Immun. *56*, 3046-3051.

Abou-Zeid, C., Smith, I., Grange, J.M., Ratliff, T.L., Steele, J., and Rook, G.A. (1988b). The secreted antigens of *Mycobacterium tuberculosis* and their relationship to those recognized by the available antibodies. J. Gen. Microbiol. *134*, 531-538.

Ahmad, S., Akbar, P.K., Wiker, H.G., Harboe, M., and Mustafa, A.S. (1999). Cloning, expression and immunological reactivity of two mammalian cell entry proteins encoded by the *mce1* operon of *Mycobacterium tuberculosis*. Scand. J. Immunol. *50*, 510-518.

Alami, M., Luke, I., Deitermann, S., Eisner, G., Koch, H.G., Brunner, J., and Muller, M. (2003). Differential interactions between a twin-arginine signal peptide and its translocase in *Escherichia coli*. Mol. Cell *12*, 937-946.

Alami, M., Trescher, D., Wu, L.F., and Muller, M. (2002). Separate analysis of twin-arginine translocation (Tat)-specific membrane binding and translocation in *Escherichia coli*. J. Biol. Chem. *277*, 20499-20503.

Al-Attiyah, R., and Mustafa, A.S. (2004). Computer-assisted prediction of HLA-DR binding and experimental analysis for human promiscuous Th1-cell peptides in the 24 kDa secreted lipoprotein (LppX) of *Mycobacterium tuberculosis*. Scand. J. Immunol. *59*, 16-24.

Andersen, A.B., and Hansen, E.B. (1989). Structure and mapping of antigenic domains of protein antigen b, a 38,000-molecular-weight protein of *Mycobacterium tuberculosis*. Infect. Immun. *57*, 2481-2488.

Andersen, A.B., Worsaae, A., and Chaparas, S.D. (1988). Isolation and characterization of recombinant lambda gt11 bacteriophages expressing eight different mycobacterial antigens of potential immunological relevance. Infect. Immun. *56*, 1344-1351.

Andersen, P. (1994). Effective vaccination of mice against *Mycobacterium tuberculosis* infection with a soluble mixture of secreted mycobacterial proteins. Infect. Immun. *62*, 2536-2544.

Andersen, P., Andersen, A.B., Sorensen, A.L., and Nagai, S. (1995). Recall of long-lived immunity to *Mycobacterium tuberculosis* infection in mice. J. Immunol. *154*, 3359-3372.

Andersen, P., Askgaard, D., Ljungqvist, L., Bennedsen, J., and Heron, I. (1991a). Proteins released from *Mycobacterium tuberculosis* during growth. Infect. Immun. *59*, 1905-1910.

Andersen, P., Askgaard, D., Ljungqvist, L., Bentzon, M.W., and Heron, I. (1991b). T-cell proliferative response to antigens secreted by *Mycobacterium tuberculosis*. Infect. Immun. *59*, 1558-1563.

Arend, S.M., Andersen, P., van Meijgaarden, K.E., Skjot, R.L., Subronto, Y.W., van Dissel, J. T., and Ottenhoff, T. H. (2000). Detection of active tuberculosis infection by T cell responses to early-secreted antigenic target 6-kDa protein and culture filtrate protein 10. J. Infect. Dis. *181*, 1850-1854.

Armitige, L.Y., Jagannath, C., Wanger, A.R., and Norris, S.J. (2000). Disruption of the genes encoding antigen 85A and antigen 85B of *Mycobacterium tuberculosis* H37Rv: effect on growth in culture and in macrophages. Infect. Immun. *68*, 767-778.

Arruda, S., Bomfim, G., Knights, R., Huima-Byron, T., and Riley, L.W. (1993). Cloning of an *M. tuberculosis* DNA fragment associated with entry and survival inside cells. Science *261*, 1454-1457.

Ashbridge, K.R., Booth, R.J., Watson, J.D., and Lathigra, R.B. (1989). Nucleotide sequence of the 19 kDa antigen gene from *Mycobacterium tuberculosis*. Nucleic Acids Res. *17*, 1249.

Av-Gay, Y., Jamil, S., and Drews, S.J. (1999). Expression and characterization of the *Mycobacterium tuberculosis* serine/threonine protein kinase PknB. Infect. Immun. *67*, 5676-5682.

Azad, A.K., Sirakova, T.D., Fernandes, N.D., and Kolattukudy, P.E. (1997). Gene knockout reveals a novel gene cluster for the synthesis of a class of cell wall lipids unique to pathogenic mycobacteria. J. Biol. Chem. *272*, 16741-16745.

Baird, P.N., Hall, L.M., and Coates, A.R. (1988). A major antigen from *Mycobacterium tuberculosis* which is homologous to the heat shock proteins *groES* from E. coli and the *htpA* gene product of *Coxiella burneti*. Nucleic Acids Res. *16*, 9047.

Baldwin, S.L., D'Souza, C., Roberts, A.D., Kelly, B.P., Frank, A.A., Lui, M.A., Ulmer, J.B., Huygen, K., McMurray, D.M., and Orme, I.M. (1998). Evaluation of new vaccines in the mouse and guinea pig model of tuberculosis. Infect. Immun. *66*, 2951-2959.

Bao, L., Chen, W., Zhang, H., and Wang, X. (2003). Virulence, immunogenicity, and protective efficacy of two recombinant *Mycobacterium bovis* bacillus Calmette-Guerin strains expressing the antigen ESAT-6 from *Mycobacterium tuberculosis*. Infect. Immun. *71*, 1656-1661.

Barry, C.E., 3rd (2001). Interpreting cell wall 'virulence factors' of *Mycobacterium tuberculosis*. Trends Microbiol. *9*, 237-241.

Beatty, W.L., Rhoades, E.R., Ullrich, H.J., Chatterjee, D., Heuser, J.E., and Russell, D.G. (2000). Trafficking and release of mycobacterial lipids from infected macrophages. Traffic *1*, 235-247.

Beatty, W.L., and Russell, D.G. (2000). Identification of mycobacterial surface proteins released into subcellular compartments of infected macrophages. Infect. Immun. *68*, 6997-7002.

Behar, S.M., Dascher, C.C., Grusby, M.J., Wang, C.R., and Brenner, M.B. (1999). Susceptibility of mice deficient in CD1D or TAP1 to infection with *Mycobacterium tuberculosis*. J. Exp. Med. *189*, 1973-1980.

Behr, M.A., Wilson, M.A., Gill, W.P., Salamon, H., Schoolnik, G.K., Rane, S., and Small, P.M. (1999). Comparative genomics of BCG vaccines by whole-genome DNA microarray. Science *284*, 1520-1523.

Belisle, J.T., Braunstein, M., Rosenkrands, I., and Andersen, P. (2004). The proteome of *Mycobacterium tuberculosis*. In Tuberculosis, S.T. Cole, K. Eisenach, B. Gicquel, W.R.J. Jacobs, and D. McMurray, eds. (Washington, D.C., ASM Press).

Belisle, J.T., Vissa, V.D., Sievert, T., Takayama, K., Brennan, P.J., and Besra, G.S. (1997). Role of the major antigen of *Mycobacterium tuberculosis* in cell wall biogenesis. Science *276*, 1420-1422.

Bendtsen, J.D., Nielsen, H., von Heijne, G., and Brunak, S. (2004). Improved prediction of signal peptides: SignalP 3.0. J. Mol. Biol. *340*, 783-795.

Bensing, B.A., and Sullam, P.M. (2002). An accessory *sec* locus of *Streptococcus gordonii* is required for export of the surface protein GspB and for normal levels of binding to human platelets. Mol. Micro *44*, 1081-1094.

Bentley-Hibbert, S.I., Quan, X., Newman, T., Huygen, K., and Godfrey, H.P. (1999). Pathophysiology of antigen 85 in patients with active tuberculosis: antigen 85 circulates as complexes with fibronectin and immunoglobulin G. Infect. Immun. *67*, 581-588.

Berks, B.C., Sargent, F., and Palmer, T. (2000). The Tat protein export pathway. Mol. Microbiol. *35*, 260-274.

Berthet, F.X., Lagranderie, M., Gounon, P., LaurentWinter, C., Ensergueix, D., Chavarot, P., Thouron, F., Maranghi, E., Pelicic, V., Portnoi, D., et al. (1998a). Attenuation of virulence by disruption of the *Mycobacterium tuberculosis erp* gene. Science *282*, 759-762.

Berthet, F.X., Rasmussen, P.B., Rosenkrands, I., Andersen, P., and Gicquel, B. (1998b). A *Mycobacterium tuberculosis* operon encoding ESAT-6 and a novel low- molecular-mass culture filtrate protein (CFP-10). Microbiology *144*, 3195-3203.

Bigi, F., Espitia, C., Alito, A., Zumarraga, M., Romano, M.I., Cravero, S., and Cataldi, A. (1997). A novel 27 kDa lipoprotein antigen from *Mycobacterium bovis*. Microbiology *143*, 3599-3605.

Bigi, F., Gioffre, A., Klepp, L., de la Paz Santangelo, M., Alito, A., Caimi, K., Meikle, V., Zumarraga, M., Taboga, O., Romano, M.I., and Cataldi, A. (2004). The knockout of the *lprG*-Rv1410 operon produces strong attenuation of *Mycobacterium tuberculosis*. Microbes Infect. *6*, 182-187.

Blander, J.M., and Medzhitov, R. (2004). Regulation of phagosome maturation by signals from toll-like receptors. Science *304*, 1014-1018.

Boesen, H., Jensen, B.N., Wilcke, T., and Andersen, P. (1995). Human T-cell responses to secreted antigen fractions of *Mycobacterium tuberculosis*. Infect. Immun. *63*, 1491-1497.

Borremans, M., de Wit, L., Volckaert, G., Ooms, J., de Bruyn, J., Huygen, K., van Vooren, J.P., Stelandre, M., Verhofstadt, R., and Content, J. (1989). Cloning, sequence determination, and expression of a 32-kilodalton-protein gene of *Mycobacterium tuberculosis*. Infect. Immun. *57*, 3123-3130.

Brandt, L., Elhay, M., Rosenkrands, I., Lindblad, E.B., and Andersen, P. (2000). ESAT-6 subunit vaccination against *Mycobacterium tuberculosis*. Infect. Immun. *68*, 791-795.

Branger, J., Leemans, J.C., Florquin, S., Weijer, S., Speelman, P., and Van Der Poll, T. (2004). Toll-like receptor 4 plays a protective role in pulmonary tuberculosis in mice. Int. Immunol. *16*, 509-516.

Braunstein, M., and Belisle, J.T. (2000). Genetics of protein secretion. In Molecular genetics of mycobacteria, G. F. Hatfull, and W. R. J. Jacobs, eds. (ASM Press), pp. 203-220.

Braunstein, M., Brown, A.M., Kurtz, S., and Jacobs, W.R., Jr. (2001). Two nonredundant SecA homologues function in mycobacteria. J. Bacteriol. *183*, 6979-6990.

Braunstein, M., Espinosa, B., Chan, J., Belisle, J.T., and Jacobs, W.R.J. (2003). SecA2 functions in the secretion of superoxide dismutase A and in the virulence of *Mycobacterium tuberculosis*. Mol. Microbiol. *48*, 453-464.

Braunstein, M., Griffin, T. I., Kriakov, J.I., Friedman, S.T., Grindley, N.D., and Jacobs, W.R., Jr. (2000). Identification of genes encoding exported *Mycobacterium tuberculosis* proteins using a Tn*552'phoA in vitro* transposition system. J. Bacteriol. *182*, 2732-2740.

Brennan, P.J., and Nikaido, H. (1995). The envelope of mycobacteria. Annu. Rev. Biochem. *64*, 29-63.

Brightbill, H.D., Libraty, D.H., Krutzik, S.R., Yang, R.B., Belisle, J.T., Bleharski, J.R., Maitland, M., Norgard, M.V., Plevy, S.E., Smale, S.T., et al. (1999). Host defense mechanisms triggered by microbial lipoproteins through toll-like receptors. Science *285*, 732-736.

Brodin, P., Majlessi, L., Brosch, R., Smith, D., Bancroft, G., Clark, S., Williams, A., Leclerc, C., and Cole, S.T. (2004). Enhanced protection against tuberculosis by vaccination with recombinant *Mycobacterium microti* vaccine that induces T cell immunity against region of difference 1 antigens. J. Infect. Dis. *190*, 115-122.

Brookes, R.H., Pathan, A.A., McShane, H., Hensmann, M., Price, D.A., and Hill, A.V. (2003). CD8+ T cell-mediated suppression of intracellular *Mycobacterium tuberculosis* growth in activated human macrophages. Eur. J. Immunol. *33*, 3293-3302.

Brooks, J.V., Frank, A.A., Keen, M.A., Bellisle, J.T., and Orme, I.M. (2001). Boosting vaccine for tuberculosis. Infect. Immun. *69*, 2714-2717.

Broome-Smith, J.K., Tadayyon, M., and Zhang, Y. (1990). Beta-lactamase as a probe of membrane protein assembly and protein export. Mol. Microbiol. *4*, 1637-1644.

Brusasca, P.N., Colangeli, R., Lyashchenko, K.P., Zhao, X., Vogelstein, M., Spencer, J.S., McMurray, D.N., and Gennaro, M.L. (2001). Immunological characterization of antigens encoded by the RD1 region of the *Mycobacterium tuberculosis* genome. Scand. J. Immunol. *54*, 448-452.

Brusasca, P.N., Peters, R.L., Motzel, S.L., Klein, H.J., and Gennaro, M.L. (2003). Antigen recognition by serum antibodies in non-human primates experimentally infected with *Mycobacterium tuberculosis*. Comp Med. *53*, 165-172.

Buchanan, G., Sargent, F., Berks, B.C., and Palmer, T. (2001). A genetic screen for suppressors of *Escherichia coli* Tat signal peptide mutations establishes a critical role for the second arginine within the twin-arginine motif. Arch. Microbiol. *177*, 107-112.

Bucher, P., and Bairoch, A. (1994). A generalized profile syntax for biomolecular sequence motifs and its function in automatic sequence interpretation. Proc. Int. Conf Intell Syst Mol. Biol. *2*, 53-61.

Buchmeier, N., Blanc-Potard, A., Ehrt, S., Piddington, D., Riley, L., and Groisman, E.A. (2000). A parallel intraphagosomal survival strategy shared by *Mycobacterium tuberculosis* and *Salmonella enterica*. Mol. Microbiol. *35*, 1375-1382.

Buttner, D., and Bonas, U. (2002). Port of entry--the type III secretion translocon. Trends Microbiol. *10*, 186-192.

Camacho, L.R., Constant, P., Raynaud, C., Laneelle, M.A., Triccas, J.A., Gicquel, B., Daffe, M., and Guilhot, C. (2001). Analysis of the phthiocerol dimycocerosate locus of *Mycobacterium tuberculosis*. Evidence that this lipid is involved in the cell wall permeability barrier. J. Biol. Chem. *276*, 19845-19854.

Camacho, L.R., Ensergueix, D., Perez, E., Gicquel, B., and Guilhot, C. (1999). Identification of a virulence gene cluster of *Mycobacterium tuberculosis* by signature-tagged transposon mutagenesis. Mol. Microbiol. *34*, 257-267.

Carroll, J.D., Wallace, R.C., Keane, J., Remold, H.G., and Arbeit, R.D. (2000). Identification of *Mycobacterium avium* DNA sequences that encode exported proteins by using *phoA* gene fusions. Tuber Lung Dis. *80*, 117-130.

Caruso, A.M., Serbina, N., Klein, E., Triebold, K., Bloom, B.R., and Flynn, J.L. (1999). Mice deficient in CD4 T cells have only transiently diminished levels of IFN-gamma, yet succumb to tuberculosis. J. Immunol. *162*, 5407-5416.

Chakhaiyar, P., Nagalakshmi, Y., Aruna, B., Murthy, K.J., Katoch, V.M., and Hasnain, S.E. (2004). Regions of High Antigenicity within the Hypothetical PPE Major Polymorphic Tandem Repeat Open-Reading Frame, Rv2608, Show a Differential Humoral Response and a Low T Cell Response in Various Categories of Patients with Tuberculosis. J. Infect. Dis. *190*, 1237-1244.

Chen, Q., Wu, H., and Fives-Taylor, P.M. (2004). Investigating the role of *secA2* in secretion and glycosylation of a fimbrial adhesin in *Streptococcus parasanguis* FW213. Mol. Microbiol. *53*, 843-856.

Chubb, A.J., Woodman, Z.L., da Silva Tatley, F.M., Hoffmann, H.J., Scholle, R.R., and Ehlers, M.R. (1998). Identification of *Mycobacterium tuberculosis* signal sequences that direct the export of a leaderless beta-lactamase gene product in *Escherichia coli*. Microbiology *144*, 1619-1629.

Ciaramella, A., Cavone, A., Santucci, M.B., Garg, S.K., Sanarico, N., Bocchino, M., Galati, D., Martino, A., Auricchio, G., D'Orazio, M., et al. (2004). Induction of Apoptosis and Release of Interleukin-1 beta by Cell Wall-Associated 19-kDa Lipoprotein during the Course of Mycobacterial Infection. J. Infect. Dis. *190*, 1167-1176.

Closs, O., Harboe, M., Axelsen, N.H., Bunch-Christensen, K., and Magnusson, M. (1980). The antigens of *Mycobacterium bovis*, strain BCG, studied by crossed immunoelectrophoresis: a reference system. Scand. J. Immunol. *12*, 249-263.

Coates, A.R., Nicolai, H., Pallen, M.J., Guy, A., Chaparas, S.D., and Mitchison, D.A. (1989). The 45 kilodalton molecule of *Mycobacterium tuberculosis* identified by immunoblotting and monoclonal antibodies as antigenic in patients with tuberculosis. Br. J. Exp. Pathol. *70*, 215-225.

Cockle, P.J., Gordon, S.V., Lalvani, A., Buddle, B.M., Hewinson, R.G., and Vordermeier, H.M. (2002). Identification of novel *Mycobacterium tuberculosis* antigens with potential as diagnostic reagents or subunit vaccine candidates by comparative genomics. Infect. Immun. *70*, 6996-7003.

Cole, S.T., Brosch, R., Parkhill, J., Garnier, T., Churcher, C., Harris, D., Gordon, S.V., Eiglmeier, K., Gas, S., Barry, C. E., 3rd, et al. (1998). Deciphering the biology of *Mycobacterium tuberculosis* from the complete genome sequence. Nature *393*, 537-544.

Collins, F.M., Lamb, J.R., and Young, D.B. (1988). Biological activity of protein antigens isolated from *Mycobacterium tuberculosis* culture filtrate. Infect. Immun. *56*, 1260-1266.

Converse, S.E., Mougous, J.D., Leavell, M.D., Leary, J.A., Bertozzi, C.R., and Cox, J.S. (2003). MmpL8 is required for sulfolipid-1 biosynthesis and *Mycobacterium tuberculosis* virulence. Proc. Natl. Acad. Sci. U.S.A. *100*, 6121-6126.

Covert, B.A., Spencer, J.S., Orme, I.M., and Belisle, J.T. (2001). The application of proteomics in defining the T cell antigens of *Mycobacterium tuberculosis*. Proteomics *1*, 574-586.

Cowley, S., Ko, M., Pick, N., Chow, R., Downing, K.J., Gordhan, B.G., Betts, J.C., Mizrahi, V., Smith, D.A., Stokes, R.W., and Av-Gay, Y. (2004). The *Mycobacterium tuberculosis* protein serine/threonine kinase PknG is linked to cellular glutamate/glutamine levels and is important for growth *in vivo*. Mol. Microbiol. *52*, 1691-1702.

Cox, J.S., Chen, B., McNeil, M., and Jacobs, W.R., Jr. (1999). Complex lipid determines tissue-specific replication of *Mycobacterium tuberculosis* in mice. Nature *402*, 79-83.

Daffe, M., and Draper, P. (1998). The envelope layers of mycobacteria with reference to their pathogenicity. Adv Microb Physiol *39*, 131-203.

Daffe, M., and Etienne, G. (1999). The capsule of *Mycobacterium tuberculosis* and its implications for pathogenicity. Tuber Lung Dis. *79*, 153-169.

Dahl, J.L., Wei, J., Moulder, J.W., Laal, S., and Friedman, R.L. (2001). Subcellular localization of the Intracellular survival-enhancing Eis protein of *Mycobacterium tuberculosis*. Infect. Immun. *69*, 4295-4302.

Davenport, M.P., Ho Shon, I.A., and Hill, A.V. (1995). An empirical method for the prediction of T-cell epitopes. Immunogenetics *42*, 392-397.

Day, J.B., Ferracci, F., and Plano, G.V. (2003). Translocation of YopE and YopN into eukaryotic cells by *Yersinia pestis yopN, tyeA, sycN, yscB* and *lcrG* deletion mutants measured using a phosphorylatable peptide tag and phosphospecific antibodies. Mol. Microbiol. *47*, 807-823.

De Bruyn, J., Bosmans, R., Nyabenda, J., and Van Vooren, J.P. (1989). Effect of zinc deficiency on the appearance of two immunodominant protein antigens (32 kDa and 65 kDa) in culture filtrates of mycobacteria. J. Gen. Microbiol. *135 (Pt 1)*, 79-84.

de Gier, J.W., and Luirink, J. (2003). The ribosome and YidC. New insights into the biogenesis of *Escherichia coli* inner membrane proteins. EMBO Rep *4*, 939-943.

De Groot, A.S., Bosma, A., Chinai, N., Frost, J., Jesdale, B.M., Gonzalez, M.A., Martin, W., and Saint-Aubin, C. (2001). From genome to vaccine: *in silico* predictions, *ex vivo* verification. Vaccine *19*, 4385-4395.

de Keyzer, J., van der Does, C., and Driessen, A.J. (2003). The bacterial translocase: a dynamic protein channel complex. Cell Mol. Life Sci. *60*, 2034-2052.

de Mendonca-Lima, L., Bordat, Y., Pivert, E., Recchi, C., Neyrolles, O., Maitournam, A., Gicquel, B., and Reyrat, J. M. (2003). The allele encoding the mycobacterial Erp protein affects lung disease in mice. Cell Microbiol. *5*, 65-73.

De Voss, J.J., Rutter, K., Schroeder, B.G., Su, H., Zhu, Y., and Barry, C.E., 3rd (2000). The salicylate-derived mycobactin siderophores of *Mycobacterium tuberculosis* are essential for growth in macrophages. Proc. Natl. Acad. Sci. U.S.A. *97*, 1252-1257.

DeLisa, M.P., Samuelson, P., Palmer, T., and Georgiou, G. (2002). Genetic analysis of the twin arginine translocator secretion pathway in bacteria. J. Biol. Chem. *277*, 29825-29831.

Delogu, G., and Brennan, M.J. (2001). Comparative immune response to PE and PE_PGRS antigens of *Mycobacterium tuberculosis*. Infect. Immun. *69*, 5606-5611.

Demangel, C., Brodin, P., Cockle, P.J., Brosch, R., Majlessi, L., Leclerc, C., and Cole, S.T. (2004). Cell envelope protein PPE68 contributes to *Mycobacterium tuberculosis* RD1 immunogenicity independently of a 10-kilodalton culture filtrate protein and ESAT-6. Infect. Immun. *72*, 2170-2176.

Dilks, K., Rose, R.W., Hartmann, E., and Pohlschroder, M. (2003). Prokaryotic utilization of the twin-arginine translocation pathway: a genomic survey. J. Bacteriol. *185*, 1478-1483.

Dillon, D.C., Alderson, M.R., Day, C.H., Bement, T., Campos-Neto, A., Skeiky, Y.A., Vedvick, T., Badaro, R., Reed, S.G., and Houghton, R. (2000). Molecular and immunological characterization of *Mycobacterium tuberculosis* CFP-10, an immunodiagnostic antigen missing in *Mycobacterium bovis* BCG. J. Clin. Microbiol. *38*, 3285-3290.

Ding, Z., and Christie, P.J. (2003). *Agrobacterium tumefaciens* twin-arginine-dependent translocation is important for virulence, flagellation, and chemotaxis but not type IV secretion. J. Bacteriol. *185*, 760-771.

Doherty, T.M., and Andersen, P. (2002). Tuberculosis vaccine development. Curr. Opin Pulm Med. *8*, 183-187.

Domenech, P., Reed, M.B., Dowd, C.S., Manca, C., Kaplan, G., and Barry, C.E., 3rd (2004). The role of MmpL8 in sulfatide biogenesis and virulence of *Mycobacterium tuberculosis*. J. Biol. Chem. *279*, 21257-21265.

Downing, K.J., McAdam, R.A., and Mizrahi, V. (1999). *Staphylococcus aureus* nuclease is a useful secretion reporter for mycobacteria. Gene *239*, 293-299.

Draper, P. (1998). The outer parts of the mycobacterial envelope as permeability barriers. Front Biosci *3*, D1253-1261.

Drennan, M.B., Nicolle, D., Quesniaux, V.J., Jacobs, M., Allie, N., Mpagi, J., Fremond, C., Wagner, H., Kirschning, C., and Ryffel, B. (2004). Toll-like receptor 2-deficient mice succumb to *Mycobacterium tuberculosis* infection. Amer. J. Pathol. *164*, 49-57.

Dubnau, E., Chan, J., Raynaud, C., Mohan, V.P., Laneele, M.A., Yu, K., Quemard, A., Smith, I., and Daffee, M. (2000). Oxygenated mycolic acids are necessary for virulence of *Mycobacterium tuberculosis* in mice, Mol. Microbiol. *36*, 630-637.

Duong, F., and Wickner, W. (1997). Distinct catalytic roles of the SecYE, SecG and SecDFyajC subunits of preprotein translocase holoenzyme. Embo J. *16*, 2756-2768.

Economou, A. (1998). Bacterial preprotein translocase: mechanism and conformational dynamics of a processive enzyme. Mol. Microbiol. *27*, 511-518.

Economou, A., and Wickner, W. (1994). SecA promotes preprotein translocation by undergoing ATP-driven cycles of membrane insertion and deinsertion. Cell *78*, 835-843.

Edwards, K.M., Cynamon, M.H., Voladri, R.K., Hager, C.C., DeStefano, M.S., Tham, K.T., Lakey, D.L., Bochan, M.R., and Kernodle, D.S. (2001). Iron-cofactored superoxide dismutase inhibits host responses to *Mycobacterium tuberculosis*. Amer. J. Respir Crit Care Med. *164*, 2213-2219.

Espitia, C., Laclette, J.P., Mondragon-Palomino, M., Amador, A., Campuzano, J., Martens, A., Singh, M., Cicero, R., Zhang, Y., and Moreno, C. (1999). The PE-PGRS glycine-rich proteins of *Mycobacterium tuberculosis*: a new family of fibronectin-binding proteins? Microbiology *145 (Pt 12)*, 3487-3495.

Espitia, C., Sciutto, E., Bottasso, O., Gonzalez-Amaro, R., Hernandez-Pando, R., and Mancilla, R. (1992). High antibody levels to the mycobacterial fibronectin-binding antigen of 30-31 kD in tuberculosis and lepromatous leprosy. Clin. Exp. Immunol. *87*, 362-367.

Ewann, F., Jackson, M., Pethe, K., Cooper, A., Mielcarek, N., Ensergueix, D., Gicquel, B., Locht, C., and Supply, P. (2002). Transient requirement of the PrrA-PrrB two-component system for early intracellular multiplication of *Mycobacterium tuberculosis*. Infect. Immun. *70*, 2256-2263.

Faith, A., Moreno, C., Lathigra, R., Roman, E., Fernandez, M., Brett, S., Mitchell, D.M., Ivanyi, J., and Rees, A.D. (1991). Analysis of human T-cell epitopes in the 19,000 MW antigen of *Mycobacterium tuberculosis*: influence of HLA-DR. Immunology *74*, 1-7.

Falla, J.C., Parra, C.A., Mendoza, M., Franco, L.C., Guzman, F., Forero, J., Orozco, O., and Patarroyo, M.E. (1991). Identification of B- and T-cell epitopes within the MTP40 protein of *Mycobacterium tuberculosis* and their correlation with the disease course. Infect. Immun. *59*, 2265-2273.

Fekkes, P., and Driessen, A.J. (1999). Protein targeting to the bacterial cytoplasmic membrane. Microbiol. Mol. Biol. Rev. *63*, 161-173.

Fekkes, P., van der Does, C., and Driessen, A.J. (1997). The molecular chaperone SecB is released from the carboxy-terminus of SecA during initiation of precursor protein translocation. Embo J. *16*, 6105-6113.

Finlay, B.B., and Falkow, S. (1997). Common themes in microbial pathogenicity revisited. Microbiol. Mol. Biol. Rev. *61*, 136-169.

Fischer, K., Chatterjee, D., Torrelles, J., Brennan, P.J., Kaufmann, S.H., and Schaible, U.E. (2001). Mycobacterial lysocardiolipin is exported from phagosomes upon cleavage of cardiolipin by a macrophage-derived lysosomal phospholipase A2. J. Immunol. *167*, 2187-2192.

Flesselles, B., Anand, N.N., Remani, J., Loosmore, S.M., and Klein, M.H. (1999). Disruption of the mycobacterial cell entry gene of *Mycobacterium bovis* BCG results in a mutant that exhibits a reduced invasiveness for epithelial cells. FEMS Microbiol. Lett. *177*, 237-242.

Flint, J.L., Kowalski, J.C., Karnati, P.K., and Derbyshire, K.M. (2004). The RD1 virulence locus of *Mycobacterium tuberculosis* regulates DNA transfer in *Mycobacterium smegmatis*. Proc. Natl. Acad. Sci. U.S.A. *101*, 12598-12603.

Florio, W., Freer, G., Daila Casa, B., Batoni, G., Maisetta, G., Senesi, S., and Campa, M. (1997). Comparative analysis of subcellular distribution of protein antigens in *Mycobacterium bovis* bacillus Calmette-Guerin. Can J. Microbiol. *43*, 744-750.

Flynn, J.L., Goldstein, M.M., Triebold, K.J., Koller, B., and Bloom, B.R. (1992). Major histocompatibility complex class I-restricted T cells are required for resistance to *Mycobacterium tuberculosis* infection. Proc. Natl. Acad. Sci. U.S.A. *89*, 12013-12017.

Fortune, S.M., Solache, A., Jaeger, A., Hill, P.J., Belisle, J.T., Bloom, B.R., Rubin, E. J., and Ernst, J.D. (2004). *Mycobacterium tuberculosis* inhibits macrophage responses to IFN-gamma through myeloid differentiation factor 88-dependent and -independent mechanisms. J. Immunol. *172*, 6272-6280.

Fremond, C.M., Nicolle, D.M., Torres, D.S., and Quesniaux, V.F. (2003). Control of *Mycobacterium bovis* BCG infection with increased inflammation in TLR4-deficient mice. Microbes Infect. *5*, 1070-1081.

Fulton, S.A., Reba, S.M., Pai, R.K., Pennini, M., Torres, M., Harding, C.V., and Boom, W.H. (2004). Inhibition of major histocompatibility complex II expression and antigen processing in murine alveolar macrophages by *Mycobacterium bovis* BCG and the 19-kilodalton mycobacterial lipoprotein. Infect. Immun. *72*, 2101-2110.

Gao, L.Y., Guo, S., McLaughlin, B., Morisaki, H., Engel, J.N., and Brown, E.J. (2004). A mycobacterial virulence gene cluster extending RD1 is required for cytolysis, bacterial spreading and ESAT-6 secretion. Mol. Microbiol. *53*, 1677-1693.

Garg, S.K., Tiwari, R.P., Tiwari, D., Singh, R., Malhotra, D., Ramnani, V.K., Prasad, G.B., Chandra, R., Fraziano, M., Colizzi, V., and Bisen, P.S. (2003). Diagnosis of tuberculosis: available technologies, limitations, and possibilities. J. Clin. Lab. Anal. *17*, 155-163.

Garsia, R.J., Hellqvist, L., Booth, R.J., Radford, A.J., Britton, W.J., Astbury, L., Trent, R.J., and Basten, A. (1989). Homology of the 70-kilodalton antigens from *Mycobacterium leprae* and *Mycobacterium bovis* with the *Mycobacterium tuberculosis* 71-kilodalton antigen and with the conserved heat shock protein 70 of eucaryotes. Infect. Immun. *57*, 204-212.

Gehring, A.J., Dobos, K.M., Belisle, J.T., Harding, C.V., and Boom, W.H. (2004). *Mycobacterium tuberculosis* LprG (Rv1411c): a novel TLR-2 ligand that inhibits human macrophage class II MHC antigen processing. J. Immunol. *173*, 2660-2668.

Gehring, A.J., Rojas, R.E., Canaday, D.H., Lakey, D.L., Harding, C.V., and Boom, W.H. (2003). The *Mycobacterium tuberculosis* 19-kilodalton lipoprotein inhibits gamma interferon-regulated HLA-DR and Fc gamma R1 on human macrophages through Toll-like receptor 2. Infect. Immun. *71*, 4487-4497.

Geluk, A., Taneja, V., van Meijgaarden, K.E., Zanelli, E., Abou-Zeid, C., Thole, J.E., de Vries, R.R., David, C.S., and Ottenhoff, T.H. (1998). Identification of HLA class II-restricted determinants of *Mycobacterium tuberculosis*-derived proteins by using HLA-transgenic, class II-deficient mice. Proc. Natl. Acad. Sci. U.S.A. *95*, 10797-10802.

Geluk, A., van Meijgaarden, K.E., Franken, K.L., Drijfhout, J.W., D'Souza, S., Necker, A., Huygen, K., and Ottenhoff, T.H. (2000). Identification of major epitopes of *Mycobacterium tuberculosis* AG85B that are recognized by HLA-A*0201-restricted CD8+ T cells in HLA-transgenic mice and humans. J. Immunol. *165*, 6463-6471.

Genetics Computer Group (1997). SPScan (Madison, WI, Genetics Computer Group).

Gennaro, M.L. (2000). Immunologic diagnosis of tuberculosis. Clin. Infect. Dis. *30 Suppl 3*, S243-246.

Gey Van Pittius, N.C., Gamieldien, J., Hide, W., Brown, G.D., Siezen, R.J., and Beyers, A.D. (2001). The ESAT-6 gene cluster of *Mycobacterium tuberculosis* and other high G+C Gram-positive bacteria. Genome Biol. *2*, RESEARCH0044.1-0044.18.

Godfrey, H.P., Feng, Z., Mandy, S., Mandy, K., Huygen, K., De Bruyn, J., Abou-Zeid, C., Wiker, H.G., Nagai, S., and Tasaka, H. (1992). Modulation of expression of delayed hypersensitivity by mycobacterial antigen 85 fibronectin-binding proteins. Infect. Immun. *60*, 2522-2528.

Gomez, M., Johnson, S., and Gennaro, M.L. (2000). Identification of secreted proteins of *Mycobacterium tuberculosis* by a bioinformatic approach. Infect. Immun. *68*, 2323-2327.

Goonetilleke, N.P., McShane, H., Hannan, C.M., Anderson, R.J., Brookes, R.H., and Hill, A.V. (2003). Enhanced immunogenicity and protective efficacy against *Mycobacterium tuberculosis* of bacille Calmette-Guerin vaccine using mucosal administration and boosting with a recombinant modified vaccinia virus Ankara. J. Immunol. *171*, 1602-1609.

Gu, S., Chen, J., Dobos, K.M., Bradbury, E.M., Belisle, J.T., and Chen, X. (2003). Comprehensive Proteomic Profiling of the Membrane Constituents of a *Mycobacterium tuberculosis* Strain. Mol. Cell Proteomics 2, 1284-1296.

Guinn, K.M., Hickey, M.J., Mathur, S.K., Zakel, K.L., Grotzke, J.E., Lewinsohn, D.M., Smith, S., and Sherman, D.R. (2004). Individual RD1-region genes are required for export of ESAT-6/CFP-10 and for virulence of *Mycobacterium tuberculosis*. Mol. Microbiol. *51*, 359-370.

Hakansson, S., Galyov, E.E., Rosqvist, R., and Wolf-Watz, H. (1996). The Yersinia YpkA Ser/Thr kinase is translocated and subsequently targeted to the inner surface of the HeLa cell plasma membrane. Mol. Microbiol. *20*, 593-603.

Harboe, M., Oettinger, T., Wiker, H.G., Rosenkrands, I., and Andersen, P. (1996). Evidence for occurrence of the ESAT-6 protein in *Mycobacterium tuberculosis* and virulent *Mycobacterium bovis* and for its absence in *Mycobacterium bovis* BCG. Infect. Immun. *64*, 16-22.

Harris, D.P., Vordermeier, H.M., Friscia, G., Roman, E., Surcel, H.M., Pasvol, G., Moreno, C., and Ivanyi, J. (1993). Genetically permissive recognition of adjacent epitopes from the 19-kDa antigen of *Mycobacterium tuberculosis* by human and murine T cells. J. Immunol. *150*, 5041-5050.

Harris, D.P., Vordermeier, H.M., Roman, E., Lathigra, R., Brett, S.J., Moreno, C., and Ivanyi, J. (1991). Murine T cell-stimulatory peptides from the 19-kDa antigen of *Mycobacterium tuberculosis*. Epitope-restricted homology with the 28-kDa protein of *Mycobacterium leprae*. J. Immunol. *147*, 2706-2712.

Harth, G., and Horwitz, M.A. (1999a). Export of recombinant *Mycobacterium tuberculosis* superoxide dismutase is dependent upon both information in the protein and mycobacterial export machinery. A model for studying export of leaderless proteins by pathogenic mycobacteria. J. Biol. Chem. *274*, 4281-4292.

Harth, G., and Horwitz, M.A. (1999b). An inhibitor of exported *Mycobacterium tuberculosis* glutamine synthetase selectively blocks the growth of pathogenic mycobacteria in axenic culture and in human monocytes: extracellular proteins as potential novel drug targets. J. Exp. Med. *189*, 1425-1436.

Harth, G., Lee, B.Y., Wang, J., Clemens, D.L., and Horwitz, M.A. (1996). Novel insights into the genetics, biochemistry, and immunocytochemistry of the 30-kilodalton major extracellular protein of *Mycobacterium tuberculosis*. Infect. Immun. *64*, 3038-3047.

Harth, G., Zamecnik, P.C., Tang, J.Y., Tabatadze, D., and Horwitz, M.A. (2000). Treatment of *Mycobacterium tuberculosis* with antisense oligonucleotides to glutamine synthetase mRNA inhibits glutamine synthetase activity, formation of the poly-L-glutamate/glutamine cell wall structure, and bacterial replication. Proc. Natl. Acad. Sci. U.S.A. *97*, 418-423.

Hartl, F.U., Lecker, S., Schiebel, E., Hendrick, J.P., and Wickner, W. (1990). The binding cascade of SecB to SecA to SecY/E mediates preprotein targeting to the *E. coli* plasma membrane. Cell *63*, 269-279.

Hatzixanthis, K., Palmer, T., and Sargent, F. (2003). A subset of bacterial inner membrane proteins integrated by the twin-arginine translocase. Mol. Microbiol. *49*, 1377-1390.

Havlir, D.V., Wallis, R.S., Boom, W.H., Daniel, T.M., Chervenak, K., and Ellner, J.J. (1991). Human immune response to *Mycobacterium tuberculosis* antigens. Infect. Immun. *59*, 665-670.

Heldwein, K.A., and Fenton, M.J. (2002). The role of Toll-like receptors in immunity against mycobacterial infection. Microbes Infect. *4*, 937-944.

Heldwein, K.A., Liang, M.D., Andresen, T.K., Thomas, K.E., Marty, A.M., Cuesta, N., Vogel, S. N., and Fenton, M.J. (2003). TLR2 and TLR4 serve distinct roles in the host immune response against *Mycobacterium bovis* BCG. J. Leukoc Biol. *74*, 277-286.

Hendrickson, R.C., Douglass, J.F., Reynolds, L.D., McNeill, P.D., Carter, D., Reed, S.G., and Houghton, R.L. (2000). Mass spectrometric identification of mtb81, a novel serological marker for tuberculosis. J. Clin. Microbiol. *38*, 2354-2361.

Hewinson, R.G., Michell, S.L., Russell, W.P., McAdam, R.A., and Jacobs, W.J. (1996). Molecular characterization of MPT83: a seroreactive antigen of *Mycobacterium tuberculosis* with homology to MPT70. Scand. J. Immunol. *43*, 490-499.

Hinsley, A.P., Stanley, N.R., Palmer, T., and Berks, B.C. (2001). A naturally occurring bacterial Tat signal peptide lacking one of the 'invariant' arginine residues of the consensus targeting motif. FEBS Lett. *497*, 45-49.

Hisert, K.B., Kirksey, M.A., Gomez, J.E., Sousa, A.O., Cox, J.S., Jacobs, W.R., Jr., Nathan, C. F., and McKinney, J.D. (2004). Identification of *Mycobacterium tuberculosis* counterimmune (cim) mutants in immunodeficient mice by differential screening. Infect. Immun. *72*, 5315-5321.

Hofmann, K., and Stoffel, W. (1993). TMbase - A database of membrane spanning proteins segments. Biol. Chem. Hoppe-Seyler *374,*, 166-170.

Horwitz, M.A., and Harth, G. (2003). A new vaccine against tuberculosis affords greater survival after challenge than the current vaccine in the guinea pig model of pulmonary tuberculosis. Infect. Immun. *71*, 1672-1679.

Horwitz, M.A., Harth, G., Dillon, B.J., and Maslesa-Galic, S. (2000). Recombinant bacillus calmette-guerin (BCG) vaccines expressing the *Mycobacterium tuberculosis* 30-kDa major secretory protein induce greater protective immunity against tuberculosis than conventional BCG vaccines in a highly susceptible animal model. Proc. Natl. Acad. Sci. U.S.A. *97*, 13853-13858.

Horwitz, M.A., Lee, B.W., Dillon, B.J., and Harth, G. (1995). Protective immunity against tuberculosis induced by vaccination with major extracellular proteins of *Mycobacterium tuberculosis*. Proc. Natl. Acad. Sci. U.S.A. *92*, 1530-1534.

Hovav, A.H., Mullerad, J., Davidovitch, L., Fishman, Y., Bigi, F., Cataldi, A., and Bercovier, H. (2003). The *Mycobacterium tuberculosis* recombinant 27-kilodalton lipoprotein induces a strong Th1-type immune response deleterious to protection. Infect. Immun. *71*, 3146-3154.

Hsu, T., Hingley-Wilson, S.M., Chen, B., Chen, M., Dai, A.Z., Morin, P.M., Marks, C.B.,

Padiyar, J., Goulding, C., Gingery, M., et al. (2003). The primary mechanism of attenuation of bacillus Calmette-Guerin is a loss of secreted lytic function required for invasion of lung interstitial tissue. Proc. Natl. Acad. Sci. U.S.A. *100*, 12420-12425.

Huard, R.C., Chitale, S., Leung, M., Lazzarini, L.C., Zhu, H., Shashkina, E., Laal, S., Conde, M. B., Kritski, A.L., Belisle, J.T., et al. (2003). The *Mycobacterium tuberculosis* complex-restricted gene *cfp32* encodes an expressed protein that is detectable in tuberculosis patients and is positively correlated with pulmonary interleukin-10. Infect. Immun. *71*, 6871-6883.

Hubbard, R.D., Flory, C.M., and Collins, F.M. (1992). Immunization of mice with mycobacterial culture filtrate proteins. Clin. Exp. Immunol. *87*, 94-98.

Huygen, K., Content, J., Denis, O., Montgomery, D.L., Yawman, A.M., Deck, R.R., DeWitt, C. M., Orme, I.M., Baldwin, S., D'Souza, C., et al. (1996). Immunogenicity and protective efficacy of a tuberculosis DNA vaccine. Nat. Med. *2*, 893-898.

Huygen, K., Drowart, A., Harboe, M., ten Berg, R., Cogniaux, J., and Van Vooren, J.P. (1993). Influence of genes from the major histocompatibility complex on the antibody repertoire against culture filtrate antigens in mice infected with live *Mycobacterium bovis* BCG. Infect. Immun. *61*, 2687-2693.

Huygen, K., Ljungqvist, L., ten Berg, R., and Van Vooren, J.P. (1990a). Repertoires of antibodies to culture filtrate antigens in different mouse strains infected with *Mycobacterium bovis* BCG. Infect. Immun. *58*, 2192-2197.

Huygen, K., Palfliet, K., Jurion, F., Lenoir, C., and van Vooren, J.P. (1990b). Antibody repertoire against culture filtrate antigens in wild house mice infected with *Mycobacterium bovis* BCG. Clin. Exp. Immunol. *82*, 369-372.

Ignatova, Z., Hornle, C., Nurk, A., and Kasche, V. (2002). Unusual signal peptide directs penicillin amidase from *Escherichia coli* to the Tat translocation machinery. Biochem. Biophys Res. Commun *291*, 146-149.

Ize, B., Gerard, F., Zhang, M., Chanal, A., Voulhoux, R., Palmer, T., Filloux, A., and Wu, L.F. (2002). In vivo dissection of the Tat translocation pathway in *Escherichia coli*. J. Mol. Biol. *317*, 327-335.

Jackett, P.S., Bothamley, G.H., Batra, H.V., Mistry, A., Young, D.B., and Ivanyi, J. (1988). Specificity of antibodies to immunodominant mycobacterial antigens in pulmonary tuberculosis. J. Clin. Microbiol. *26*, 2313-2318.

Jackson, M., Portnoi, D., Catheline, D., Dumail, L., Rauzier, J., Legrand, P., and Gicquel, B.
(1997). *Mycobacterium tuberculosis* Des protein: an immunodominant target for the humoral response of tuberculous patients. Infect. Immun. *65*, 2883-2889.

Jackson, M., Raynaud, C., Laneelle, M.A., Guilhot, C., Laurent-Winter, C., Ensergueix, D., Gicquel, B., and Daffe, M. (1999). Inactivation of the antigen 85C gene profoundly affects the mycolate content and alters the permeability of the *Mycobacterium tuberculosis* cell envelope. Mol. Microbiol. *31*, 1573-1587.

Johansen, K.A., Gill, R.E., and Vasin, M.L. (1996). Biochemical and molecular analysis of phospholipase C and phospholipase D activity in mycobacteria. Infect. Immun. *64*, 3259-3266.

Johnson, C.M., Cooper, A.M., Frank, A.A., Bonorino, C.B., Wysoki, L.J., and Orme, I.M. (1997). *Mycobacterium tuberculosis* aerogenic rechallenge infections in B cell-deficient mice. Tuber Lung Dis. *78*, 257-261.

Jungblut, P.R., Schaible, U.E., Mollenkopf, H., Zimny-Arndt, U., Raupach, B., Mattow, J., Halada, P., Lamer, S., Hagens, K., and Kaufmann, S.H. (1999). Comparative proteome analysis of *Mycobacterium tuberculosis* and *Mycobacterium bovis* BCG strains: towards functional genomics of microbial pathogens. Mol. Microbiol. *33*, 1103-1117.

Kamath, A.B., Alt, J., Debbabi, H., and Behar, S.M. (2003). Toll-like receptor 4-defective C3H/HeJ mice are not more susceptible than other C3H substrains to infection with *Mycobacterium tuberculosis*. Infect. Immun. *71*, 4112-4118.

Kamath, A.T., Feng, C.G., Macdonald, M., Briscoe, H., and Britton, W.J. (1999a). Differential protective efficacy of DNA vaccines expressing secreted proteins of *Mycobacterium tuberculosis*. Infect. Immun. *67*, 1702-1707.

Kamath, A.T., Hanke, T., Briscoe, H., and Britton, W.J. (1999b). Co-immunization with DNA vaccines expressing granulocyte-macrophage colony-stimulating factor and mycobacterial secreted proteins enhances T-cell immunity, but not protective efficacy against *Mycobacterium tuberculosis*. Immunology *96*, 511-516.

Kasik, J.E. (1979). Mycobacterial Beta-lactamases. In Beta-lactamases, J.M.T. Hamilton-Miller, and J.T. Smith, eds. (New York, NY, Academic Press), pp. 339-350.

Klein, M.R., and Fox, A. (2001). Mycobacterium-specific human CD8 T cell responses. Arch. Immunol. Ther Exp. (Warsz) *49*, 379-389.

Klein, M.R., Smith, S.M., Hammond, A.S., Ogg, G.S., King, A.S., Vekemans, J., Jaye, A., Lukey, P.T., and McAdam, K.P. (2001). HLA-B*35-restricted CD8 T cell epitopes in the antigen 85 complex of *Mycobacterium tuberculosis*. J. Infect. Dis. *183*, 928-934.

Kolattukudy, P.E., Fernandes, N.D., Azad, A.K., Fitzmaurice, A.M., and Sirakova, T.D. (1997). Biochemistry and molecular genetics of cell-wall lipid biosynthesis in mycobacteria. Mol. Microbiol. *24*, 263-270.

Koul, A., Choidas, A., Tyagi, A.K., Drlica, K., Singh, Y., and Ullrich, A. (2001). Serine/threonine protein kinases PknF and PknG of *Mycobacterium tuberculosis*: characterization and localization. Microbiology *147*, 2307-2314.

Leao, S.C., Rocha, C.L., Murillo, L.A., Parra, C.A., and Patarroyo, M.E. (1995). A species-specific nucleotide sequence of *Mycobacterium tuberculosis* encodes a protein that exhibits hemolytic activity when expressed in *Escherichia coli*. Infect. Immun. *63*, 4301-4306.

Lee, B.Y., Hefta, S.A., and Brennan, P.J. (1992). Characterization of the major membrane protein of virulent *Mycobacterium tuberculosis*. Infect. Immun. *60*, 2066-2074.

Lee, B.Y., and Horwitz, M.A. (1999). T-cell epitope mapping of the three most abundant extracellular proteins of *Mycobacterium tuberculosis* in outbred guinea pigs. Infec Immunity *67*, 2665-2670.

Lefevre, P., Denis, O., De Wit, L., Tanghe, A., Vandenbussche, P., Content, J., and Huygen, K. (2000). Cloning of the gene encoding a 22-kilodalton cell surface antigen of *Mycobacterium bovis* BCG and analysis of its potential for DNA vaccination against tuberculosis. Infect. Immun. *68*, 1040-1047.

Lenz, L.L., Mohammadi, S., Geissler, A., and Portnoy, D.A. (2003). SecA2-dependent secretion of autolytic enzymes promotes *Listeria monocytogenes* pathogenesis. Proc. Natl. Acad. Sci. U.S.A. *100*, 12432-12437.

Lenz, L.L., and Portnoy, D.A. (2002). Identification of a second *Listeria secA* gene associated with protein secretion and the rough phenotype. Mol. Microbiol. *45*, 1043-1056.

Lewis, K.N., Liao, R., Guinn, K.M., Hickey, M.J., Smith, S., Behr, M.A., and Sherman, D.R. (2003). Deletion of RD1 from *Mycobacterium tuberculosis* mimics bacille Calmette-Guerin attenuation. J. Infect. Dis. *187*, 117-123.

Lewthwaite, J.C., Coates, A.R., Tormay, P., Singh, M., Mascagni, P., Poole, S., Roberts, M.,

Sharp, L., and Henderson, B. (2001). *Mycobacterium tuberculosis* chaperonin 60.1 is a more potent cytokine stimulator than chaperonin 60.2 (Hsp 65) and contains a CD14-binding domain. Infect. Immun. *69*, 7349-7355.

Lim, E.M., Rauzier, J., Timm, J., Torrea, G., Murray, A., Gicquel, B., and Portnoi, D. (1995). Identification of *Mycobacterium tuberculosis* DNA sequences encoding exported proteins by using *phoA* gene fusions. J. Bacteriol. *177*, 59-65.

Lim, J.H., Kim, H.J., Lee, K.S., Jo, E.K., Song, C.H., Jung, S.B., Kim, S.Y., Lee, J.S., Paik, T. H., and Park, J.K. (2004). Identification of the new T-cell-stimulating antigens from *Mycobacterium tuberculosis* culture filtrate. FEMS Microbiol. Lett. *232*, 51-59.

Ljungqvist, L., Worsaae, A., and Heron, I. (1988). Antibody responses against *Mycobacterium tuberculosis* in 11 strains of inbred mice: novel monoclonal antibody specificities generated by fusions, using spleens from BALB.B10 and CBA/J mice. Infect. Immun. *56*, 1994-1998.

Lodes, M.J., Dillon, D.C., Mohamath, R., Day, C.H., Benson, D.R., Reynolds, L.D., McNeill, P., Sampaio, D.P., Skeiky, Y.A., Badaro, R., et al. (2001). Serological expression cloning and immunological evaluation of MTB48, a novel *Mycobacterium tuberculosis* antigen. J. Clin. Microbiol. *39*, 2485-2493.

Lopez, M., Sly, L.M., Luu, Y., Young, D., Cooper, H., and Reiner, N.E. (2003). The 19-kDa *Mycobacterium tuberculosis* protein induces macrophage apoptosis through Toll-like receptor-2. J. Immunol. *170*, 2409-2416.

Lowrie, D.B., Silva, C.L., and Tascon, R.E. (1997). DNA vaccines against tuberculosis. Immunol. Cell Biol. *75*, 591-594.

Lozes, E., Huygen, K., Content, J., Denis, O., Montgomery, D.L., Yawman, A.M.,

Vandenbussche, P., Van Vooren, J.P., Drowart, A., Ulmer, J.B., and Liu, M.A. (1997). Immunogenicity and efficacy of a tuberculosis DNA vaccine encoding the components of the secreted antigen 85 complex. Vaccine *15*, 830-833.

Mahairas, G.G., Sabo, P.J., Hickey, M.J., Singh, D.C., and Stover, C.K. (1996). Molecular analysis of genetic differences between *Mycobacterium bovis* BCG and virulent *M. bovis*. J. Bacteriol. *178*, 1274-1282.

Majlessi, L., Rojas, M.J., Brodin, P., and Leclerc, C. (2003). CD8+-T-cell responses of Mycobacterium-infected mice to a newly identified major histocompatibility complex class I-restricted epitope shared by proteins of the ESAT-6 family. Infect. Immun. *71*, 7173-7177.

Malhotra, V., Sharma, D., Ramanathan, V.D., Shakila, H., Saini, D.K., Chakravorty, S., Das, T. K., Li, Q., Silver, R.F., Narayanan, P.R., and Tyagi, J.S. (2004). Disruption of response regulator gene, *devR*, leads to attenuation in virulence of *Mycobacterium tuberculosis*. FEMS Microbiol. Lett. *231*, 237-245.

Malin, A.S., Huygen, K., Content, J., Mackett, M., Brandt, L., Andersen, P., Smith, S.M., and Dockrell, H.M. (2000). Vaccinia expression of *Mycobacterium tuberculosis*-secreted proteins: tissue plasminogen activator signal sequence enhances expression and immunogenicity of *M. tuberculosis* Ag85. Microbes Infect. *2*, 1677-1685.

Manca, C., Lyashchenko, K., Colangeli, R., and Gennaro, M.L. (1997a). MTC28, a novel 28-kilodalton proline-rich secreted antigen specific for the *Mycobacterium tuberculosis* complex. Infect. Immun. *65*, 4951-4957.

Manca, C., Lyashchenko, K., Wiker, H.G., Usai, D., Colangeli, R., and Gennaro, M.L. (1997b). Molecular cloning, purification, and serological characterization of MPT63, a novel antigen secreted by *Mycobacterium tuberculosis*. Infect. Immun. *65*, 16-23.

Manoil, C., and Beckwith, J. (1985). Tn*phoA*: a transposon probe for protein export signals. Proc. Natl. Acad. Sci. U.S.A. *82*, 8129-8133.

Mattow, J., Schaible, U.E., Schmidt, F., Hagens, K., Siejak, F., Brestrich, G., Haeselbarth, G., Muller, E.C., Jungblut, P.R., and Kaufmann, S.H. (2003). Comparative proteome analysis of culture supernatant proteins from virulent

Mycobacterium tuberculosis H37Rv and attenuated *M. bovis* BCG Copenhagen. Electrophoresis *24*, 3405-3420.

Mazmanian, S.K., Ton-That, H., and Schneewind, O. (2001). Sortase-catalysed anchoring of surface proteins to the cell wall of *Staphylococcus aureus*. Mol. Microbiol. *40*, 1049-1057.

McShane, H., Brookes, R., Gilbert, S.C., and Hill, A.V. (2001). Enhanced immunogenicity of CD4(+) t-cell responses and protective efficacy of a DNA-modified vaccinia virus Ankara prime-boost vaccination regimen for murine tuberculosis. Infect. Immun. *69*, 681-686.

Meister, G.E., Roberts, C.G., Berzofsky, J.A., and De Groot, A.S. (1995). Two novel T cell epitope prediction algorithms based on MHC-binding motifs; comparison of predicted and published epitopes from *Mycobacterium tuberculosis* and HIV protein sequences. Vaccine *13*, 581-591.

Menozzi, F.D., Bischoff, R., Fort, E., Brennan, M.J., and Locht, C. (1998). Molecular characterization of the mycobacterial heparin-binding hemagglutinin, a mycobacterial adhesin. Proc. Natl. Acad. Sci. USA. *95*, 12625-12630.

Menozzi, F.D., Rouse, J.H., Alavi, M., Laude-Sharp, M., Muller, J., Bischoff, R., Brennan, M.J., and Locht, C. (1996). Identification of a heparin-binding hemagglutinin present in mycobacteria. J. Exp. Med. *184*, 993-1001.

Mogues, T., Goodrich, M.E., Ryan, L., LaCourse, R., and North, R.J. (2001). The relative importance of T cell subsets in immunity and immunopathology of airborne *Mycobacterium tuberculosis* infection in mice. J. Exp. Med. *193*, 271-280.

Mohagheghpour, N., Gammon, D., Kawamura, L.M., vanVollenhoven, A., Benike, C.J., and Engleman, E.G. (1998). CTL response to *Mycobacterium tuberculosis*: Identification of an immunogenic epitope in the 19-kDa lipoprotein. J. Immunol. *161*, 2400-2406.

Mollenkopf, H.J., Groine-Triebkorn, D., Andersen, P., Hess, J., and Kaufmann, S.H. (2001). Protective efficacy against tuberculosis of ESAT-6 secreted by a live *Salmonella typhimurium* vaccine carrier strain and expressed by naked DNA. Vaccine *19*, 4028-4035.

Mollenkopf, H.J., Jungblut, P.R., Raupach, B., Mattow, J., Lamer, S., Zimny-Arndt, U., Schaible, U.E., and Kaufmann, S.H. (1999). A dynamic two-dimensional polyacrylamide gel electrophoresis database: the mycobacterial proteome via Internet. Electrophoresis *20*, 2172-2180.

Monahan, I.M., Betts, J., Banerjee, D.K., and Butcher, P.D. (2001). Differential expression of mycobacterial proteins following phagocytosis by macrophages. Microbiology *147*, 459-471.

Morris, S., Kelley, C., Howard, A., Li, Z., and Collins, F. (2000). The immunogenicity of single and combination DNA vaccines against tuberculosis. Vaccine *18*, 2155-2163.

Mueller-Ortiz, S.L., Sepulveda, E., Olsen, M.R., Jagannath, C., Wanger, A.R., and Norris, S.J. (2002). Decreased infectivity despite unaltered C3 binding by a Delta*hbhA* mutant of *Mycobacterium tuberculosis*. Infect. Immun. *70*, 6751-6760.

Mustafa, A.S. (2002). Development of new vaccines and diagnostic reagents against tuberculosis. Mol. Immunol. *39*, 113-119.

Mustafa, A.S., Shaban, F.A., Al-Attiyah, R., Abal, A.T., El-Shamy, A.M., Andersen, P., and Oftung, F. (2003). Human Th1 cell lines recognize the *Mycobacterium tuberculosis* ESAT-6 antigen and its peptides in association with frequently expressed HLA class II molecules. Scand. J. Immunol. *57*, 125-134.

Nagai, S., Wiker, H.G., Harboe, M., and Kinomoto, M. (1991). Isolation and partial characterization of major protein antigens in the culture fluid of *Mycobacterium tuberculosis*. Infect. Immun. *59*, 372-382.

Naito, M., Izumi, S., and Yamada, T. (1998). Two-dimensional electrophoretic analysis of humoral responses to culture filtrate of *Mycobacterium bovis* BCG in patients with leprosy and tuberculosis. Int. J. Lepr Mycobact Dis. *66*, 208-213.

Nakai, K., and Kanehisa, M. (1991). Expert system for predicting protein localization sites in gram-negative bacteria. Proteins *11*, 95-110.

Nau, G.J., Richmond, J.F., Schlesinger, A., Jennings, E.G., Lander, E.S., and Young, R.A. (2002). Human macrophage activation programs induced by bacterial pathogens. Proc. Natl. Acad. Sci. U.S.A. *99*, 1503-1508.

Neyrolles, O., Gould, K., Gares, M.P., Brett, S., Janssen, R., O'Gaora, P., Herrmann, J.L., Prevost, M.C., Perret, E., Thole, J.E., and Young, D. (2001). Lipoprotein access to MHC class I presentation during infection of murine macrophages with live mycobacteria. J. Immunol. *166*, 447-457.

Ng, V.H., Cox, J.S., Sousa, A.O., MacMicking, J.D., and McKinney, J.D. (2004). Role of KatG catalase-peroxidase in mycobacterial pathogenesis: countering the phagocyte oxidative burst. Mol. Microbiol. *52*, 1291-1302.

Noss, E.H., Pai, R.K., Sellati, T.J., Radolf, J.D., Belisle, J., Golenbock, D.T., Boom, W.H., and Harding, C.V. (2001). Toll-like receptor 2-dependent inhibition of macrophage class II MHC expression and antigen processing by 19-kDa lipoprotein of *Mycobacterium tuberculosis*. J. Immunol. *167*, 910-918.

Ochsner, U.A., Snyder, A., Vasil, A.I., and Vasil, M.L. (2002). Effects of the twin-arginine translocase on secretion of virulence factors, stress response, and pathogenesis. Proc. Natl. Acad. Sci. U.S.A. *99*, 8312-8317.

Okkels, L.M., and Andersen, P. (2004). Protein-protein interactions of proteins from the ESAT-6 family of *Mycobacterium tuberculosis*. J. Bacteriol. *186*, 2487-2491.

Olsen, A.W., and Andersen, P. (2003). A novel TB vaccine; strategies to combat a complex pathogen. Immunol. Lett. *85*, 207-211.

Ortalo-Magne, A., Dupont, M.A., Lemassu, A., Andersen, A.B., Gounon, P., and Daffe, M. (1995). Molecular composition of the outermost capsular material of the tubercle bacillus. Microbiology *141*, 1609-1620.

Ostroff, R.M., Vasil, A.I., and Vasil, M.L. (1990). Molecular comparison of a nonhemolytic and a hemolytic phospholipase C from *Pseudomonas aeruginosa*. J. Bacteriol. *172*, 5915-5923.

Owens, M.U., Swords, W.E., Schmidt, M.G., King, C.H., and Quinn, F.D. (2002). Cloning, expression, and functional characterization of the *Mycobacterium tuberculosis secA* gene. FEMS Microbiol. Lett. *211*, 133-141.

Pai, R.K., Convery, M., Hamilton, T.A., Boom, W.H., and Harding, C.V. (2003). Inhibition of IFN-gamma-induced class II transactivator expression by a 19-kDa lipoprotein from *Mycobacterium tuberculosis*: a potential mechanism for immune evasion. J. Immunol. *171*, 175-184.

Pal, P.G., and Horwitz, M.A. (1992). Immunization with extracellular proteins of *Mycobacterium tuberculosis* induces cell-mediated immune responses and substantial protective immunity in a guinea pig model of pulmonary tuberculosis. Infect. Immun. *60*, 4781-4792.

Pallen, M.J. (2002). The ESAT-6/WXG100 superfamily -- and a new Gram-positive secretion system? Trends Microbiol. *10*, 209-212.

Palmer, T., and Berks, B.C. (2003). Moving folded proteins across the bacterial cell membrane. Microbiology *149*, 547-556.

Panigada, M., Sturniolo, T., Besozzi, G., Boccieri, M.G., Sinigaglia, F., Grassi, G.G., and Grassi, F. (2002). Identification of a promiscuous T-cell epitope in *Mycobacterium tuberculosis* Mce proteins. Infect. Immun. *70*, 79-85.

Parish, T., Smith, D.A., Roberts, G., Betts, J., and Stoker, N.G. (2003). The *senX3-regX3* two-component regulatory system of *Mycobacterium tuberculosis* is required for virulence. Microbiology *149*, 1423-1435.

Parish, T., and Stoker, N.G. (2000). Use of a flexible cassette method to generate a double unmarked *Mycobacterium tuberculosis tlyA plcABC* mutant by gene replacement. Microbiology *146 (Pt 8)*, 1969-1975.

Parra, C.A., Londono, L.P., Del Portillo, P., and Patarroyo, M.E. (1991). Isolation, characterization, and molecular cloning of a specific *Mycobacterium tuberculosis* antigen gene: identification of a species-specific sequence. Infect. Immun. *59*, 3411-3417.

Perez, E., Samper, S., Bordas, Y., Guilhot, C., Gicquel, B., and Martin, C. (2001). An essential role for *phoP* in *Mycobacterium tuberculosis* virulence. Mol. Microbiol. *41*, 179-187.

Pethe, K., Alonso, S., Biet, F., Delogu, G., Brennan, M.J., Locht, C., and Menozzi, F.D. (2001). The heparin binding haemagglutinin of *M. tuberculosis* is required for extrapulmonary dissemination. Nature *412*, 190-194.

Piddington, D.L., Fang, F.C., Laessig, T., Cooper, A.M., Orme, I.M., and Buchmeier, N.A. (2001). Cu,Zn superoxide dismutase of *Mycobacterium tuberculosis* contributes to survival in activated macrophages that are generating an oxidative burst. Infect. Immun. *69*, 4980-4987.

Plano, G.V., Day, J.B., and Ferracci, F. (2001). Type III export: new uses for an old pathway.

Mol Micro *40*, 284-293.

Pollock, J.M., and Andersen, P. (1997). The potential of the ESAT-6 antigen secreted by virulent mycobacteria for specific diagnosis of tuberculosis. J. Infect. Dis. *175*, 1251-1254.

Pope, R.M., Wallis, R.S., Sailer, D., Buchanan, T.M., and Pahlavani, M.A. (1991). T cell activation by mycobacterial antigens in inflammatory synovitis. Cell Immunol. *133*, 95-108.

Poquet, I., Ehrlich, S.D., and Gruss, A. (1998). An export-specific reporter designed for gram-positive bacteria: application to *Lactococcus lactis*. J. Bacteriol. *180*, 1904-1912.

Porcelli, I., de Leeuw, E., Wallis, R., van den Brink-van der Laan, E., de Kruijff, B., Wallace, B. A., Palmer, T., and Berks, B.C. (2002). Characterization and membrane assembly of the TatA component of the *Escherichia coli* twin-arginine protein transport system. Biochemistry *41*, 13690-13697.

Post, F.A., Manca, C., Neyrolles, O., Ryffel, B., Young, D.B., and Kaplan, G. (2001). *Mycobacterium tuberculosis* 19-kilodalton lipoprotein inhibits *Mycobacterium smegmatis*-induced cytokine production by human macrophages *in vitro*. Infect. Immun. *69*, 1433-1439.

Pradel, N., Ye, C., Livrelli, V., Xu, J., Joly, B., and Wu, L.F. (2003). Contribution of the Twin Arginine Translocation System to the Virulence of Enterohemorrhagic *Escherichia coli* O157:H7. Infect. Immun. *71*, 4908-4916.

Pugsley, A.P. (1993). The complete general secretory pathway in gram-negative bacteria. Microbiol. Rev. *57*, 50-108.

Pym, A.S., Brodin, P., Brosch, R., Huerre, M., and Cole, S.T. (2002). Loss of RD1 contributed to the attenuation of the live tuberculosis vaccines *Mycobacterium bovis* BCG and

Mycobacterium microti. Mol. Microbiol. *46*, 709-717.

Pym, A.S., Brodin, P., Majlessi, L., Brosch, R., Demangel, C., Williams, A., Griffiths, K.E., Marchal, G., Leclerc, C., and Cole, S.T. (2003). Recombinant BCG exporting ESAT-6 confers enhanced protection against tuberculosis. Nat. Med. *9*, 533-539.

Quesniaux, V., Fremond, C., Jacobs, M., Parida, S., Nicolle, D., Yeremeev, V., Bihl, F., Erard, F., Botha, T., Drennan, M., et al. (2004). Toll-like receptor pathways in the immune responses to mycobacteria. Microbes Infect. *6*, 946-959.

Raynaud, C., Etienne, G., Peyron, P., Laneelle, M. A., and Daffe, M. (1998). Extracellular enzyme activities potentially involved in the pathogenicity of *Mycobacterium tuberculosis*. Microbiology *144*, 577-587.

Raynaud, C., Guilhot, C., Rauzier, J., Bordat, Y., Pelicic, V., Manganelli, R., Smith, I., Gicquel, B., and Jackson, M. (2002). Phospholipases C are involved in the virulence of *Mycobacterium tuberculosis*. Mol. Microbiol. *45*, 203-217.

Recchi, C., Rauzier, J., Gicquel, B., and Reyrat, J. M. (2002). Signal-sequence-independent secretion of the staphylococcal nuclease in *Mycobacterium smegmatis*. Microbiology *148*, 529-536.

Reiling, N., Holscher, C., Fehrenbach, A., Kroger, S., Kirschning, C.J., Goyert, S., and Ehlers, S. (2002). Cutting edge: Toll-like receptor (TLR)2- and TLR4-mediated pathogen recognition in resistance to airborne infection with *Mycobacterium tuberculosis*. J. Immunol. *169*, 3480-3484.

Renshaw, P.S., Panagiotidou, P., Whelan, A., Gordon, S.V., Hewinson, R.G., Williamson, R.A., and Carr, M.D. (2002). Conclusive evidence that the major T-cell antigens of the *Mycobacterium tuberculosis* complex ESAT-6 and CFP-10 form a tight, 1:1 complex and characterization of the structural properties of ESAT-6, CFP-10, and the ESAT-6*CFP-10 complex. Implications for pathogenesis and virulence. J. Biol. Chem. *277*, 21598-21603.

Roberts, A.D., Sonnenberg, M.G., Ordway, D.J., Furney, S.K., Brennan, P.J., Belisle, J.T., and Orme, I.M. (1995). Characteristics of protective immunity engendered by vaccination of mice with purified culture filtrate protein antigens of *Mycobacterium tuberculosis*. Immunology *85*, 502-508.

Rodrigue, A., Chanal, A., Beck, K., Muller, M., and Wu, L.F. (1999). Co-translocation of a periplasmic enzyme complex by a hitchhiker mechanism through the bacterial tat pathway. J. Biol. Chem. *274*, 13223-13228.

Rosenkrands, I., King, A., Weldingh, K., Moniatte, M., Moertz, E., and Andersen, P. (2000a). Towards the proteome of *Mycobacterium tuberculosis*. Electrophoresis *21*, 3740-3756.

Rosenkrands, I., Rasmussen, P.B., Carnio, M., Jacobsen, S., Theisen, M., and Andersen, P. (1998). Identification and characterization of a 29-kilodalton protein from *Mycobacterium tuberculosis* culture filtrate recognized by mouse memory effector cells. Infect. Immun. *66*, 2728-2735.

Rosenkrands, I., Weldingh, K., Jacobsen, S., Hansen, C.V., Florio, W., Gianetri, I., and Andersen, P. (2000b). Mapping and identification of *Mycobacterium tuberculosis* proteins by two-dimensional gel electrophoresis, microsequencing and immunodetection. Electrophoresis *21*, 935-948.

Rousseau, C., Winter, N., Pivert, E., Bordat, Y., Neyrolles, O., Ave, P., Huerre, M., Gicquel, B., and Jackson, M. (2004). Production of phthiocerol dimycocerosates protects *Mycobacterium tuberculosis* from the cidal activity of reactive nitrogen intermediates produced by macrophages and modulates the early immune response to infection. Cell Microbiol. *6*, 277-287.

Rumschlag, H.S., Shinnick, T.M., and Cohen, M.L. (1988). Serological responses of patients with lepromatous and tuberculoid leprosy to 30-, 31-, and 32-kilodalton antigens of *Mycobacterium tuberculosis*. J. Clin. Microbiol. *26*, 2200-2202.

Saleh, M.T., Fillon, M., Brennan, P.J., and Belisle, J.T. (2001). Identification of putative exported/secreted proteins in prokaryotic proteomes. Gene *269*, 195-204.

Samanich, K.M., Belisle, J.T., Sonnenberg, M.G., Keen, M.A., Zolla-Pazner, S., and Laal, S. (1998). Delineation of human antibody responses to culture filtrate antigens of *Mycobacterium tuberculosis*. J. Infect. Dis. *178*, 1534-1538.

Samuelson, J.C., Chen, M., Jiang, F., Moller, I., Wiedmann, M., Kuhn, A., Phillips, G.J., and Dalbey, R.E. (2000). YidC mediates membrane protein insertion in bacteria. Nature *406*, 637-641.

Sander, P., Rezwan, M., Walker, B., Rampini, S.K., Kroppenstedt, R.M., Ehlers, S., Keller, C., Keeble, J.R., Hagemeier, M., Colston, M.J., et al. (2004). Lipoprotein processing is required for virulence of *Mycobacterium tuberculosis*. Mol. Microbiol. *52*, 1543-1552.

Sanger Institute (2003). *Corynebacterium diphtheriae* genome sequence. http://www.sanger.ac.uk/Projects/C_diphtheriae/

Santini, C.L., Ize, B., Chanal, A., Muller, M., Giordano, G., and Wu, L.F. (1998). A novel *sec*-independent periplasmic protein translocation pathway in *Escherichia coli*. Embo J. *17*, 101-112.

Sargent, F., Bogsch, E.G., Stanley, N.R., Wexler, M., Robinson, C., Berks, B.C., and Palmer, T. (1998). Overlapping functions of components of a bacterial Sec-independent protein export pathway. Embo J. *17*, 3640-3650.

Sargent, F., Gohlke, U., De Leeuw, E., Stanley, N.R., Palmer, T., Saibil, H.R., and Berks, B.C. (2001). Purified components of the *Escherichia coli* Tat protein transport system form a double-layered ring structure. Eur. J. Biochem. *268*, 3361-3367.

Sassetti, C.M., Boyd, D.H., and Rubin, E.J. (2003). Genes required for mycobacterial growth defined by high density mutagenesis. Mol. Microbiol. *48*, 77-84.

Sassetti, C.M., and Rubin, E.J. (2003). Genetic requirements for mycobacterial survival during infection. Proc. Natl. Acad. Sci. U.S.A. *100*, 12989-12994.

Schafer, J.R., Jesdale, B.M., George, J.A., Kouttab, N.M., and De Groot, A.S. (1998). Prediction of well-conserved HIV-1 ligands using a matrix-based algorithm, EpiMatrix. Vaccine *16*, 1880-1884.

Schiebel, E., Driessen, A.J., Hartl, F.U., and Wickner, W. (1991). Delta mu H+ and ATP function at different steps of the catalytic cycle of preprotein translocase. Cell *64*, 927-939.

Schmidt, M.G., and Kiser, K.B. (1999). SecA: the ubiquitous component of preprotein translocase in prokaryotes. Microbes Infect. *1*, 993-1004.

Schneider, G. (1999). How many potentially secreted proteins are contained in a bacterial genome? Gene *237*, 1113-1121.

Schwarzer, N., Nost, R., Seybold, J., Parida, S.K., Fuhrmann, O., Krull, M., Schmidt, R., Newton, R., Hippenstiel, S., Domann, E., et al. (1998). Two distinct phospholipases C of *Listeria monocytogenes* induce ceramide generation, nuclear factor-kappa B activation, and E-selectin expression in human endothelial cells. J. Immunol. *161*, 3010-3018.

Serek, J., Bauer-Manz, G., Struhalla, G., van den Berg, L., Kiefer, D., Dalbey, R., and Kuhn, A. (2004). *Escherichia coli* YidC is a membrane insertase for Sec-independent proteins. Embo J. *23*, 294-301.

Sherman, D.R., Guinn, K.M., Hickey, M.J., Mathur, S.K., Zakel, K.L., and Smith, S. (2004). *Mycobacterium tuberculosis* H37Rv: Delta RD1 is more virulent than *M. bovis* bacille Calmette-Guerin in long-term murine infection. J. Infect. Dis. *190*, 123-126.

Shim, T.S., Turner, O.C., and Orme, I.M. (2003). Toll-like receptor 4 plays no role in susceptibility of mice to *Mycobacterium tuberculosis* infection. Tuberculosis (Edinb) *83*, 367-371.

Shimono, N., Morici, L., Casali, N., Cantrell, S., Sidders, B., Ehrt, S., and Riley, L.W. (2003). Hypervirulent mutant of *Mycobacterium tuberculosis* resulting from disruption of the mce1 operon. Proc. Natl. Acad. Sci. U.S.A. *100*, 15918-15923.

Shinnick, T.M., Sweetser, D., Thole, J., van Embden, J., and Young, R.A. (1987). The etiologic agents of leprosy and tuberculosis share an immunoreactive protein antigen with the vaccine strain *Mycobacterium bovis* BCG. Infect. Immun. *55*, 1932-1935.

Silva, V.M., Kanaujia, G., Gennaro, M.L., and Menzies, D. (2003). Factors associated with humoral response to ESAT-6, 38 kDa and 14 kDa in patients with a spectrum of tuberculosis. Int. J. Tuberc Lung Dis. *7*, 478-484.

Singh, B., Singh, G., Trajkovic, V., Sharma, P. (2003a). Intracellular expression of *Mycobacterium tuberculosis* specific 10-kDa antigen down-regulates macrophage B7.1 expression and nitric oxide release. Clin. Epx Immunol. *134*, 70-77.

Singh, H., and Raghava, G.P. (2001). ProPred: prediction of HLA-DR binding sites. Bioinformatics *17*, 1236-1237.

Singh, K.K., Zhang, X., Patibandla, A.S., Chien, P., Jr., and Laal, S. (2001). Antigens of *Mycobacterium tuberculosis* expressed during preclinical tuberculosis: serological immunodominance of proteins with repetitive amino acid sequences. Infect. Immun. *69*, 4185-4191.

Singh, R., Rao, V., Shakila, H., Gupta, R., Khera, A., Dhar, N., Singh, A., Koul, A., Singh, Y., Naseema, M., et al. (2003b). Disruption of *mptpB* impairs the ability of *Mycobacterium tuberculosis* to survive in guinea pigs. Mol. Microbiol. *50*, 751-762.

Skeiky, Y.A., Lodes, M.J., Guderian, J.A., Mohamath, R., Bement, T., Alderson, M.R., and Reed, S.G. (1999). Cloning, expression, and immunological evaluation of two putative secreted serine protease antigens of *Mycobacterium tuberculosis*. Infect. Immun. *67*, 3998-4007.

Skjot, R.L., Brock, I., Arend, S.M., Munk, M.E., Theisen, M., Ottenhoff, T.H., and Andersen, P. (2002). Epitope mapping of the immunodominant antigen TB10.4 and the two homologous proteins TB10.3 and TB12.9, which constitute a subfamily of the *esat-6* gene family. Infect. Immun. *70*, 5446-5453.

Skjot, R.L., Oettinger, T., Rosenkrands, I., Ravn, P., Brock, I., Jacobsen, S., and Andersen, P. (2000). Comparative evaluation of low-molecular-mass proteins from *Mycobacterium tuberculosis* identifies members of the ESAT-6 family as immunodominant T-cell antigens. Infect. Immun. *68*, 214-220.

Sonnenberg, M.G., and Belisle, J.T. (1997). Definition of *Mycobacterium tuberculosis* culture filtrate proteins by two-dimensional polyacrylamide gel electrophoresis, N-terminal amino acid sequencing, and electrospray mass spectrometry. Infect. Immun. *65*, 4515-4524.

Sorensen, A.L., Nagai, S., Houen, G., Andersen, P., and Andersen, A.B. (1995). Purification and characterization of a low-molecular-mass T-cell antigen secreted by *Mycobacterium tuberculosis*. Infect. Immun. *63*, 1710-1717.

Sousa, A.O., Mazzaccaro, R.J., Russell, R.G., Lee, F.K., Turner, O.C., Hong, S., Van Kaer, L., and Bloom, B.R. (2000). Relative contributions of distinct MHC class I-dependent cell populations in protection to tuberculosis infection in mice. Proc. Natl. Acad. Sci. U.S.A. *97*, 4204-4208.

Stanley, N.R., Palmer, T., and Berks, B.C. (2000). The twin arginine consensus motif of Tat signal peptides is involved in Sec-independent protein targeting in *Escherichia coli*. J. Biol. Chem. *275*, 11591-11596.

Stanley, S.A., Raghavan, S., Hwang, W.W., and Cox, J.S. (2003). Acute infection and macrophage subversion by *Mycobacterium tuberculosis* require a specialized secretion system. Proc. Natl. Acad. Sci. U.S.A. *100*, 13001-13006.

Stenger, S., and Modlin, R.L. (2002). Control of *Mycobacterium tuberculosis* through mammalian Toll-like receptors. Curr. Opin Immunol. *14*, 452-457.

Sturniolo, T., Bono, E., Ding, J., Raddrizzani, L., Tuereci, O., Sahin, U., Braxenthaler, M., Gallazzi, F., Protti, M.P., Sinigaglia, F., and Hammer, J. (1999). Generation of tissue-specific and promiscuous HLA ligand databases using DNA microarrays and virtual HLA class II matrices. Nat. Biotechnol *17*, 555-561.

Sugawara, I., Yamada, H., Li, C., Mizuno, S., Takeuchi, O., and Akira, S. (2003). Mycobacterial infection in TLR2 and TLR6 knockout mice. Microbiol. Immunol. *47*, 327-336.

Sutcliffe, I.C., and Russell, R.R. (1995). Lipoproteins of gram-positive bacteria. J. Bacteriol. *177*, 1123-1128.

Tascon, R.E., Stavropoulos, E., Lukacs, K.V., and Colston, M.J. (1998). Protection against *Mycobacterium tuberculosis* infection by CD8+ T cells requires the production of gamma interferon. Infect. Immun. *66*, 830-834.

Tekaia, F., Gordon, S.V., Garnier, T., Brosch, R., Barrell, B.G., and Cole, S.T. (1999). Analysis of the proteome of *Mycobacterium tuberculosis in silico*. Tuber Lung Dis. *79*, 329-342.

Temmerman, S., Pethe, K., Parra, M., Alonso, S., Rouanet, C., Pickett, T., Drowart, A., Debrie, A.S., Delogu, G., Menozzi, F.D., et al. (2004). Methylation-dependent T cell immunity to *Mycobacterium tuberculosis* heparin-binding hemagglutinin. Nat. Med. *10*, 935-941.

Terada, L.S., Johansen, K.A., Nowbar, S., Vasil, A.I., and Vasil, M.L. (1999). *Pseudomonas aeruginosa* hemolytic phospholipase C suppresses neutrophil respiratory burst activity. Infect. Immun. *67*, 2371-2376.

Terasaka, K., Yamaguchi, R., Matsuo, K., Yamazaki, A., Nagai, S., and Yamada, T. (1989). Complete nucleotide sequence of immunogenic protein MPB70 from *Mycobacterium bovis* BCG. FEMS Microbiol. Lett. *49*, 273-276.

The Institute for Genomic Research (2003). *Bacillus anthracis* sequence. http://www.tigr.org/tdb/b_anthracis/faq.shtml

Thoma-Uszynski, S., Kiertscher, S.M., Ochoa, M.T., Bouis, D.A., Norgard, M.V., Miyake, K., Godowski, P.J., Roth, M.D., and Modlin, R.L. (2000). Activation of toll-like receptor 2 on human dendritic cells triggers induction of IL-12, but not IL-10. J. Immunol. *165*, 3804-3810.

Timm, J., Perilli, M.G., Duez, C., Trias, J., Orefici, G., Fattorini, L., Amicosante, G., Oratore, A., Joris, B., Frere, J.M., and et al. (1994). Transcription and expression analysis, using *lacZ* and *phoA* gene fusions, of *Mycobacterium fortuitum* beta-lactamase genes cloned from a natural isolate and a high-level beta-lactamase producer. Mol. Microbiol. *12*, 491-504.

Tobian, A.A., Potter, N.S., Ramachandra, L., Pai, R.K., Convery, M., Boom, W.H., and Harding, C.V. (2003). Alternate class I MHC antigen processing is inhibited by Toll-like receptor signaling pathogen-associated molecular patterns: *Mycobacterium tuberculosis* 19-kDa lipoprotein, CpG DNA, and lipopolysaccharide. J. Immunol. *171*, 1413-1422.

Tokuda, H., and Matsuyama, S. (2004). Sorting of lipoproteins to the outer membrane in *E. coli*. Biochim Biophys Acta *1693*, 5-13.

Trajkovic, V., Singh, G., Singh, B., Singh, S., and Sharma, P. (2002). Effect of *Mycobacterium tuberculosis*-specific 10-kilodalton antigen on macrophage release of tumor necrosis factor alpha and nitric oxide. Infect. Immun. *70*, 6558-6566.

Tullius, M.V., Harth, G., and Horwitz, M.A. (2001). High Extracellular Levels of *Mycobacterium tuberculosis* Glutamine Synthetase and Superoxide Dismutase in Actively Growing Cultures Are Due to High Expression and Extracellular Stability Rather than to a Protein-Specific Export Mechanism. Infect. Immun. *69*, 6348-6363.

Tullius, M.V., Harth, G., and Horwitz, M.A. (2003). Glutamine synthetase GlnA1 is essential for growth of *Mycobacterium tuberculosis* in human THP-1 macrophages and guinea pigs. Infect. Immun. *71*, 3927-3936.

Ulrichs, T., Munk, M.E., Mollenkopf, H., Behr-Perst, S., Colangeli, R., Gennaro, M.L., and Kaufmann, S.H. (1998). Differential T cell responses to *Mycobacterium tuberculosis* ESAT6 in tuberculosis patients and healthy donors. Eur. J. Immunol. *28*, 3949-3958.

Valent, Q.A., Scotti, P.A., High, S., de Gier, J.W., von Heijne, G., Lentzen, G., Wintermeyer, W., Oudega, B., and Luirink, J. (1998). The *Escherichia coli* SRP and SecB targeting pathways converge at the translocon. Embo J. *17*, 2504-2512.

Valle, M.T., Megiovanni, A.M., Merlo, A., Li Pira, G., Bottone, L., Angelini, G., Bracci, L., Lozzi, L., Huygen, K., and Manca, F. (2001). Epitope focus, clonal composition and Th1 phenotype of the human CD4 response to the secretory mycobacterial antigen Ag85. Clin. Exp. Immunol. *123*, 226-232.

van der Laan, M., Bechtluft, P., Kol, S., Nouwen, N., and Driessen, A. J. (2004). F1F0 ATP synthase subunit c is a substrate of the novel YidC pathway for membrane protein biogenesis. J. Cell Biol. *165*, 213-222.

Vazquez-Boland, J.A., Kocks, C., Dramsi, S., Ohayon, H., Geoffroy, C., Mengaud, J., and Cossart, P. (1992). Nucleotide sequence of the lecithinase operon of *Listeria monocytogenes* and possible role of lecithinase in cell-to-cell spread. Infect. Immun. *60*, 219-230.

Verbon, A., Hartskeerl, R.A., Schuitema, A., Kolk, A.H., Young, D.B., and Lathigra, R. (1992a). The 14,000-molecular-weight antigen of *Mycobacterium tuberculosis* is related to the alpha-crystallin family of low-molecular-weight heat shock proteins. J. Bacteriol. *174*, 1352-1359.

Verbon, A., Kuijper, S., Jansen, H.M., Speelman, P., and Kolk, A.H. (1992b). Antibodies against secreted and non-secreted antigens in mice after infection with live *Mycobacterium tuberculosis*. Scand. J. Immunol. *36*, 371-384.

Vordermeier, H.M., Harris, D.P., Lathigra, R., Roman, E., Moreno, C., and Ivanyi, J. (1993). Recognition of peptide epitopes of the 16,000 MW antigen of *Mycobacterium tuberculosis* by murine T cells. Immunology *80*, 6-12.

Vordermeier, H.M., Zhu, X., and Harris, D.P. (1997). Induction of CD8+ CTL recognizing mycobacterial peptides. Scand. J. Immunol. *45*, 521-526.

Vordermeier, M., Whelan, A.O., and Hewinson, R.G. (2003). Recognition of mycobacterial epitopes by T cells across mammalian species and use of a program that predicts human HLA-DR binding peptides to predict bovine epitopes. Infect. Immun. *71*, 1980-1987.

Voulhoux, R., Ball, G., Ize, B., Vasil, M.L., Lazdunski, A., Wu, L.F., and Filloux, A. (2001). Involvement of the twin-arginine translocation system in protein secretion via the type II pathway. Embo J. *20*, 6735-6741.

Walburger, A., Koul, A., Ferrari, G., Nguyen, L., Prescianotto-Baschong, C., Huygen, K., Klebl, B., Thompson, C., Bacher, G., and Pieters, J. (2004). Protein kinase G from pathogenic mycobacteria promotes survival within macrophages. Science *304*, 1800-1804.

Wang, J., Li, C., Yang, H., Mushegian, A., and Jin, S. (1998). A novel serine/threonine protein kinase homologue of *Pseudomonas aeruginosa* is specifically inducible within the host infection site and is required for full virulence in neutropenic mice. J. Bacteriol. *180*, 6764-6768.

Wards, B.J., de Lisle, G.W., and Collins, D.M. (2000). An *esat6* knockout mutant of *Mycobacterium bovis* produced by homologous recombination will contribute to the development of a live tuberculosis vaccine. Tuber Lung Dis. *80*, 185-189.

Webb, J.R., Vedvick, T.S., Alderson, M.R., Guderian, J. A., Jen, S.S., Ovendale, P.J., Johnson, S.M., Reed, S.G., and Skeiky, Y.A.W. (1998). Molecular cloning, expression, and immunogenicity of MTB12, a novel low-molecular-weight antigen secreted by *Mycobacterium tuberculosis*. Infect. Immun. *66*, 4208-4214.

Weiner, J.H., Bilous, P.T., Shaw, G.M., Lubitz, S.P., Frost,L., Thomas, G.H., Cole, J.A., and Turner, R.J. (1998). A novel and ubiquitous system for membrane targeting and secretion of cofactor-containing proteins. Cell *93*, 93-101.

Weinrich Olsen, A., van Pinxteren, L.A., Meng Okkels, L., Birk Rasmussen, P., and Andersen, P. (2001). Protection of mice with a tuberculosis subunit vaccine based on a fusion protein of antigen 85b and esat-6. Infect. Immun. *69*, 2773-2778.

Weldingh, K., Rosenkrands, I., Jacobsen, S., Rasmussen, P.B., Elhay, M.J., and Andersen, P. (1998). Two-dimensional electrophoresis for analysis of *Mycobacterium tuberculosis* culture filtrate and purification and characterization of six novel proteins. Infect. Immun. *66*, 3492-3500.

Wiker, H.G., Harboe, M., Bennedsen, J., and Closs, O. (1988). The antigens of *Mycobacterium tuberculosis*, H37Rv, studied by crossed immunoelectrophoresis. Comparison with a reference system for *Mycobacterium bovis*, BCG. Scand. J. Immunol. *27*, 223-239.

Wiker, H.G., Harboe, M., and Lea, T.E. (1986). Purification and characterization of two protein antigens from the heterogeneous BCG85 complex in *Mycobacterium bovis* BCG. Int. Arch. Allergy Appl Immunol. *81*, 298-306.

Wiker, H.G., Sletten, K., Nagai, S., and Harboe, M. (1990). Evidence for three separate genes encoding the proteins of the mycobacterial antigen 85 complex. Infect. Immun. *58*, 272-274.

Wiker, H.G., Wilson, M.A., and Schoolnik, G.K. (2000). Extracytoplasmic proteins of *Mycobacterium tuberculosis* - mature secreted proteins often start with aspartic acid and proline. Microbiology *146 (Pt 7)*, 1525-1533.

Worsaae, A., Ljungqvist, L., Haslov, K., Heron, I., and Bennedsen, J. (1987). Allergenic and blastogenic reactivity of three antigens from *Mycobacterium tuberculosis* in sensitized guinea pigs. Infect. Immun. *55*, 2922-2927.

Worsaae, A., Ljungqvist, L., and Heron, I. (1988). Monoclonal antibodies produced in BALB.B10 mice define new antigenic determinants in culture filtrate preparations of *Mycobacterium tuberculosis*. J. Clin. Microbiol. *26*, 2608-2614.

Wu, C.H. (1996). Biosynthesis of lipoproteins. In *Escherichia coli* and *Salmonella typhimurium*, F. C. Neidhart, ed. (Washington, DC, ASM Press), pp. 1005-1014.

Yahr, T.L., and Wickner, W.T. (2001). Functional reconstitution of bacterial Tat translocation *in vitro*. Embo J. *20*, 2472-2479.

Yen, M.R., Tseng, Y.H., Nguyen, E.H., Wu, L.F., and Saier, M.H., Jr. (2002). Sequence and phylogenetic analyses of the twin-arginine targeting (Tat) protein export system. Arch. Microbiol. *177*, 441-450.

Yeremeev, V.V., Kondratieva, T.K., Rubakova, E.I., Petrovskaya, S.N., Kazarian, K.A., Telkov, M.V., Biketov, S.F., Kaprelyants, A.S., and Apt, A.S. (2003). Proteins of the Rpf family: immune cell reactivity and vaccination efficacy against tuberculosis in mice. Infect. Immun. *71*, 4789-4794.

Yeremeev, V.V., Lyadova, I.V., Nikonenko, B.V., Apt, A.S., Abou-Zeid, C., Inwald, J., and

Young, D.B. (2000). The 19-kD antigen and protective immunity in a murine model of tuberculosis. Clin. Exp. Immunol. *120*, 274-279.

Yoneda, M., and Fukui, Y. (1965). Isolation, purification, and characterization of extracellular antigens of *Mycobacterium tuberculosis*. Amer. Rev. Respir Dis. *92*, 9-18.

Young, D., Garbe, T., Lathigra, R., Abou-Zeid, C., and Zhang, Y. (1991). Characterization of prominent protein antigens from mycobacteria. Bull Int. Union Tuberc Lung Dis. *66*, 47-51.

Young, D.B., Kaufmann, S.H., Hermans, P.W., and Thole, J.E. (1992). Mycobacterial protein antigens: a compilation. Mol. Microbiol. *6*, 133-145.

Yuan, Y., Crane, D.D., Simpson, R.M., Zhu, Y. Q., Hickey, M.J., Sherman, D.R., and Barry, C. E., 3rd (1998). The 16-kDa alpha-crystallin (Acr) protein of *Mycobacterium tuberculosis* is required for growth in macrophages. Proc. Natl. Acad. Sci. U.S.A. *95*, 9578-9583.

Zahrt, T.C., and Deretic, V. (2001). *Mycobacterium tuberculosis* signal transduction system required for persistent infections. Proc. Natl. Acad. Sci. U.S.A. *98*, 12706-12711.

Zhang, Y., Lathigra, R., Garbe, T., Catty, D., and Young, D. (1991). Genetic analysis of *superoxide dismutase*, the 23 kilodalton antigen of *Mycobacterium tuberculosis*. Mol. Microbiol. *5*, 381-391.

Chapter 4

Vaccine Strategies

D.M. Collins, B.M. Buddle and G.W. de Lisle

ABSTRACT

A wider appreciation in developed countries of the size and intractability of the global tuberculosis problem has resulted in intensive efforts being devoted over the last 15 years towards producing new tuberculosis vaccines. These efforts have been aided by new insights into mycobacterial pathogenesis gained from developments in genomics, molecular genetics and immunology. A wide range of different vaccine candidates have recently been produced, including live attenuated mycobacteria, and protein and DNA subunits. An increasingly advocated concept in the last few years has been the use of different priming and boosting vaccines rather than a single all purpose vaccine, although the single vaccine approach has not been abandoned. Already, a few of the new candidate vaccines tested, have induced protection in animal models that is equivalent or better than that of the current vaccine, bacille Calmette and Guérin (BCG). Several of the candidates that may have moderate advantages over BCG are already in early human trials, but it will be many years before a replacement for BCG is available. Attempts to produce better vaccines than those in the current trials are continuing. It is expected that in some cases these attempts will be successful due to the continual improvements that are occurring in understanding host protection and mycobacterial pathogenesis, and also because some classes of vaccine candidates take longer to optimise than others. Improved animal models are being developed that more closely mimic conditions in human populations, so that a more accurate assessment of candidate vaccines can be made before they enter human trials. There is now almost universal acceptance that much better tuberculosis vaccines than BCG can be made, and considerable attention is being devoted towards designing clinical trials and considering operational issues for their deployment.

GENERAL INTRODUCTION

The size of the statistics about tuberculosis continue to both shock and depress; shock because a third of the world's population is infected and nearly two million people die of the disease each year (Dye et al., 1999), and depress because despite enormous efforts, the incidence of tuberculosis continues to increase (World Health Organisation, 2004). A major factor contributing to this increase is the HIV/AIDS pandemic, as co-infection with HIV is the strongest risk factor for progression from tuberculosis infection to disease (Zumla et al., 2000). Effective tuberculosis drugs are available and their widespread implementation in many high-incidence countries using the directly observed therapy short-course (DOTS) initiative has led to greatly increased numbers of patients being treated (Frieden et al., 2003). However, the current detection rate for new cases of pulmonary disease in these countries is too low for DOTS treatment to control the epidemic, and improvement of the detection rate will be difficult (Dye et al., 2003). A tuberculosis vaccine, bacille Calmette and Guérin (BCG), is widely used but it appears to have had little overall effect in limiting the epidemic. In the face of this appalling situation, and in fact largely because of it, there has been an upsurge in interest over the last 15 years in performing research directed at improving our understanding of tuberculosis pathogenesis with the goal of developing new and improved control strategies. Much of this research has been specifically aimed at producing a new vaccine or has been broadly justified in terms of this goal. This is because of the inadequacies of BCG and because most experts believe that an effective tuberculosis vaccine will be essential if tuberculosis is ever to be controlled (McMurray, 2003).

Compared to more acute diseases such as salmonellosis, tuberculosis research suffers from the requirement to culture slow-growing organisms and perform protracted pathogenesis studies and animal vaccination trials. Nevertheless, the amount of high quality tuberculosis research over the last 15 years, coupled with spectacular developments in relevant disciplines, have enabled substantial progress towards new vaccines to be made. This is best exemplified by the plethora of reviews on tuberculosis vaccine development that have appeared recently (Agger and Andersen, 2002; Brandt and Orme, 2002; Brennan et al., 2004; Britton and Palendira, 2003; Collins and Kaufmann, 2001; Dietrich et al., 2003; Doherty, 2004, Flynn, 2004; Ginsberg, 2002; Kana and Mizrahi, 2004; Kumar et al., 2003; McMurray, 2003; Orme et al., 2001; Reed et al., 2003; von Reyn and Vuola, 2002; Young and Stewart, 2002; Xing, 2001) and the fact that a few new vaccines are already in early human trials (Hoag, 2004). Taken together, these reviews cover all the areas of tuberculosis vaccine development that have been the mainstream of research for the last 15 years. This chapter will summarise common themes that are in many of these reviews as well as highlight particularly relevant areas that have made recent progress. Most research has been concerned with the development of vaccines for treating naive populations or those who may be infected but have no evidence of disease and this will be the subject of the chapter. These vaccines are generally referred to as prophylactic, although it can be argued that vaccines employed for treatment of the latter group could be considered in a sense as therapeutic vaccines. While the development of therapeutic vaccines to treat clinical disease will not be further discussed, it should be noted that there is continuing interest in research on this approach (Bonato et al., 2004). The first human trials in which killed *Mycobacterium vaccae* was used as an adjunct to standard chemotherapy were unsuccessful (de Bruyn and Garner, 2002; Mwinga et al., 2002), but further trials are likely to be conducted (Stanford et al., 2004).

At critical stages in the development of all new tuberculosis vaccines, comparisons are always made to BCG, an attenuated strain of *Mycobacterium bovis*. *M. bovis* itself is the cause of tuberculosis in cattle and other mammals and causes a small but significant percentage of human tuberculosis; it is very closely related to *Mycobacterium tuberculosis,* the most common human agent of tuberculosis. These two species, together with *Mycobacterium africanum* are essentially the cause of all human tuberculosis and together with a few less important species are collectively referred to as the *M. tuberculosis* complex. Over three billion people have been vaccinated with BCG since the 1920s (Agger and Andersen, 2002), and there is general agreement that it prevents serious forms of the disease in children. In contrast, its ability to protect teenagers and adults is questionable, based on the results of a range of BCG trials where protection varied from 0-80% (Fine, 1989). Its proponents take the view that BCG is an inherently good vaccine of proven safety that is clearly affected adversely by unknown host and/or environmental variables which alter the ability of certain groups of vaccinees to respond successfully (Smith et al., 2000), and take heart from a meta-analysis of 26 vaccination trials that showed there was an average protection of 50% (Brewer, 2000). This belief inspires a number of approaches to improve BCG, that are described in a later section. In this context, it is interesting to note a recent report of a BCG vaccine trial followed for 60 years that showed little waning of BCG protective efficacy over all that time (Aronson et al., 2004). Detractors of BCG claim that its protection diminishes in teenagers and that in general it does not appear to protect adults (Brandt and Orme, 2002), and point out that it provides no protection in high-incidence countries such as India (Fine, 1989) which is precisely where protection is most needed. This negative assessment of BCG inspires those developing a wide range of new tuberculosis vaccines. In recent years, while hundreds of new candidate vaccines have been developed (Fruth and Young, 2004), only a minority have induced protection comparable to BCG and very few have so far provided better protection (Ginsberg, 2002). In large part, this is because although there have been major advances in understanding *M. tuberculosis* itself and the host's response to infection and vaccination, both the exact features that a good vaccine requires and the detailed host response that constitutes protective immunity have remained elusive. Against this background, there have been two approaches towards vaccine development. On the one hand, there has been a concerted effort to dissect the fundamental mechanisms by which the immune system responds to infection with *M. tuberculosis* (Kaufmann, 2001). In parallel, a wide range of potential vaccine candidates have been generated and subjected to empirical testing in animal vaccination models (Young and Stewart, 2002). Some of these vaccine candidates have been made randomly, others have been engineered with the expectation that some feature being incorporated will stimulate a particular facet of the immune response that is deemed important. In practice, there are many links between these different approaches and an iterative development in understanding of what is required for a good vaccine is occurring.

This chapter will cover the different classes of vaccine candidates that are being investigated and the rationale for their development. Since much of this work has been heavily dependent on advances made in molecular genetics and genomics of *M. tuberculosis* (Kana and Mizrahi, 2004), and the evaluation of the candidates is critically dependent on the use of animal models (McMurray, 2003), there will be a focus on these aspects. Where appropriate, a few salient immunological terms have been included, but the field of tuberculosis immunology is too complex to summarise briefly and is the subject of regular review elsewhere (Flynn, 2004; Orme, 2004; North and Jung, 2004). Many other important aspects of candidate vaccines, including their biological and formulation stability, their ability to persist in the host, their immunogenicity and their potential for reversion, are the subject of intense scrutiny by regulatory authorities but are not dealt with in the confines of this chapter.

VACCINE CANDIDATES

The range of vaccine candidates being developed can be categorised in different ways. In this chapter, a division into living and non-living vaccines has been used but for some purposes it can be convenient to classify candidates as either whole mycobacteria or subunits. While the early expectation was that these vaccines would normally be used independently, there has been increasing interest from those developing subunit vaccines, to use them sequentially with a whole mycobacterial vaccine in a prime-boost scenario. In part, this trend is driven by a desire to better stimulate all relevant parts of the immune system, but it also reflects the reality that there are already large populations of previously BCG-vaccinated individuals whose immunity is unlikely to be boosted by repeat BCG vaccination (Karonga trial prevention group, 1996; Leung et al., 2001) but might be boosted by a subunit vaccine (Doherty, 2004). In any case, the current inability of most vaccine candidates when used alone, to provide longer lasting protection against tuberculosis than BCG, makes the prime-boost scenario more attractive. It is important to note that, while the vaccine candidates referred to below include members of all the main types of vaccines being investigated, they are by no means a comprehensive list. The examples given should therefore be seen as exemplifying that particular type of vaccine and in some cases giving an indication of the possible future prospects for that type rather than necessarily for the precise example quoted.

BCG AND ITS MODIFICATIONS

BCG ITSELF

It is evident from the slowness with which clinical trials of tuberculosis vaccines can be performed and the results then adopted across many national boundaries, that even if one of the vaccines that are currently in early human trials were to provide significantly better protection than BCG, it would be many years before use of BCG would be widely supplanted. With this in mind, various researchers are still attempting to employ BCG in a more effective way by taking account of improved understanding of this vaccine. It has been known for a considerable time that BCG strains used in different countries differ from each other in a number of features (Collins and de Lisle, 1987), probably because they were maintained apart in continuous culture for many years (Behr, 2002). These differences have been proposed as one reason to explain the discordant results of vaccine trials (ten Dam and Pio, 1982). Wide differences in production procedures, variable doses and different immunisation procedures for BCG have also received more attention and all these factors are now being taken into account in proposing and implementing BCG vaccination programmes (Comstock, 2000; Hussey et al., 2002; Dietrich et al., 2003). While these sorts of approaches do not use modern molecular techniques directly, they are guided by the knowledge that these techniques have revealed. In one sense they are the most important innovations in the area of tuberculosis vaccines, as they offer the only likely improvements in tuberculosis vaccination that will be available for human use in the next few years.

BCG RECOMBINANTS

The modern genomic era has enabled the genomic sequencing of three strains of the *M. tuberculosis* complex and identification of the major genetic deletions (Brosch et al., 2002; Behr, 2002) and some of the minor differences (Collins et al., 2003a) occurring in BCG relative to virulent *M. bovis*. This, together with an improved understanding of tuberculosis pathogenesis, has led to four different recombinant methods being investigated for improving BCG: 1) addition of an antigen that is present in virulent strains of the *M. tuberculosis* complex but deleted in BCG, 2) over expression of a gene, 3) down regulation or inactivation of a gene, 4) expression of heterologous (foreign) genes that modulate immunity. In the approach based on replacing deleted antigens, much interest has focused on the small highly immunogenic protein ESAT-6 and its likely function (Wards et al., 2000; Hsu et al., 2003). The genes responsible for encoding and secreting ESAT-6 are absent from all BCG strains. It was proposed that production of a recombinant BCG that expressed ESAT-6 might increase its vaccine efficacy, and such an increase was observed (Pym et al., 2003). In contrast, Bao et al. (2003) also produced a recombinant BCG that secreted ESAT-6, but found no such improvement in vaccine efficacy. More detailed analysis of both papers shows that in one study (Bao et al., 2003) only ESAT-6 was produced and not the associated protein CFP-10 which may affect the activity of ESAT-6 in the host (Renshaw et al., 2002). In the study by Pym et al. (2003) both ESAT-6 and CFP-10 as well as other nearby genes required for their secretion were expressed, but the large recombinant fragment also contained additional genes around the *esat-6* locus some of them of unknown function that duplicated genes already present in BCG. Since it is known that gene dosage can have major phenotypic effects that could easily influence vaccine efficacy, the increased efficacy of this recombinant may be accounted for by the products of these genes. It is interesting to note in this context that removal of *esat-6* from two live attenuated *M. bovis* strains that already had vaccine efficacy comparable to BCG had no significant effect on efficacy (Collins et al., 2003b). In summary, it is not possible from these studies to be completely certain of the effect on efficacy of a BCG expressing ESAT-6 and more studies are warranted.

The best known example of a BCG recombinant over-expressing an antigen is in the case of the antigen 85B secreted protein. This recombinant showed increased protection relative to BCG in a guinea pig model (Horwitz and Harth, 2003) and by now will have completed phase 1 human trials (Hoag, 2004). This protein is responsible for attaching mycolic acids to the cell wall of *M. tuberculosis*, and is not only its major secretory protein in broth culture but is also among the major proteins of all *M. tuberculosis* proteins expressed in human macrophages. The rationale for its original choice is that among the extracellular proteins of intracellular pathogens, the ones released in greatest abundance will probably be among the most effective in inducing immunoprotection. By virtue of their abundance in the phagosome they will be processed and presented more frequently and therefore be likely to induce a particularly strong cellular response (Horwitz et al., 1995). This is a common rationale not only for live vaccines but also for vaccine candidates based on naked DNA or protein sub-units. In this particular case, the expression of the protein in BCG seems to have been a logical progression from a simpler approach, as the same group had been unsuccessful in inducing better protection than BCG when using subunit vaccines administered together with various experimental adjuvants and immunostimulatory molecules such as interleukin-12 and CpG motifs (Nor and Musa, 2004). Over expression of other antigens in BCG is also being investigated (Rao et al., 2003), and it is possible that a better vaccine of this type may ultimately be achieved by over expressing two or more proteins in a single recombinant. Two BCG recombinants in which the immunomodulatory 19-kDa lipoprotein (Lopez et al., 2003) was either deleted or over expressed have been produced (Yeremeev et al., 2000). While in this case, both these recombinants had very similar vaccine efficacy to their BCG parent, the concept of altering the expression of immunomodulatory proteins of BCG to produce a better vaccine remains attractive and is likely to be further pursued.

A third method of modifying BCG involves down regulating or inactivating genes. This has been done either to increase its immunogenicity, for example by reducing its superoxide dismutase activity to enhance apoptosis and antigen presentation (von Reyn and Vuola, 2002), or to improve its safety for immune-compromised hosts. While the possibility of BCG being more virulent for compromised hosts is a concern, it should be noted that untoward effects of BCG have not been widely observed in studies of HIV-positive children (Dietrich et al., 2003). Auxotrophic mutants of BCG with deletion of an amino acid biosynthesis gene were shown to be much less virulent in severe-combined-immunodeficient (SCID) mice than BCG and gave comparable protection (Guleria et al., 1996). On the other hand, in guinea pig vaccination models, BCG auxotrophs have provided somewhat less protection than BCG (Jackson et al., 1999; Chambers et al., 2000), and the auxotrophic approach may offer more promise when used as a method of attenuating virulent *M. tuberculosis* or *M. bovis* strains.

A clear example of where the enormous growth in knowledge of the immune system has impacted on production of candidate tuberculosis vaccines is in the range of BCG recombinants expressing heterologous immunomodulatory genes. BCG recombinants expressing a number of different murine cytokines were some of the earliest candidate tuberculosis vaccines of the molecular era (Murray et al., 1996). In particular, BCG modified to secrete interleukin-2, interferon-γ, or granulocyte–macrophage colony-stimulating factor were shown to be substantially more potent stimulators of the cell-mediated immune response than BCG itself. Unfortunately, this stimulation induced more rapid clearance in animal models leading to reduced protection and these vaccines have not been pursued further (Brandt and Orme, 2002). A very different approach to stimulating host immunity has been to produce a recombinant BCG that expresses listeriolysin (Hess and Kaufmann, 2001). The rationale for this was that BCG is deficient in its ability to stimulate a strong $CD8^+$ T cell response; that *Listeria monocytogenes* produces a lysin that enables the organisms to escape from the phagosome into the cytosol of infected macrophages resulting in a level of MHC class I presentation that leads to a strong $CD8^+$ T cell response; and that incorporating the lysin gene into BCG might cause a similar effect. In fact, BCG expressing listeriolysin did not escape more quickly from the phagosome, but there were indications of a stronger $CD8^+$ T cell response and this approach has now being refined by incorporating listeriolysin into BCG which has a deleted urease gene.

ATTENUATED STRAINS OF THE *M. TUBERCULOSIS* COMPLEX

The rationale for production of a new vaccine by attenuation of strains of the *M. tuberculosis* complex, is that many good vaccines against other diseases are attenuated strains of the pathogenic organism, that BCG was formed from its original virulent parent *M. bovis* by an accumulation of undirected mutations, and that different mutations may produce a better vaccine. Compared to virulent strains of the *M. tuberculosis* complex, BCG has lost many functional genes that are known (Brosch et al., 2002; Behr, 2002; Collins et al., 2003a) and probably some which will be revealed when the sequencing of the genome of BCG is completed. Known mutated genes in BCG, such as *esat-6* (Wards et al., 2000; Hsu et al., 2003) and *phoT* (Collins et al., 2003a), have already been shown to play a role in virulence mechanisms. By selecting mutations that interfere with different stages of the infectious process, it may be possible to optimise immunogenicity and induction of memory, and to inactivate genes encoding immunosuppressive components. There is potential that mutations which alter the ability of *M. tuberculosis* to interfere with events in the infected cell might generate strains that are more immunogenic than the natural infection which itself appears to provide no greater protection than

BCG (Mollenkopf et al., 2004b). With this in mind, a large number of mutants of *M. tuberculosis* and *M. bovis* have been produced both by random mutagenesis techniques and by specific mutation, and many of these have been tested for their attenuation in animals. In a fair proportion of cases, those mutants that have been found to be attenuated have been tested in animal models to determine their vaccine potential. When attenuated mutants with good vaccine efficacy are discovered, the mutations can be re-engineered in other strains or further genetic deletions or additions can be made using well-established allelic exchange and genetic incorporation techniques (Kana and Mizrahi, 2004).

RANDOM MUTANTS

In the random mutagenesis approach, large numbers of mutants of a random nature are produced by transposon insertion or illegitimate recombination and then tested in animals either directly or after an *in vitro* screen that is used to reduce the numbers down to a manageable size for animal testing. Mutants that have become sufficiently attenuated to be used as live vaccines are then tested for vaccine efficacy in an animal model. The benefit of this approach is that it makes no prior assumptions about which gene or genes will be inactivated. This is a very powerful advantage, as it is not known which genes should be inactivated in *M. tuberculosis* to produce a better vaccine than BCG, and random mutagenesis implemented on a wide enough scale can be used to inactivate any non-essential gene. While ultimately this approach may be decisive and result in vaccines with greatly improved efficacy, the development of such vaccines lags behind those developed in other ways. This is because of the time required to perform the processes involved in discovering and characterising such genes, followed by the time required to induce, and determine the effect on vaccine efficacy, of further mutations designed to give assurance of stable attenuation and to incorporate other desirable features. Transposon mutagenesis has been most used (McAdam et al., 2002; Sasetti and Rubin, 2003) but illegitimate recombination has also been employed (Collins et al., 2002). In order to mutate as wide a range of genes as possible, large numbers of mutants are generated by these techniques and screened in some way to limit the number of animals that are eventually used to assess the virulence of individual mutants. The first approach of this type was to screen mutants for auxotrophy (McAdam et al., 1995) which is the inability to grow without the addition of intermediary metabolites such as amino acids. There was good reason for this; auxotrophs of other bacterial pathogens have frequently been found to be avirulent and sometimes to have good vaccine efficacy. While this first auxotroph screening was actually carried out on BCG and not *M. tuberculosis* itself, it is included in this section because it identified a leucine auxotroph of BCG that was found to be attenuated in SCID mice (Guleria et al., 1996). This result led to production of a leucine auxotroph of *M. tuberculosis* by allelic exchange with the specific aim of investigating its vaccine potential (Hondalus et al., 2000). Although this *M. tuberculosis* mutant itself was not as good a vaccine as BCG, the introduction of one or more selected mutations in auxotrophs of this type is being used to produce an improved vaccine (Sampson et al., 2004). This outline of the pathway used to develop an auxotroph of *M. tuberculosis* also illustrates the way in which different molecular genetic approaches are employed to take best advantage of discoveries as they occur. While the divisions of the chapter into sections based on techniques enables a large number of developments to be presented in an orderly way, to some extent it cuts across the pragmatic progress of tuberculosis vaccine development, where different techniques are often employed at different stages of progress.

Another mutant screen is based on the isolation of mutants that cannot be recultured after being grown to stationary phase in minimal medium (Collins et al., 2002). While this method of selection involves steps that are technically similar to the selection of auxotrophs, the types of mutants selected are genetically more varied and include many genes of unknown function. Mutants with protective efficacy at least comparable to BCG have been reported using this approach (de Lisle et al., 1999; Collins et al., 2002).

An alternative technique to *in vitro* screening has been to use signature tag mutagenesis which avoids the need for *in vitro* screening. In this technique, pools of individually tagged mutants of *M. tuberculosis* are inoculated into animals and the absence of any particular attenuated mutant is identified by comparing the recovered tags to the inoculated tags. In two separate studies, a restricted set of attenuating mutants was identified, probably reflecting a combination of limitations in the model of virulence used and the relatively small number of mutants examined (Camacho et al., 1999; Cox et al., 1999). Lipid genes important for production and secretion of phthiocerols were identified in both studies and, importantly, one of these mutants was found to provide better efficacy than BCG when tested in a mouse vaccination model (Pinto et al., 2004). The technique has also been applied to *M. bovis*, resulting in identification of different attenuating mutations, including many mutants in which different parts of the *esat-6* region were deleted. (Collins et al., 2005). In future, more sophisticated applications of signature tag mutagenesis (Hisert et al., 2004) could be used to identify genes that are important for particular parts of the infection cycle and inactivation or over expression of such genes might be used to improve a live *M. tuberculosis* vaccine. A major attraction of inactivating virulence genes to produce a vaccine is the notion that such strains will replicate normally in the host, at least for a time, producing a vaccine that is superior to auxotrophic vaccines that have mutations in biosynthetic genes and are likely to be affected in their ability to colonize and replicate (Kochi et al., 2003).

SPECIFIC MUTANTS

The two key steps in producing a mutant of the *M. tuberculosis* complex in which a specific gene is inactivated are selection of the gene and implementation of a method for allelic exchange. A range of allelic exchange methods are available and, at least in principle and usually in practice, any non-essential gene can now be easily deleted (Kana and Mizrahi, 2004). Selection of the gene to be inactivated can be carried out in a large number of different ways (Collins and Gicquel, 2000; Clark-Curtiss and Haydel, 2003), all of which involve assumptions about the possible role of that gene in virulence or its role in encoding some other phenotype whose removal will result in an improved vaccine. One approach is to inactivate a gene in *M. tuberculosis* that is homologous to genes whose inactivation in other pathogens had led to attenuated strains with good vaccine potential. This rationale was used to select the regulatory gene *phoP* for inactivation in *M. tuberculosis* (Perez et al., 2001). At this stage however, the most well-known candidates of this type are auxotrophs of *M. tuberculosis*. Inactivation of the *purC* gene of *M. tuberculosis* has been used to produce a purine auxotroph (Jackson et al, 1999), inactivation of *panCD* to produce a pantothenate (vitamin B5) auxotroph (Sambandamurthy et al., 2002) and inactivation of *proC* and *trpD* to produce auxotrophs of proline and tryptophan respectively (Parish et al., 1999). Some of these auxotrophs appear at least as protective as BCG and are less virulent in SCID mice (Sambandamurthy et al., 2002; Smith et al., 2001), so they may be potentially safer vaccines than BCG for use in immunocompromised patients such as those with HIV infection. The safety of the *panCD* auxotroph has been further enhanced by incorporating a second auxotrophic mutation, this time in a leucine biosynthetic gene (*leuD*), and the combination of these two attenuating mutations did not appear to affect vaccine efficacy (Sampson et al., 2004).

From a genetic perspective, BCG is very similar to virulent strains of the *M. tuberculosis* complex, and approaches used to modify BCG can also be applied to these virulent strains themselves or to the incorporation of additional mutations into auxotrophs or other attenuated strains of the *M. tuberculosis* complex. Thus, in continuation of work in which the 19 kDa antigen was deleted from BCG (Yeremeev et al., 2000), it has now been deleted from *M. tuberculosis* in a study that investigated the role of this protein in apoptosis with the aim that such a mutation might be incorporated as part of a future rationally-designed vaccine (Ciaramella et al., 2004). Inactivation of this protein in a future vaccine might also contribute to vaccine safety as its natural inactivation in some strains of *M. tuberculosis* causes moderate attenuation in mice (Lathigra et al., 1996) and its deletion from *M. bovis* causes moderate attenuation in guinea pigs (D.M. Collins, unpublished results). There is no practical reason preventing genes being added or upregulated in attenuated strains of the *M. tuberculosis* complex by following the examples in BCG with cytokines and antigen 85B but it has been too soon for much progress in this area to have occurred. One advantage of producing such strains would be to closely compare their effects to equivalent BCG recombinants in animal vaccination models in order to gain a better understanding of protective immune responses.

While most of the candidates being developed in this category are recombinants of *M. tuberculosis* and *M. bovis*, there is still continuing interest in *Mycobacterium microti*. This is a member of the *M. tuberculosis* complex that causes tuberculosis in voles but is avirulent in most other mammals including humans. In a large trial in the

United Kingdom, it induced protection comparable to BCG (Hart and Sutherland, 1977). The recent application of genomic techniques to the *M. tuberculosis* complex has revealed that *M. microti* is a natural *esat-6* mutant and restoration of the *esat-6* gene region has been shown to improve its vaccine efficacy in mice (Brodin et al., 2004).

One technical disadvantage of implementing a vaccine based on an attenuated strain of the *M. tuberculosis* complex, is that such vaccination is likely to sensitise vaccinees to tuberculin based tests. This could severely limit the usefulness of these tests in establishing whether infection with *M. tuberculosis* had occurred. One approach to overcoming this disadvantage, is to implement tests based on dominant antigens such as ESAT-6 and MPT-64 (Haslov et al., 1995; Johnson et al., 1999) and to inactivate their encoding genes in a live vaccine candidate. Genes encoding both ESAT-6 (Collins et al., 2003b) and MPB-64 (D.M. Collins, unpublished results) have been deleted from *M. bovis* with this purpose in mind. A further advantage of deleting *esat-6*, but not the gene encoding MPB-64, is that this causes moderate attenuation of *M. bovis* and would add to the safety of using such a vaccine.

OTHER LIVE VACCINES

The concept of producing a tuberculosis vaccine by expressing immunogenic proteins of the *M. tuberculosis* complex in other live vectors such as bacteria and viruses is included in this section because the ability of these live vectors to stimulate a cell mediated immune response is a critical factor in their effectiveness. In addition, although they do not cause the same degree of concern as live mycobacterial vaccines, there are increased regulatory concerns about their safety and persistence that do not apply to protein and DNA vaccines. Nevertheless, these candidates are sub-unit vaccines, as they contain only one, or potentially a few, mycobacterial antigens. Ultimately, this may be their Achilles heel if it transpires that a very limited set of mycobacterial antigens is unable to induce good protection, or at least effectively boost a previous vaccination, in a large percentage of an outbred human population that has diverse responses to particular antigens. The candidate currently receiving most attention is a vaccinia virus that expresses *M. tuberculosis* antigen 85A (Goonetilleke et al., 2003). This candidate has been included here because, although it is a disabled virus that does not replicate in the host, it induces cellular immune responses similar to a replicating virus (Ramirez et al., 2000). This recombinant vaccinia was the first new tuberculosis vaccine both to enter (Ginsberg, 2002) and to complete (McShane et al., 2004) phase I human trials but it will be some years before even preliminary results on its effectiveness in humans can be determined. It is proposed to use it is as a booster vaccine to a prior BCG vaccination. Although, when used in this way in mice, the vaccinia recombinant did not provide any better protection than boosting with BCG itself, its proponents claim a number of potential advantages (Goonetilleke et al., 2003). These include it not being compromised by exposure to environmental mycobacteria, and its safety for immune compromised individuals. Until results of testing these attributes in animal models are available, it is difficult to comment on the potential usefulness of this vaccine. The expression of antigen 85A as a vaccine is also being investigated in adenonvirus (Wang et al., 2004) and poxvirus (Ginsberg, 2002).

Avirulent bacterial vectors are also being developed as tuberculosis vaccines, including attenuated *Salmonella typhimurium* expressing antigen 85B (Hess and Kaufmann, 2001), and *Listeria monocytogenes* expressing antigen 85A, 85B and MPB/MPT51 (Miki et al., 2004). Encouraging protection has already been reported for both vectors in mouse challenge studies. As with viral based vectors, vaccines of this type could be used as boosters and are not likely to be compromised by previous exposure to environmental mycobacteria.

SUBUNIT VACCINES

Most of the candidate tuberculosis vaccines tested have been subunits. This category includes primarily proteins and DNA but there is some interest in investigating other components of mycobacteria such as lipids. Various approaches have been used to determine which proteins or which epitopes within them are likely to stimulate $CD4^+$ and $CD8^+$ T cells (Okkels et al., 2003) but, as with other types of candidate vaccines, good stimulation does not necessarily lead to induction of protection (Britton and Palendira, 2003). In practice, the best known candidates were largely selected on their ability to induce a strong TH1 response in mice (Okkels et al., 2003). Major attractions of subunit vaccines are the ease of producing them, at least in the quantities required for extensive experimentation, and avoidance of the need to address the reversion-to-virulence safety issue of attenuated live vaccines. Counterbalancing these advantages are the requirements for subunit vaccines to be formulated with adjuvants and to be inoculated several times to achieve a good immune response. Until recently, there were no good adjuvants that could be used to stimulate T cells in humans, but a number of formulations including monophosphoryl lipid A have now been developed for protein subunits; and lipids as well as CpG oligonucleotides are being extensively used as adjuvants in DNA vaccines (Brandt and Orme, 2002; Okkels et al., 2003; McMurray, 2003; Reed et al., 2003). Other adjuvants such as cytokines are also being investigated (Britton and Palendira, 2003), and it is likely that further optimisation of adjuvants will improve the efficacy of these vaccines. The degree to which these improvements will add to the cost of subunit vaccines and the requirement for multiple inoculations will present considerable economic and logistic challenges which will need to be overcome for widespread implementation of such vaccines to occur in developing countries.

PROTEIN SUBUNIT VACCINES

Half the vaccines tested in the large National Institutes of Health vaccine screening programme up to 2001 were subunits (Orme et al., 2001). This reflected knowledge gained in previous years of the proteome of the *M. tuberculosis* complex as well as the availability of techniques to help identify the most immunogenic proteins. Culture filtrates of *M. tuberculosis* contain a wide range of proteins and have been shown to provide considerable protection in animal models, but the difficulty of standardising these preparations and of attaining better efficacy than BCG makes it unlikely that they will ever be used for human vaccination. At the other extreme, a large number of single protein antigens of *M. tuberculosis* have been tested as candidates (Okkels et al., 2003; Xing, 2001). While most of these have provided little protection and are often not recognised immunologically by a majority of an outbred population, a few immunodominant proteins such as ESAT-6 and the Antigen 85 complex have given more encouraging results (Britton and Palendira, 2003; Young and Stewart, 2002). The protection is even better when both proteins are used together and a fusion protein of antigen 85B and ESAT-6 induced protection equal to that of BCG in mice (Olsen et al., 2001) and induced nearly as good a survival curve as BCG in guinea pigs (Olsen et al., 2004). When mice were pre-sensitised with environmental mycobacteria, this fusion protein continued to give good protection, whereas BCG replication was affected under these conditions and it induced less protection than in naive animals (Brandt et al., 2002). This is potentially important because one of the favoured explanations for the lack of protective efficacy of BCG in some human trials has been that those populations were exposed to high levels of environmental mycobacteria and that this masked or interfered with the ability of BCG to induce a good protective response (Fine et al., 2001; Anon., 2004).

The most well-known of the fusion proteins being used as a candidate vaccine is antigen 72f which has already reached phase I clinical trials in humans (Skeiky et al., 2004). This combines two proteins (a serine proteinase and a PPE protein) that between them give good stimulation of $CD4^+$, and $CD8^+$ T cells. The proteins were initially identified from their ability to stimulate T cells in healthy humans who were positive in the standard skin test for tuberculosis infection (Dillon et al., 1999; Skeiky et al., 2004). While this does not of course necessarily mean that their T cell responsiveness to these antigens is critical for good protection, it intuitively seems a reasonable approach to try to stimulate some of the immune reactions that are present in a majority of humans who remain well despite likely infection with *M. tuberculosis*. Individually, both proteins induced some protection in mice but when inoculated together the protection was better than with either separately. Fusion of the two proteins did not increase the protection but enabled the proteins to be produced more cheaply, an important consideration for any vaccine proposed for use in developing countries. The 72f vaccine also gave good protection in a guinea pig model. The most likely future use of protein subunit vaccines is as booster vaccines following BCG or some other primary vaccine and when 72f is used in this way in monkeys it provides better protection than BCG alone (Reed et al., 2003). An alternative approach might be to co-administer a protein sub-unit vaccine with a live mycobacterial vaccine because, when compared to BCG alone, co-administration of 72f and BCG in guinea pigs substantially increased survival after low-dose aerosol challenge with *M. tuberculosis* (Brandt et al., 2004).

LIPID VACCINES

While most tuberculosis vaccine development has concentrated on stimulating good $CD4^+$ and $CD8^+$ T cell responses, unconventional T cell sub-sets such as γδ and double-negative αβ sub-sets have also been implicated in protective immunity against tuberculosis. These T cells recognise mycobacterial non-protein antigens, including glycolipids, via the CD1 antigen-presentation pathway and have important effector functions for controlling infections (Moody et al., 2000; Gilleron et al., 2004). Patients recently infected with *M. tuberculosis* had increased CD1-restricted T cell responses to a lipid antigen (Moody et al., 2000). Support for the use of lipid antigens in vaccines against tuberculosis has recently been provided by a study where immunisation with a mycobacterial lipid improved pulmonary pathology in the guinea pig model of tuberculosis (Dascher et al., 2003). The rationale for using mycobacterial lipids would be to include them as part of a multi-subunit formulation in order to extend the range of T cells that were stimulated.

DNA VACCINES

While protein vaccines were the most popular vaccines tested up till 2001, DNA vaccines were the most popular type by 2003 (McMurray, 2003). This was partly because of the ease of producing them, and does not reflect a widespread belief that other vaccines are less likely to be successful. Most of the proteins that have been tested as subunit vaccines have also been tested as DNA vaccines, usually administered as a plasmid with the coding sequence under the control of a strong eukaryotic promoter such as the tissue plasminogen activator sequence (McMurray, 2003; Britton and Palendira, 2003). A claimed advantage of such vaccines is their ability to stimulate a strong $CD8^+$ T cell response (Agger and Andersen, 2002), but in practice the protective effect they induce rarely rivals that of BCG (Britton and Palendira, 2003) and is usually comparable or worse than that of the corresponding protein candidate. DNA vaccines have in many cases given disappointing results (Doherty, 2004) and have worked much better in mice than larger animals including humans (von Reyn and Vuola, 2002), primarily because of the difficulties of delivering DNA vaccines to sufficient antigen presenting cells in larger animals to get a good immune response (van Drunen Littel-van den Hurk et al., 2004). Other issues being addressed include altering the codon composition of the vaccine so that it is similar to preferred codons in the host, and testing different adjuvant formulations. A secondary reason for DNA vaccines providing more protection in mice than other animals is that the CpG oligonucleotides first used as adjuvants were developed using mouse models and it appears that different oligonucleotide sequences may be optimum in other animal species (Bauer et al., 2001). Partly as a result of the difficulties of improving DNA vaccines, the emphasis on using them has switched away from independent use to incorporation in various prime-boost combinations with BCG and/or protein subunits, and this sometimes induces better protection than either of the prime or boost vaccines alone (Agger and Andersen, 2002; Britton and Palendira, 2003; Doherty, 2004; Mollenkopf at al., 2004a; Skinner et al., 2003).

ANIMAL MODELS FOR ASSESSING TUBERCULOSIS VACCINES

Animal models are essential for evaluating new tuberculosis vaccines because *in vitro* correlates of protection still remain elusive and cannot be used as a reliable substitute. It is important that the key events in the pathogenesis of tuberculosis be taken into account when designing or choosing an animal model, as misleading results are obtained from using poorly designed models (Wiegeshaus and Smith, 1989). The last ten years has seen an almost universal adoption of more appropriate models, with key features including the provision of sufficient time between vaccination and challenge for induction of an adaptive immune response, and the use of low dose aerosol challenge which mimics the natural process of infection. The vast majority of the hundreds of vaccine candidates that have been produced have first been tested in mice as these are relatively inexpensive. Most of the comments relating to efficacy of vaccines in earlier sections of this chapter were of necessity based on such murine studies. Promising candidates have then been more fully examined in guinea pigs whose tuberculosis pathology more closely resembles that in humans (Young and Stewart, 2002). Relatively few candidates have been tested in larger animal models such as cattle or non-human primates as the cost of such tests can only be justified for particularly promising candidates. The different stages of animal testing that a promising candidate might proceed through are summarised in Table 1.

In the short-term mouse model used in the programme funded by The National Institutes of Health, C57BL/6 mice are vaccinated at 3 week intervals for testing subunit vaccines (Orme et al. 2001). A BCG control group is inoculated at the time of the last subunit vaccination or at the same time as the single vaccination of a candidate live vaccine. All groups are challenged by aerosol infection 4 weeks after the last vaccination. At approximately 30 days post challenge, all animals are sacrificed and the bacterial load is determined. With group sizes of 4-5 mice, a 0.7 log reduction in bacterial counts is usually significant. Most of the vaccines tested have not achieved this level of protection. Vaccines that show promise in the mouse vaccination model are usually re-tested in a guinea pig model. Tuberculous guinea pigs, like humans but unlike mice, develop high levels of delayed type hypersensitivity to tuberculin and produce progressive lung pathology where necrosis is a prominent feature. A similar short-term vaccination-challenge model has been developed for the guinea-pig. When BCG-vaccinated guinea pigs are examined 4-5 weeks after challenge with a low-dose aerosol, the differences in bacterial numbers and pathology from unvaccinated controls are much larger than

Table 1. Animal models for different stages of tuberculosis vaccine development.		
Stage	Purpose	Type
1	Identify candidates at least equivalent to BCG in laboratory animals	a. Short-term trial in mice b. Short-term trial in guinea pigs
2	Identify candidates safer than BCG in immunocompromised hosts	a. Survival of SCID mice b. Survival of mice with more limited immune deficiencies
3	Identify candidates better than BCG in laboratory animals	a. Long-term survival of mice b. Long-term survival of guinea pigs
4	Identify candidates with characteristics required for success in human trials	a. Animal previously vaccinated with BCG b. Prior sensitisation of animals with environmental mycobacteria c. Animals latently infected with *M. tuberculosis.*
5	Identify candidates more likely to be efficacious in humans, particularly in neonates	a. Non-human primates b. Cattle
6	Determine if candidates protect humans	Human vaccine trials

those seen in the mouse. In this system, BCG induces high levels of protection where the lung bacterial load is usually 2 log or more lower than those of controls (McMurray et al. 1999). By increasing group sizes up to 24, Horwitz et al. (2000) was able to demonstrate that a recombinant BCG expressing the *M. tuberculosis* 30-kDa major secretory protein induced better protection than conventional BCG.

In the short-term animal models, it has been difficult to identify vaccines that are more effective than BCG and progress has been made towards adapting the models to identify situations where BCG is less-effective. Over 30 years ago, Wiegeshaus et al. (1970) showed that many guinea pigs vaccinated with BCG and subsequently challenged by aerosol with virulent *M. tuberculosis* died from a progressive infection. Survival has recently been acknowledged as an important read-out of the efficacy of new vaccines (Orme et al., 2001) and is related to the type of changes in lung pathology. Early deaths in guinea pigs are associated with extensive lung changes including consolidation, an influx of macrophages and necrosis. Guinea pigs which are protected have granulomas in which lymphocytes are a prominent cellular component of the lesions, and there is minimal consolidation and necrosis. In survival experiments, guinea pigs are weighed regularly and sacrificed after they have lost a significant amount of weight. Survival times depend on the virulence of the challenge organism and the challenge dose. Unvaccinated guinea pigs challenged by aerosol with *M. tuberculosis* will survive 8 - 28 weeks, whereas BCG vaccinated animals will survive 40 weeks or more post challenge (Baldwin et al., 1998; Horwitz et al., 2003; Skeiky et al., 2004). Importantly, a few of the vaccines inoculated have been shown to result in significantly longer survival times than achieved with BCG. While survival studies in guinea pigs clearly identify new vaccines that induce significantly greater protection than BCG, these experiments may last over a year and are expensive.

It is important to identify tuberculosis vaccines which can be used synergistically with BCG or are more effective than BCG in real-life situations. The strategy of vaccinating neonates with BCG in developing countries has well-recognised health benefits and is likely to be continued for many years pending the demonstration in phase III trials of a better primary vaccine. It is generally acknowledged that BCG vaccine efficacy against pulmonary tuberculosis is highly variable. In the Chingleput trial, there was a gradual loss of effectiveness of BCG as individuals reached 10-15 years of age (Fine, 2001) and in Malawi the effectiveness of BCG was not significantly improved by repeated BCG vaccination (Karonga prevention trial group, 1996). Hence, there is a need to develop animal models that can be used to identify vaccines which can be utilised in conjunction with BCG to enhance protection. Such a model was implemented by Brooks et al. (2001) who observed that mice vaccinated with BCG gradually lost their capacity to resist an aerosol *M. tuberculosis* challenge infection as they aged. They found that when these mice were given a booster vaccination with antigen 85A protein in mid-life, resistance to an aerosol challenge with *M. tuberculosis* was improved compared to BCG alone.

Evidence indicates that BCG is less efficacious where the human populations have been exposed to environmental mycobacteria prior to vaccination with BCG (Fine et al., 2001). Brandt et al., (2002) developed a mouse model where prior sensitisation of mice with several strains of environmental mycobacteria, blocked both subsequent *in vivo* multiplication of BCG and protection against *M. tuberculosis* infection. In contrast, the efficacy of a vaccine based on a mycobacterial fusion protein (antigen 85B/ESAT-6) was unaffected by prior exposure to the environmental mycobacteria. We developed a related model in guinea pigs and found that BCG was significantly less protective after pre-sensitisation of guinea pigs with an *M. avium* strain containing IS901, whereas a newly developed attenuated strain of *M. bovis* was unaffected by pre-sensitisation and provided significantly better protection than BCG under those conditions (de Lisle et al., 2005).

One-third of the world's population has been infected with *M. tuberculosis* (Dye et al., 1999). Since at least 10% of these people are expected to develop clinical tuberculosis within their lifetime and HIV co-infection increases this likelihood to 10% per year (de Jong et al., 2004), it would be of great value to have available a vaccine that could be used to prevent latent tuberculosis reactivating. Models have been developed to assess such vaccines against tuberculosis (Lowrie et al., 1999; Repique et al., 2002; Scanga et al., 1999), based on the Cornell mouse model (McCune et al., 1966). Mice are infected by intravenous or aerosol routes and after 4 weeks, antibiotic therapy is administered for up to 3 months. At a later time point, glucocorticoid or some more specific immune suppressor is used to reactivate latent tuberculosis infections, which have been reduced to non-detectable levels by the antibiotic therapy. At specified intervals after the administration of the immune suppressor, mice are sacrificed to assess the mycobacterial burden in relevant organs. Vaccination of mice in such a model with a DNA construct expressing the hsp60 protein of *Mycobacterium leprae* completely prevented bacterial reactivation (Lowrie et al., 1999). However, these results have not been repeated by others and in some situations the immunised mice developed classical Koch reactions characterised by multifocal necrosis throughout the lung granulomas (Taylor et al., 2003). From a broader perspective, widely varying spontaneous reactivation rates are often observed in the Cornell model even without immune suppression (Scanga et al., 1999; Lenaerts et al., 2004) and this probably explains why no version of the model has become the accepted standard.

There is general agreement that non-human primates have significant advantages over mice and guinea pigs for modelling human pulmonary tuberculosis for purposes of vaccine evaluation, but for cost and other reasons they have been rarely used (McMurray 2000). With a number of candidate vaccines in early human trials and others approaching this stage, interest in using these animals has increased, the relative advantages of one monkey species over another is being evaluated (Langermans et al., 2001; Reed et al., 2003), and they will undoubtedly be used more in the future. There are no current requirements from regulatory agencies that monkey trials should be performed before human trials are commenced but whether such trials will come to be regarded as either advisable or mandatory before human phase III trials are approved remains to be determined. Other large animal models of vaccination, particularly cattle, have been more extensively evaluated (Buddle et al., 2003a). While the main reason for these models is to develop a veterinary vaccine against bovine tuberculosis, they have several features which attract more general interest. Unlike laboratory animals, cattle are natural hosts of tuberculosis and their clinical disease can take years to develop. Because of the size of the animals, the kinetics of infection can be easily followed in individual animals and, unlike guinea pigs, a large array of immunological reagents are available. Finally, cows are immunologically competent at birth allowing for the use of neonatal vaccination. These features have enabled the use of calves for modelling vaccination in neonatal humans (Buddle et al., 2003b).

CONCLUSION

It is now over 50 years since inactivated whole cell mycobacterial preparations and live *M. microti* were tested as replacement vaccines for BCG (von Reyn and Vuola, 2002). While fresh enthusiasm has been redirected very recently towards both these types of vaccine (Haile et al., 2004; Brodin et al., 2004), most tuberculosis vaccine development of the last 15 years has involved the vaccines described earlier in this chapter. This has resulted in a small number of new vaccines entering phase I human trials and others approaching this stage. One of these, BCG over-expressing antigen 85B, showed moderately increased protection relative to BCG in animal models (Horwitz and Harth, 2003), and might replace BCG. The others are subunit vaccines that induce similar protection to BCG in animal models and could be used as booster vaccines, perhaps delivered many years after a vaccination with BCG. The concept of different priming and boosting vaccines rather than a single all purpose vaccine has emerged in the last few years to become a dominant theme in tuberculosis vaccine development.

Most of the work referred to in this review was of necessity from published reports. The number of such reports is large, but even so an expert in the field believes it is merely the tip of an iceberg and that there are perhaps hundreds of new vaccines that have been tested for which no published data is available (McMurray, 2003). It is possible that some of these candidates may also be approaching phase I human trials. One development that is apparent in reviewing the recent literature is that clinical trials and regulatory and operational issues around using new human tuberculosis vaccines have been the central focus of one review (Brennan et al., 2004) and have featured to at least some extent in many others (Doherty, 2004; Fruth and Young, 2004; von Reyn and Vuola, 2002; Young and Stewart, 2002). Ten years ago these concerns barely rated a mention. This development reflects the almost universal acceptance that much better tuberculosis vaccines than BCG can be made, and the fact that the first new candidates have already gone through the regulatory processes for early human trials. Such progress provides increased hope for those dealing on a daily basis with this dreadful disease.

FUTURE TRENDS

The protagonists for the new vaccines that are currently in early human trials, anticipate the eventual testing of their candidates in phase III trials. While results of these trials will be awaited with considerable interest, it will be 5-10 years before they are available. It is difficult to conceive that any of the current new vaccines that are in or approaching early human trials will finally prove to be the best available. This is not stated to denigrate the testing of these new vaccines in humans as this is appropriate, nor does it imply deficiencies in their development which has been intensive, but it reflects instead the current state of the art. The continuing uncertainty both about what constitutes a protective immune response, and the appropriateness of animal models to reflect the situation applying in humans, greatly hinder the development of optimal tuberculosis vaccines. Together with the intense investigations that are continuing in all areas of mycobacterial pathogenesis and functional genomics, cell biology, and immunology including adjuvant development; progress in producing better vaccines than those now available will undoubtedly be made. A more detailed understanding of how *M. tuberculosis* persists in the host (Stewart et al., 2003) and the various stages of infection (Hingley-Wilson et al., 2003) will contribute to this progress. Certain secreted antigens, particularly ESAT-6 and the antigen 85 complex, have received much more attention than others, both in live and subunit vaccines, largely because they are the major secreted proteins in culture filtrates and have strong immunogenicity in animal models. This may change in future, as the information from more recent results of broader surveys of secreted and immunostimulatory tuberculosis antigens (Mollenkopf et al., 2004a) feed through into the development of new candidate vaccines. Given the long time frame over which human trials will need to be conducted for new vaccines, particularly phase III trials, it is highly likely that candidates showing greatly improved results in animal studies will be entering phase I human trials while some of the vaccines currently being tested are still in phase II or phase III trials. At critical future time points, decisions will have to be made about whether to implement a less efficacious vaccine or to wait a year or more for a potentially better vaccine to complete human trials. By the standard of research grant allocations, decisions on clinical trials involve very large amounts of money. The vaccine candidates that are now at the early stages of human trials would not have progressed thus far without substantial support from various targeted funding sources such as AERAS in the USA, the European consortium TBVAC and the German organisation Vakzine Projekt. As candidates proceed into phase II and III human trials, a major role in facilitating or funding the work will also be played by various other organisations, including the World Health Organisation and the recently announced European and Developing Countries Clinical Trials Partnership (Olliaro and Smith, 2004). Clearly, vaccine companies will also be important participants. Ultimately, political and commercial interests will be at least as important as scientific issues in deciding which new vaccines are deployed globally.

REFERENCES

Agger, E.M., and Andersen, P. (2002). A novel TB vaccine; towards a strategy based on our understanding of BCG failure. Vaccine *21*, 7-14.

Anonymous (2004). BCG vaccine: WHO position paper. Weekly Epidemiol. Rec. 4, 27-38.

Aronson, N.E., Santosham, M., Comstock, G.W., Howard, R.S., Moulton, L.H., Rhoades, E.R., Harrison, L.H. (2004). Long-term efficacy of BCG vaccine in American Indians and Alaska Natives: A 60-year follow-up study. JAMA *291*, 2086-2091.

Baldwin, S.L., D'Souza, C., Roberts, A.D., Kelly, B.P., Frank, A.A., Lui, M.A., Ulmer, J.B., Huygen, K., McMurray, D.N., and Orme, I.M. (1998). Evaluation of new vaccines in the mouse and guinea pig models of tuberculosis. Infect. Immun. *66*, 2951-2959.

Bao, L., Chen, W., Zhang, H., and Wang, X. (2003). Virulence, immunogenicity, and protective efficacy of two recombinant *Mycobacterium bovis* bacillus Calmette-Guérin strains expressing the antigen ESAT-6 from *Mycobacterium tuberculosis*. Infect. Immun. *71*, 1656-1661.

Bauer, S., Kirschning, C.J., Hacker, H., Redecke, V., Hausmann, S., Akira, S., Wagner, H., and Lipford, G.B. (2001). Human TLR9 confers responsiveness to bacterial DNA via species-specific CpG motif recognition. Proc. Natl. Acad. Sci. USA. *98*, 9237-9242.

Behr, M.A. (2002). BCG, different strains, different vaccines? Lancet Infect. Dis. *2*, 86-92.

Bonato, V.L., Goncalves, E.D., Soares, E.G., Santos Jr., R.R., Sartori, A., Coelho-Castelo, A.A., and Silva, C.L. (2004). Immune regulatory effect of pHSP65 DNA therapy in pulmonary tuberculosis: activation of CD8+ cells, interferon-gamma recovery and reduction of lung injury. Immunology *113*, 130-138.

Brandt, L., and Orme, I. (2002). Prospects for new vaccines against tuberculosis. Biotechniques *33*, 1098-1102.

Brandt, L., Cunha J.F., Olsen A.W., Chilima, B., Hirsch, P., Appelberg, R., and Andersen, P. (2002). Failure of the *Mycobacterium bovis* BCG vaccine: some species of environmental mycobacteria block multiplication of BCG and induction of protective immunity to tuberculosis. Infect. Immun. *70*, 672-678.

Brandt, L., Skeiky, Y.A., Alderson, M.R., Lobet, Y., Dalemans, W., Turner, O.C., Basaraba, R.J., Izzo, A.A., Lasco, T.M., Chapman, P.L., Reed, S.G., and Orme, I.M. (2004). The protective effect of the *Mycobacterium bovis* BCG vaccine is increased by coadministration with the *Mycobacterium tuberculosis* 72-kilodalton fusion polyprotein Mtb72F in *M. tuberculosis*-infected guinea pigs. Infect. Immun. 72, 6622-6632.

Brennan, M.J., Morris, S.L., and Sizemore, C.F. (2004). Tuberculosis vaccine development: research, regulatory and clinical strategies. Expert Opin. Biol. Ther. *4*, 1493-1504.

Brewer, T.F. (2000). Preventing tuberculosis with bacillus Calmette-Guérin vaccine: a meta-analysis of the literature. Clin. Infect. Dis. Suppl. *3*, S64-S67.

Britton, W.J., and Palendira, U. (2003). Improving vaccines against tuberculosis. Immunol. Cell Biol. *81*, 34-45.

Brodin, P., Majlessi, L., Brosch, R., Smith, D., Bancroft, G., Clark, S., Williams, A., Leclerc, C., and Cole, S.T. (2004). Enhanced protection against tuberculosis by vaccination with recombinant *Mycobacterium microti* vaccine that induces T cell immunity against region of difference 1 antigens. J. Infect. Dis. *190*, 115-122.

Brooks, J.V., Frank, A.A., Keen, M.A., Belisle, J.T., and Orme, I.M. (2001). Boosting vaccine for tuberculosis. Infect. Immun. *69*, 2714-2717.

Brosch, R., Gordon, S.V., Marmiesse, M., Brodin, P., Buchrieser, C., Eiglmeier, K., Garnier, T., Gutierrez, C., Hewinson, G., Kremer, K., Parsons, L.M., Pym, A.S., Samper, S., van Soolingen, D., and Cole, S.T. (2002). A new evolutionary scenario for the *Mycobacterium tuberculosis* complex. Proc. Natl. Acad. Sci. USA. *99*, 3684-3689.

Buddle, B.M., Pollock, J.M., Skinner, M.A., and Wedlock, D.N. (2003a). Development of vaccines to control bovine tuberculosis in cattle and relationship to vaccine development for other intracellular pathogens. Int. J. Parasitol. *33*, 555-566.

Buddle, B.M., Wedlock, D.N., Parlane, N.A., Corner, L.A., de Lisle, G.W., and Skinner, M.A. (2003b). Revaccination of neonatal calves with *Mycobacterium bovis* BCG reduces the level of protection against bovine tuberculosis induced by a single vaccination. Infect. Immun. *71*, 6411-6419.

Camacho, L.R., Ensergueix, D., Perez, E., Gicquel, B., and Guilhot, C. (1999). Identification of a virulence gene cluster of *Mycobacterium tuberculosis* by signature-tagged transposon mutagenesis. Mol. Microbiol. *34*, 257-267.

Chambers, M.A., Williams, A., Gavier-Widen, D., Whelan, A., Hall, G., Marsh, P.D., Bloom, B.R., Jacobs, W.R., and Hewinson, R.G. (2000). Identification of a *Mycobacterium bovis* BCG auxotrophic mutant that protects guinea pigs against *M. bovis* and hematogenous spread of *Mycobacterium tuberculosis* without sensitization to tuberculin. Infect. Immun. *68*, 7094-7099.

Ciaramella, A., Cavone, A., Santucci, M.B., Garg, S.K., Sanarico, N., Bocchino, M., Galati, D., Martino, A., Auricchio, G., D'Orazio, M., Stewart, G.R., Neyrolles, O., Young, D.B., Colizzi, V., and Fraziano, M. (2004). Induction of apoptosis and release of interleukin-1 beta by cell wall-associated 19-kDa lipoprotein during the course of mycobacterial infection. J. Infect. Dis. *190*, 1167-1176.

Clark-Curtiss, J.E., and Haydel, S.E. (2003). Molecular genetics of *Mycobacterium tuberculosis* pathogenesis. Annu. Rev. Microbiol. *5*, 517-549.

Collins, D.M. and de Lisle, G.W. (1987). BCG identification by DNA restriction fragment patterns. J. Gen. Microbiol. *133*, 1431-1434.

Collins, D.M., and Gicquel, B. (2000). Genetics of mycobacterial virulence. In Molecular Genetics of Mycobacteria, G.F. Hatfull and W.R. Jacobs Jr., eds. (Washington DC: ASM Press).

Collins, H.L., and Kaufmann, S.H. (2001). Prospects for better tuberculosis vaccines. Lancet Infect. Dis. *1*, 21-28.

Collins, D.M., Wilson, T., Campbell, S., Buddle, B.M., Wards, B.J., Hotter, G., de Lisle, G.W. (2002). Production of avirulent mutants of *Mycobacterium bovis* with vaccine properties by the use of illegitimate recombination and screening of stationary phase cultures. Microbiology *148*, 3019-3027.

Collins, D.M., Kawakami, R.P., Buddle, B.M., Wards, B.J., de Lisle, G.W. (2003a). Different susceptibility of two animal species infected with isogenic mutants of *Mycobacterium bovis* enabled the identification of *phoT* with roles in tuberculosis virulence and phosphate transport. Microbiology *149*, 3203-3212.

Collins, D.M., Kawakami, R.P., Wards, B.J., Campbell, S., de Lisle, G.W. (2003b). Vaccine and skin testing properties of two avirulent *Mycobacterium bovis* mutants with and without an additional *esat-6* mutation. Tuberculosis (Edinb.) *83*, 361-366.

Collins, D.M., Skou, B., White, S., Bassett, S., Collins, L., For, R., Hurr, K., Hotter, G., and de Lisle, G.W. (2005). Generation of *Mycobacterium bovis* attenuated strains by signature-tagged mutagenesis in search of novel vaccine candidates. Infect. Immun. *73*, 2379-2386.

Comstock, G.W. (2000). Simple, practical ways to assess the protective efficacy of a new tuberculosis vaccine. Clin. Infect. Dis. Suppl. *3*, S250-S253.

Cox, J.S., Chen, B., McNeil, M., and Jacobs Jr., W.R. (1999). Complex lipid determines tissue-specific replication of *Mycobacterium tuberculosis* in mice. Nature *402*, 79-83.

Dascher, C.C., Hiromatsu, K., Xiong, X., Morehouse, C., Watts, G., Liu, G., McMurray, D.N., LeClair, K.P., Porcelli, S.A., and Brenner, M.B. (2003). Immunization with a mycobacterial lipid vaccine improves pulmonary pathology in the guinea pig model of tuberculosis. Int. Immunol. *15*, 915-925.

de Bruyn, and G., Garner, P. (2003). *Mycobacterium vaccae* immunotherapy for treating tuberculosis. Cochrane Database Syst. Rev. *1*, CD001166.

de Jong, B.C., Israelski, D.M., Corbett, E.L., and Small, P.M. (2004). Clinical management of tuberculosis in the context of HIV infection. Annu. Rev. Med. *55*, 283-301.

de Lisle, G.W., Wilson, T., Collins, D.M., and Buddle, B.M. (1999). Vaccination of guinea pigs with nutritionally impaired avirulent mutants of *Mycobacterium bovis* protects against tuberculosis. Infect. Immun. *67*, 2624-2626.

de Lisle, G.W., Wards, B.J., Buddle, B.M., and Collins, D.M. (2005). The efficacy of live tuberculosis vaccines after presensitisation with *Mycobacterium avium*. Tuberculosis (Edinb.) *85*, 73-79.

Dietrich, G., Viret, J.F., and Hess, J. (2003). *Mycobacterium bovis* BCG-based vaccines against tuberculosis: novel developments. Vaccine *21*, 667-670.

Dillon, D.C., Alderson, M.R., Day, C.H, Lewinsohn, D.M., Coler, R., Bement, T., Campos-Neto, A., Skeiky, Y.A., Orme, I.M., Roberts, A., Steen, S., Dalemans, W., Badaro, R., and Reed, S.G. (1999). Molecular characterization and human T-cell responses to a member of a novel *Mycobacterium tuberculosis* mtb39 gene family. Infect. Immun. *67*, 2941- 2950.

Doherty, T.M. (2004). New vaccines against tuberculosis. Trop. Med. Int. Health *9*, 818-826.

Dye, C., Scheele, S., Dolin, P., Pathania, V., and Raviglione, M.C. (1999). Global burden of tuberculosis. JAMA *282*, 677-686.

Dye, C., Watt, C.J., Bleed, D.M., and Williams, B.G. (2003). What is the limit to case detection under the DOTS strategy for tuberculosis control? Tuberculosis (Edinb.) *83*, 35-43.

Fine, P.E.M. (1989). The BCG story: lessons from the past and implications for the future. Rev. Infect. Dis. 11, S353-S359.

Fine, P.E. (2001). BCG: The challenge continues. Scand. J. Infect. Dis. *33*, 243-245.

Fine, P.E., Floyd, S., Stanford, J.L., Nkhosa, P., Kasunga, A., Chaguluka, S., Warndorf, D.K., Jenkins, P.A., Yates, M., and Ponnighaus, J.M. (2001). Environmental mycobacteria in northern Malawi: implications for the epidemiology of tuberculosis and leprosy. Epidemiol. Infect. *126*, 379-387.

Flynn, J.L. (2004). Immunology of tuberculosis and implications in vaccine development. Tuberculosis (Edinb.) *84*, 93-101.

Frieden, T.R., Sterling, T.R., Munsiff, S.S., Watt, C.J., and Dye, C. (2003). Tuberculosis. Lancet *362*, 887-899.

Fruth, U. and Young, D. (2004). Prospects for new TB vaccines: Stop TB Working Group on TB Vaccine Development. Int. J. Tuberc. Lung Dis. *8*, 151-155.

Gilleron, M., Stenger, S., Mazorra, Z., Wittke, F., Mariotti, S., Bohmer, G., Prandi, J., Mori, L., Puzo, G., and De Libero, G. (2004). Diacylated sulfoglycolipids are novel mycobacterial antigens stimulating CD1-restricted T cells during infection with *Mycobacterium tuberculosis*. J. Exp. Med. *199*, 649-659.

Ginsberg, A.M. (2002). What's new in tuberculosis vaccines? Bull. World Health Organ. *80*, 483-488.

Goonetilleke, N.P., McShane, H., Hannan, C.M., Anderson, R.J., Brookes, R.H., and Hill, A.V. (2003). Enhanced immunogenicity and protective efficacy against *Mycobacterium tuberculosis* of bacille Calmette-Guérin vaccine using mucosal administration and boosting with a recombinant modified vaccinia virus Ankara. J. Immunol. *171*, 1602-1609.

Guleria, I., Teitelbaum, R., McAdam, R.A., Kalpana, G., Jacobs Jr., W.R., and Bloom, B.R. (1996). Auxotrophic vaccines for tuberculosis. Nat. Med. *2*, 334-337.

Haile, M., Schroder, U., Hamasur, B., Pawlowski, A., Jaxmar, T., Kallenius, G., and Svenson, S.B. (2004). Immunization with heat-killed *Mycobacterium bovis* bacille Calmette-Guérin (BCG) in Eurocine L3 adjuvant protects against tuberculosis. Vaccine. *22*, 1498-1508.

Hart, P.D., and Sutherland, I. (1977). BCG and vole bacillus vaccines in the prevention of tuberculosis in adolescence and early adult life. Br. Med. J. *2(6082)*, 293-295.

Haslov, K., Andersen, A., Nagai, S., Gottschau, A., Sorensen, T. and Andersen, P. (1995). Guinea pig cellular immune responses to proteins secreted by *Mycobacterium tuberculosis*. Infect. Immun. *63*, 804-810.

Hess, J., and Kaufmann, S.H. (2001). Development of live recombinant vaccine candidates against tuberculosis. Scand. J. Infect. Dis. *33*, 723-724.

Hingley-Wilson, S.M., Sambandamurthy, V.K., and Jacobs Jr., W.R. (2003). Survival perspectives from the world's most successful pathogen, *Mycobacterium tuberculosis*. Nat. Immunol. *4*, 949-95.

Hisert, K.B., Kirksey, M.A., Gomez, J.E., Sousa, A.O., Cox, J.S., Jacobs Jr., W.R., Nathan, C.F., and McKinney, J.D. (2004). Identification of *Mycobacterium tuberculosis* counterimmune (cim) mutants in immunodeficient mice by differential screening. Infect. Immun. *72*, 5315-5321.

Hoag, H. (2004). New vaccines enter fray in fight against tuberculosis. (2004). Nat. Med. *10*, 6.

Hondalus, M.K., Bardarov, S., Russell, R., Chan, J., Jacobs Jr., W.R., and Bloom, B.R. (2000). Attenuation of and protection induced by a leucine auxotroph of *Mycobacterium tuberculosis*. Infect. Immun. *68*, 2888-2898.

Horwitz, M.A., Lee, B.-W.E., Dillon, B.J. and Harth, G. (1995). Protective immunity against tuberculosis induced by vaccination with major extracellular proteins of *Mycobacterium tuberculosis*. Proc. Natl. Acad. Sci. USA *92*, 1530-1534.

Horwitz, M.A., Harth, G., Dillon, B.J., and Maslesa-Galic, S. (2000). Recombinant bacillus Calmette-Guérin (BCG) vaccines expressing the *Mycobacterium tuberculosis* 30-kDa major secretory protein induce greater protective immunity against tuberculosis than conventional BCG vaccines in a highly susceptible animal model. Proc. Natl. Acad. Sci. U.S.A. *97*, 13853-13858.

Horwitz, M.A., and Harth, G. (2003). A new vaccine against tuberculosis affords greater survival after challenge than the current vaccine in the guinea pig model of pulmonary tuberculosis. Infect Immun. *71*, 1672-1679.

Hsu, T., Hingley-Wilson, S.M., Chen, B., Chen, M., Dai, A.Z. Morin, P.M., Marks, C.B., Padiyar, J. Goulding, C. Gingery, M., Eisenberg, D., Russell, R.G., Derrick, S.C., Collins, F.M., Morris, S.L., King, C.H., and Jacobs Jr., W.R. (2003). The primary mechanism of attenuation of bacillus Calmette-Guérin is a loss of secreted lytic function required for invasion of lung interstitial tissue. Proc. Natl. Acad. Sci. USA *100*, 12420-12425.

Hussey, G.D, Watkins, M.L., Goddard, E.A., Gottschalk, S., Hughes, E.J., Iloni, K., Kibel, M.A., and Ress, S.R. (2002). Neonatal mycobacterial specific cytotoxic T-lymphocyte and cytokine profiles in response to distinct BCG vaccination strategies. Immunology *105*, 314-324.

Jackson, M., Phalen, S.W., Lagranderie, M., Ensergueix, D., Chavarot, P., Marchal, G., McMurray, D.N., Gicquel, B., and Guilhot, C. (1999). Persistence and protective efficacy of a *Mycobacterium tuberculosis* auxotroph vaccine. Infect. Immun. *67*, 2867-2873.

Johnson, P.D.R., Stuart, R.L. Grayson, M.L., Olden, D., Clancy, A. Ravn, P., Andersen, P., Britton, W.J., and Rothel, J.S. (1999). Tuberculin-purified protein derivative-, MPT-64-, and ESAT-6-stimulated gamma interferon responses in medical students before and after *Mycobacterium bovis* BCG vaccination and in patients with tuberculosis. Clin. Diag. Lab. Immunol. *6*, 934-937.

Kana, B.D., and Mizrahi, V. (2004). Molecular genetics of *Mycobacterium tuberculosis* in relation to the discovery of novel drugs and vaccines. Tuberculosis (Edinb.) *84*, 63-75.

Karonga prevention trial group. (1996). Randomised controlled trial of single BCG, repeated BCG or combined BCG and killed *Mycobacterium leprae* vaccine for prevention of leprosy and tuberculosis in Malawi. Lancet. *348*, 17-24.

Kaufmann, S.H. (2001). How can immunology contribute to the control of tuberculosis? Nat. Rev. Immunol. *1*, 20-30.

Kumar, H., Malhotra, D., Goswami, S, . and Bamezai, R.N. (2003). How far have we reached in tuberculosis vaccine development? Crit. Rev. Microbiol. *29*, 297-312.

Kochi, S.K., Killeen, K.P., and Ryan, U.S. (2003). Advances in the development of bacterial vector technology. Expert Rev. Vaccines *2*, 31-43.

Langermans, J.A., Andersen, P., van Soolingen, D., Vervenne, R.A., Frost, P.A., van der Laan, T., van Pinxteren, L.A., van den Hombergh, J., Kroon, S., Peekel, I., Florquin, S., and Thomas, A.W. (2001). Divergent effect of bacillus Calmette-Guérin (BCG) vaccination on *Mycobacterium tuberculosis* infection in highly related macaque species: implications for primate models in tuberculosis vaccine research. Proc. Natl. Acad. Sci. U.S.A. *98*, 11497-11502.

Lathigra, R., Zhang, Y., Hill, M., Garcia, M.J., Jackett, P.S., and Ivanyi, J. (1996). Lack of production of the 19-kDa glycolipoprotein in certain strains of *Mycobacterium tuberculosis*. Res. Microbiol. *147*, 237-249.

Lenaerts, A.J., Chapman, P.L., and Orme, I.M. (2004). Statistical limitations to the Cornell model of latent tuberculosis infection for the study of relapse rates. Tuberculosis (Edinb.) *84*, 361-364.

Leung, C.C., Tam, C.M., Chan, S.L., Chan-Yeung, M., Chan, C.K., Chang, K.C. (2001). Efficacy of the BCG revaccination programme in a cohort given BCG vaccination at birth in Hong Kong. Int. J. Tuberc. Lung Dis. *5*, 717-723.

Lopez, M., Sly, L.M., Luu, Y., Young, D., Cooper, H., and Reiner, N.E. (2003). The 19-kDa *Mycobacterium tuberculosis* protein induces macrophage apoptosis through Toll-like receptor-2.J. Immunol. *170*, 2409-2416.

Lowrie, D.B., Tascon, R.E., Bonato, V.L., Lima, V.M., Facciol, L.H., Stavropoulos, E., Colston, M.J., Hewinson, R.G., Moelling, K., and Silva, C.L. (1999). Therapy of tuberculosis in mice by DNA vaccination. Nature *400*, 269-271.

McAdam, R.A., Weisbrod, T.R., Martin, J., Scuderi, J.D., Brown, A.M., Cirillo, J.D., Bloom, B.R., and Jacobs Jr., W.R. (1995). In vivo growth characteristics of leucine and methionine auxotrophic mutants of *Mycobacterium bovis* BCG generated by transposon mutagenesis. Infect. Immun. *63*, 1004-1012.

McAdam, R.A., Quan, S., Smith, D.A., Bardarov, S., Betts, J.C., Cook, F.C., Hooker, E.U., Lewis, A.P., Woollard, P., Everett, M.J., Lukey, P.T., Bancroft, G.J., Jacobs Jr., W.R. and Duncan, K. (2002). Characterization of a *Mycobacterium tuberculosis* H37Rv transposon library reveals insertions in 351 ORFs and mutants with altered virulence. Microbiology *148*, 2975-2986.

McCune, R.M., Feldmann, F.M., Lambert, H.P., and McDermott, W. (1966). Microbial persistence. I. The capacity of tubercle bacilli to survive sterilization in mouse tissues. J. Exp. Med. *123*, 445-468.

McMurray, D.N., Dai, G., and Phalen, S. (1999). Mechanisms of vaccine-induced resistance in a guinea pig model of pulmonary tuberculosis. Tubercle Lung Dis. *79*, 261-266.

McMurray, D.N. (2000). A nonhuman primate model for preclinical testing of new tuberculosis vaccines. Clin. Infect. Dis. *30*, S210-S212.

McMurray, D.N. (2003). Recent progress in the development and testing of vaccines against human tuberculosis. Int. J. Parasitol. *33*, 547-554.

McShane, H., Pathan, A.A., Sander, C.R., Keating, S.M., Gilbert, S.C., Huygen, K., Fletcher, H.A., and Hill, A.V. (2004). Recombinant modified vaccinia virus

Ankara expressing antigen 85A boosts BCG-primed and naturally acquired antimycobacterial immunity in humans. Nat. Med. *10*, 1240-1244.

Miki, K., Nagata, T., Tanaka, T., Kim, Y.H., Uchijima, M., Ohara, N., Nakamura, S., Okada, M., and Koide, Y. (2004). Induction of protective cellular immunity against *Mycobacterium tuberculosis* by recombinant attenuated self-destructing *Listeria monocytogenes* strains harboring eukaryotic expression plasmids for antigen 85 complex and MPB/MPT51. Infect. Immun. *72*, 2014-2021.

Mollenkopf, H.J., Grode, L., Mattow, J., Stein, M., Mann, P., Knapp, B, Ulmer, J., and Kaufmann, S.H. (2004a). Application of mycobacterial proteomics to vaccine design: improved protection by *Mycobacterium bovis* BCG prime-Rv3407 DNA boost vaccination against tuberculosis. Infect Immun. *72*, 6471-6479.

Mollenkopf, H.J., Kursar, M., and Kaufmann, S.H. (2004b). Immune response to postprimary tuberculosis in mice: *Mycobacterium tuberculosis* and M*ycobacterium bovis* bacille Calmette-Guérin induce equal protection. J. Infect. Dis. *190*, 588-597.

Moody, D.B., Ulrichs, T., Muhlecker, W., Young, D.C., Gurcha, S.S., Grant, E., Rosat, J.P., Brenner, M.B., Costello, C.E., Besra, G.S., and Porcelli, S.A. (2000). CD1c-mediated T-cell recognition of isoprenoid glycolipids in *Mycobacterium tuberculosis* infection. Nature *404*, 884-888.

Murray, P.J., Aldovini, A., and Young, R.A. (1996). Manipulation and potentiation of antimycobacterial immunity using recombinant bacille Calmette-Guérin strains that secrete cytokines. Proc. Natl. Acad. Sci. USA *93*, 934-939.

Mwinga, A., Nunn, A., Ngwira, B., Chintu, C., Warndorff, D., Fine, P., Darbyshire, J., and Zumla A, (2002). *Mycobacterium vaccae* (SRL172) immunotherapy as an adjunct to standard antituberculosis treatment in HIV-infected adults with pulmonary tuberculosis: a randomised placebo-controlled trial. Lancet *360*, 1050-1055.

Nor, N.M., and Musa, M. (2004). Approaches towards the development of a vaccine against tuberculosis: recombinant BCG and DNA vaccine. Tuberculosis (Edinb.) *84*, 102-109.

North, R.J., and Jung, Y.J. (2004). Immunity to tuberculosis. Annu. Rev. Immunol. *22*, 599-623.

Okkels, L.M., Doherty, T.M., and Anderson, P. (2003). Selecting the components for a safe and efficient tuberculosis subunit vaccine - recent progress and post-genomic insights. Curr. Pharm. Biotechnol. *4*, 69-83.

Olliaro, P., and Smith, P.G. (2004). The European and developing countries clinical trails partnership. J. HIV Ther. *9*, 53-56.

Olsen, A.W., William, A., Okkels, L.M., Hatch, G., and Andersen, P. (2004). Protective effect of a tuberculosis subunit vaccine based on a fusion of antigen 85B and ESAT-6 in the aerosol guinea pig model. Infect. Immun. *72*, 6148-6150.

Olsen, A.W., van Pinxteren, L.A.H., Okkels, L. M., Rasmussen, P.B., and Andersen, P. (2001). Protection of mice with a tuberculosis subunit vaccine based on a fusion protein of antigen 85B and ESAT-6. Infect. Immun. *69*, 277-32778.

Orme I. (2004). Adaptive immunity to mycobacteria. Curr. Opin. Microbiol. *7*, 58-61.

Orme, I.M., McMurray, D.N., and Belisle, J.T. (2001). Tuberculosis vaccine development: recent progress. Trends Microbiol. *9*, 115-118.

Parish, T., Gordhan, B.G., McAdam, R.A., Duncan, K., Mizrahi, V., and Stoker, N.G. (1999). Production of mutants in amino acid biosynthesis genes of *Mycobacterium tuberculosis* by homologous recombination. Microbiology *145*, 3497-3503.

Perez, E., Samper, S., Borda, Y., Guilhot, C., Gicquel, B., and Martin, C. (2001). An essential role for *phoP* in *Mycobacterium tuberculosis* virulence. Mol. Microbiol. *41*, 179-187.

Pinto, R., Saunders, B.M., Camacho, L.R., Britton, W.J., Gicquel, B. and Triccas, J.A. (2004). *Mycobacterium tuberculosis* defective in phthiocerol dimycocerosate translocation provides greater protective immunity against tuberculosis than the existing bacille Calmette-Guérin vaccine. J. Infect. Dis. *189*, 105-112.

Pym, A.S., Brodin, P., Majlessi, L., Brosch, R., Demangel, C., Williams, A., Griffiths, K.E., Marchal, G., Leclerc, C., Cole, S.T. (2003). Recombinant BCG exporting ESAT-6 confers enhanced protection against tuberculosis. Nat. Med. *9*, 533-539.

Ramirez, J.C., Gherardi, M.M., and Esteban, M. (2000). Biology of attenuated modified vaccinia virus Ankara recombinant vector in mice: virus fate and activation of B- and T-cell immune responses in comparison with the Western Reserve strain and advantages as a vaccine. J. Virol. *74*, 923-933.

Rao, V., Dhar, N., and Tyagi, A.K. (2003). Modulation of host immune responses by overexpression of immunodominant antigens of *Mycobacterium tuberculosis* in bacille Calmette-Guérin. Scand. J. Immunol. *58*, 449-461.

Reed, S.G., Alderson, M.R., Dalemans, W., Lobet, Y., and Skeiky, Y.A. (2003). Prospects for a better vaccine against tuberculosis. Tuberculosis (Edinb.). *83*, 213-219.

Renshaw, P.S., Panagiotidou, P., Whelan, A., Gordon, S.V., Hewinson, R.G., Williamson, R.A., and Carr, M.D. (2002). Conclusive evidence that the major T-cell antigens of the *Mycobacterium tuberculosis* complex ESAT-6 and CFP-10 form a tight, 1:1 complex and characterization of the structural properties of ESAT-6, CFP-10, and the ESAT-6*CFP-10 complex. Implications for pathogenesis and virulence. J. Biol. Chem. *277*, 21598-21603.

Repique, C.J., Li, A., Collins, F.M., and Morris, S.L. (2002). DNA immunization in a mouse model of latent tuberculosis: Effect of DNA vaccination on reactivation of disease and on reinfection with a secondary challenge. Infect. Immun. *70*, 3318-3323.

Sambandamurthy, V.K., Wang, X., Chen, B., Russell, R.G., Derrick, S., Collins, F.M., Morris, S.L., and Jacobs, W.R. (2002). A pantothenate auxotroph of *Mycobacterium tuberculosis* is highly attenuated and protects mice against tuberculosis. Nat. Med. *8*, 1171-1174.

Sampson, S.L., Dascher, C.C., Sambandamurthy, V.K., Russell, R.G., Jacobs Jr., W.R., Bloom, B.R., and Hondalus, M.K. (2004). Protection elicited by a double leucine and pantothenate auxotroph of *Mycobacterium tuberculosis* in guinea pigs. Infect. Immun. *72*, 3031-3037.

Sassetti, C.M., and Rubin, E.J. (2003). Genetic requirements for mycobacterial survival during infection. Proc. Natl. Acad. Sci. USA *100*, 12989-12994.

Scanga, C.A., Mohan, V.P., Joseph, H., Yu, K., Chan, J., and Flynn, J.L. (1999). Reactivation of latent tuberculosis: variations on the Cornell murine model. Infect. Immun. *67*, 4531-4538.

Skeiky, Y.A., Alderson, M.R., Ovendale, P.J., Guderian, J.A., Brandt, L., Dillon, D.C., Campos-Neto, A., Lobet, Y., Dalemans, W., Orme, I.M., and Reed, S.G. (2004). Differential immune responses and protective efficacy induced by components of a tuberculosis polyprotein vaccine, Mtb72F, delivered as naked DNA or recombinant protein. J. Immunol. *172*, 7618-7628.

Skinner, M.A., Buddle, B.M., Wedlock, D.N., Keen, D., de Lisle, G.W., Tascon, R.E., Ferraz, J.C., Lowrie, D.B., Cockle, P.J., Vordermeier, H.M., and Hewinson, R.G. (2003). A DNA prime-*Mycobacterium bovis* BCG boost vaccination strategy for cattle induces protection against bovine tuberculosis. Infect. Immun. *71*, 4901-4907.

Smith, D.A., Parish, T., Stoker, N.G., and Bancroft, G.J. (2001). Characterization of auxotrophic mutants of *Mycobacterium tuberculosis* and their potential as vaccine candidates. Infect. Immun. *69*, 1142-1150.

Smith, D., Wiegeshaus, E., and Balasubramanian, V. (2000). An analysis of some hypotheses related to the Chingleput bacille Calmette-Guérin trial. Clin. Infect. Dis. *31*, S77-S80.

Stanford, J., Stanford, C., and Grange, J. (2004). Immunotherapy with *Mycobacterium vaccae* in the treatment of tuberculosis. Front Biosci. *9*, 1701-1719.

Stewart, G.R., Robertson, B.D., and Young, D.B. (2003). Tuberculosis: a problem with persistence. Nat. Rev. Microbiol. *1*, 97-105.

Taylor, J.L., Turner, O.C., Basaraba, R.J., Belisle, J.T., Huygen, K., and Orme, I.M. (2003). Pulmonary necrosis resulting from DNA vaccination against tuberculosis. Infect. Immun. *71*, 2192-2198.

ten Dam, H.G., and Pio, A. (1982). Pathogenesis of tuberculosis and effectiveness of BCG vaccination. Tubercle *63*, 225-233.

van Drunen Littel-van den Hurk, S., Babiuk, S.L., and Babiuk, L.A. (2004). Strategies for improved formulation and delivery of DNA vaccines to veterinary target species. Immunol. Rev. *199*, 113-125.

von Reyn, C.F., and Vuola, J.M. (2002). New vaccines for the prevention of tuberculosis. Clin. Infect. Dis. *35*, 465-474.

Wang, J., Thorson, L., Stokes, R.W., Santosuosso, M., Huygen, K., Zganiacz, A., Hitt, M., and Xing, Z. (2004). Single mucosal, but not parenteral, immunization with recombinant adenoviral-based vaccine provides potent protection from pulmonary tuberculosis. J. Immunol. *173*, 6357-6365.

Wards, B.J., de Lisle, G.W., and Collins, D.M. (2000). An *esat6* knockout mutant of *Mycobacterium bovis* produced by homologous recombination will contribute to the development of a live tuberculosis vaccine. Tubercle Lung Dis. *80*, 185-189.

Wiegeshaus, E.H., McMurray, D.N., Grover, A.A., Harding, G.E., and Smith, D.W. (1970). Host-parasite relationships in experimental airborne tuberculosis. III. Relevance of microbial enumeration to acquired resistance in guinea pigs. Am. Rev. Resp. Dis. *102*, 422-429.

Wiegeshaus, E.H., and Smith, D.W. (1989). Evaluation of the protective potency of new tuberculosis vaccines. Rev. Infect. Dis. *11 Suppl. 2*, S484-S490.

World Health Organisation. (2004). Global tuberculosis control – surveillance, planning, financing; WHO Report WHO/HTM/TB/2004.331.

Xing Z. (2001). The hunt for new tuberculosis vaccines: anti-TB immunity and rational design of vaccines. Curr. Pharm. Des. *7*, 1015-1037.

Yeremeev, V.V., Stewart, G.R., Neyrolles, O., Skrabal, K., Avdienko, V.G., Apt, A.S., and Young, D.B. (2000). Deletion of the 19 kDa antigen does not alter the protective efficacy of BCG. Tuber. Lung Dis. *80*, 243-247.

Young, D.B., and Stewart, G.R. (2002). Tuberculosis vaccines. Br. Med. Bull. *62*, 73-86.

Zumla, A., Malon, P., Henderson, J., and Grange, J.M. (2000). Impact of HIV infection on tuberculosis. Postgrad. Med. J. *76*, 259-268.

Chapter 5

Drug Resistance in *Mycobacterium tuberculosis*

Rabia Johnson, Elizabeth M. Streicher*, Gail E. Louw, Robin M. Warren, Paul D. van Helden and Thomas C. Victor*

ABSTRACT

Anti-tuberculosis drugs are a double-edged sword. Whilst they destroy pathogenic *Mycobacterium tuberculosis,* they also select for drug resistant bacteria against which those drugs are subsequently ineffective. Global surveillance has shown that drug resistant tuberculosis is widespread and is now a threat to tuberculosis control programs in many countries. The application of molecular methods during the last decade has greatly changed our understanding of drug resistance in tuberculosis. Application of molecular epidemiological methods has also been central to the description of outbreaks of drug resistance. We describe the recommendations for tuberculosis treatment according to the WHO guidelines, the drug resistance problem world-wide, mechanisms of resistance to first and second line drugs and the application of molecular methods to detect gene mutations causing resistance. It is envisaged that molecular techniques will be important adjuncts to traditional culture-based procedures for rapid drug resistance screening. Prospective analysis and intervention to prevent transmission may be particularly helpful in areas with ongoing transmission of drug resistant strains as recent mathematical modeling indicate that the burden of multi-drug resistant (MDR) strains cannot be contained in the absence of specific efforts to limit transmission.

INTRODUCTION

DRUG RESISTANCE AND GLOBAL SURVEILLANCE: HISTORY

Shortly after the first anti-tuberculosis (TB) drugs were introduced (streptomycin (STR), para-aminosalicylic acid (PAS), isoniazid (INH)) resistance to these drugs was observed in clinical isolates of *Mycobacterium tuberculosis* (Crofton and Mitchison, 1948). This lead to the need to measure resistance accurately and easily. The Institute Pasteur introduced the critical proportion method in 1961 for drug susceptibility testing in TB and this method became the standard method of use (Espinal, 2003). Studies on drug resistance in various countries in the 1960's showed that developing countries had a much higher incidence of drug resistance than developed countries (Espinal, 2003). By the end of the 1960's rifampin (RIF)

* Equal contribution as principle author.

was introduced and with the use of combination therapy, there was a decline in drug resistant and drug susceptible TB in developed countries. This led to a decline in funding and interest in TB control programs. As a result, no concrete monitoring of drug resistance was carried out for the following 20 years (Espinal, 2003). The arrival of HIV/AIDS in the 1980's resulted in an increase in transmission of TB associated with outbreaks of multi-drug-resistant TB (MDR-TB) (Edlin et al., 1992; Fischl et al., 1992) *i.e.* resistant to INH and RIF. In the early 1990's drug resistance surveillance was resumed in developed countries, but the true incidence remained unclear in the developing world (Cohn et al., 1997).

THE WHO/IUATLD GLOBAL PROJECT ON DRUG-RESISTANCE SURVEILLANCE

In 1994 the Global Project on Drug-Resistance Surveillance was initiated to monitor the trends of resistance. The first report was published in 1997 and contained data from 35 geographical settings for the period 1994-1996 (World Health Organization, 1997; Pablos-Mendez et al., 1998). The report showed that drug resistance was present globally, and that MDR-TB ranged from 0% to 14% in new cases (median:1.4%) and 0% to 54% in previously treated cases (median:13%). A second report for the period 1996-1999, followed in 2000 and included surveillance data from 58 geographical sites (Espinal, 2003; World Health Organization, 2000). This report confirmed that drug resistant TB was a sufficient problem since MDR-TB ranged from 0-16% (median: 1%) among new cases and from 0% to 48% (median: 9%) in previously treated cases. The recently published third report has data on 77 geographical sites, collected between 1999 and 2002, representing 20% of the global total of new smear-positive TB cases (World Health Organization, 2003). Eight countries did not report any MDR-TB amongst new cases, while the highest incidence of MDR-TB amongst new cases occurred in Kazakhstan and Israel (14%). Significant increases in MDR-TB prevalence were seen in Estonia, Lithuania, Tomsk Oblast (Russian Federation) and Poland and significant decreasing trends in Hong Kong, Thailand and the USA. The highest prevalence of MDR-TB among previously treated cases was reported in Oman (58.3%, 7/12) and Kazakhstan (56.4%, 180/319). The annual incidence of MDR-TB in most Western and Central European Countries was estimated to be fewer than 10 cases each. Alarmingly, it is estimated that the annual incidence of MDR-TB for 2 provinces in China (Henan and Hubei) is 1000 and for Kazakhstan and South Africa it is more than 3000. According to the report, the most effective means to prevent the emergence of drug resistance is by implementing the direct observed therapy strategy (DOTS) (World Health Organization, 2003).

CURRENT RECOMMENDATIONS FOR TB TREATMENT BY WHO

TB persists as a global public health problem and the main focus for the twentieth century is firstly to cure the individual patient and secondly to minimize the transmission of *M. tuberculosis* to other persons (World Health Organization, 2003;

Blumberg et al., 2003). The ongoing TB problem has been due to the neglect of TB control by governments, inadequate access and infrastructure, poor patient adherence to medication, poor management of TB control programs, poverty, population growth and migration, and a significant rise in the number of TB cases in HIV infected individuals. Treatment of patients with TB is most successful within a comprehensive framework based upon the following five key components:

- Government commitment
- Case detection by sputum smear microscopy
- Standardized treatment regimen of six to eight months
- A regular, uninterrupted supply of all essential anti-TB drugs
- A standard recording and reporting system

These five key elements are the recommended approach by the World Health Organization (WHO) to TB control and are called the DOTS strategy (Walley, 1997). DOTS is an inexpensive strategy for the detection and treatment of TB. DOTS was implemented as part of an adherence strategy in which patients are observed to swallow each dose of anti-TB medication, until completion of the therapy. Monthly sputum specimens are taken until 2 consecutive specimens are negative. Currently there are four recommended regimens for treating patients with TB infection by drug–susceptible organisms. Each regiment has an initial phase of 2 months intensive phase followed by a choice of several options for the continuation phase of either 4 or 7 months. The recommended regimens together with the number of doses specified by the regimen are described in Table 1.

Since the introduction of the DOTS strategy in the early '90s by the WHO, considerable progress has been made in global TB control (Sterling et al., 2003). In 1997, the estimated average treatment success rate world wide was almost 80%. However, less than 25% of people who are sick with TB are treated through the DOTS strategy (Bastian et al., 2000). A total of 180 countries (including both developed and undeveloped countries) had adopted and implemented the DOTS strategy by the end of 2002 and 69% of the global population was living in areas covered by the DOTS strategy (Blumberg et al., 2003). However, even though DOTS programs are in place, treatment success rates are very low in developed countries due to poor management of TB control programs and patient non-compliance (Lienhardt and Ogden, 2004; Bastian et al., 2003) Furthermore, the effectiveness of DOTS is facing new challenges with respect to the spread and increase of MDR-TB and the co-epidemic of TB/HIV (World Health Organization, 2003). WHO and partners have addressed these new challenges and have developed a new strategy called DOTS-Plus for the treatment of MDR-TB and its co-epidemic TB/HIV. The goal of DOTS-plus is to prevent further development and spread of MDR-TB and is a comprehensive management initiative built upon the DOTS strategy (Table 2). It is important to note that DOTS-Plus should only be implemented in areas were

Table 1. Drug Regimen for Culture-Positive Pulmonary TB Caused by Drug-Susceptible Organisms.

Intensive Phase			Continuation Phase		
Regimen	Drugs	Doses	Regimen	Drugs	Doses
1	INH, RIF PZA, EMB	7 d /wk for 56 doses (8wk) or 5 d/wk for 40 doses (8wk)	1	INH/ RIF	7 d/wk for 126 doses (18 wk) or 5d/wk for 90 doses (18 wk)
			1	INH/ RIF	2d/wk for 36 doses (18wk)
			1	INH/ RPT	1 wk for 18 doses (18wk)
2	INH, RIF PZA, EMB	7 d/wk for 14 doses (2wks), then 2 d/wk for 12 doses (6wks) or 5 d/wk for 10 doses (2wk), then 2 d/wk for 12 doses (6 wk)	2	INH/ RIF	2d/ wk for 36 doses
			2	INH/ RPT	1 wk for 18 doses (18wk)
3	INH, RIF PZA, EMB	3d/wk for 24 doses (8wk)	3	INH/ RIF	3 wk for 54 doses (18wk)
4	INH, RIF, EMB	7 d/wk for 56 doses (8wk) or 5 d/wk for 40 doses (8 wk)	4	INH/ RIF	7 d/wk for 217 doses (31 wk) or 2d/wk for 62 doses (31 wk)

INH-isoniazid; RIF-rifampicin; RPT-rifapentine; PZA-pyrazinamide.
Note: Streptomycin (STR) efficiency is equal to that of EMB and was use as an interchangeable drug with EMB in the initial phase of treatment. Due to the increase of resistance the drug is rendered less useful. Thus, STR is not recommended to be interchangeable with EMB unless the organism is known to be susceptible to the drug or the patient is from a community in which STR resistance is unlikely. Extracted from Blumberg et al. (2003).

the DOTS strategy is in place as there can be no DOTS-plus without an effective DOTS program.

DRUG SUSCEPTIBILITY TESTING

Drug susceptibility testing is carried out on sub-cultured bacteria after the initial positive culture is obtained for diagnosis. It usually takes 3-6 weeks to obtain the initial positive culture with an additional 3 weeks for susceptibility testing (reduced to about 15 days when using the BACTEC system) (Rastogi et al., 1989; Siddiqi et al., 1985; Snider, Jr. et al., 1981; Tarrand and Groschel, 1985). Thus, susceptibility testing is time consuming and costly, and there are numerous problems associated with the standardization of tests and the stability of the drugs in different culture media (Martin-Casabona et al., 1997; Victor et al., 1997). The slow diagnosis of

Table 2. DOTS Compared to DOTS-Plus Strategy.	
DOTS	DOTS-plus
• DOTS prevent emergence of drug resistant TB and MDR-TB • Make primarily use of 1st line drugs that are less expensive	• DOTS-plus design to cure MDR-TB using second line drugs. • Make use of 2nd line drugs that are more toxic and expensive,difficult to treat less effective to administrate and often poorly tolerated • DOTS-plus needed in areas where MDR-TB has emerged due to previous inadequate TB control • DOTS-plus only recommended in settings where DOTS strategy is fully in place to prevent against the development of further drug resistance

drug resistance may be a major contributor to the transmission of MDR-TB (Victor et al., 2002). The WHO recommended that drug susceptibility testing is done by the proportion method on Löwenstein-Jensen medium, but other media, such as Middlebrook 7H10, 7H11, 7H12 (BACTEC460TB) and other methods, including the absolute concentration and resistance ratio methods, may also be used (World Health Organization, 2001). For the ratio method, serial dilutions are cultured on 2 control media (without the drug) and 2 test media (with two different drug concentrations). The colonies on the different slants are counted after 21 and 40 days of growth. The proportion of resistant bacilli is calculated by comparing colony counts on drug free and drug containing media. For a resistant isolate the calculated proportion is higher and for a susceptible strain the calculated proportion is lower than the critical proportion (World Health Organization, 2001).

MOLECULAR MECHANISMS OF DRUG RESISTANCE

In order to control the drug resistance epidemic it is necessary to gain insight into how *M. tuberculosis* develops drug resistance. This knowledge will help us to understand how to prevent the occurrence of drug resistance as well as identifying genes associated with drug resistance of new drugs. The development of clinical drug resistance in TB is summarized in Fig. 1 and is classified as acquired resistance when drug resistant mutants are selected as a result of ineffective treatment or as primary resistance when a patient is infected with a resistant strain. Mutations in the genome of *M. tuberculosis* that can confer resistance to anti-TB drugs occur spontaneously with an estimated frequency of 3.5×10^{-6} for INH and 3.1×10^{-8} for RIF. Because the chromosomal loci responsible for resistance to various drugs are not linked, the risk of a double spontaneous mutation is extremely low: 9×10^{-14} for both INH and RIF (Dooley and Simone, 1994). MDR-TB defined as resistance to at least INH and RIF will thus occur mainly in circumstances where sequential

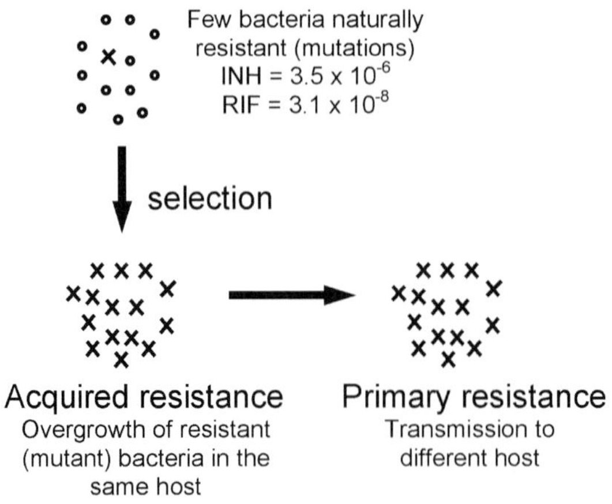

Fig. 1. Acquired resistance develops due to natural selection which is a function of ineffective treatment and non-compliance.

drug resistance follows sustained treatment failure. Treatment can be divided into first line and second line drugs according to the WHO TB treatment regimen and the mechanisms of these will be discussed separately.

FIRST LINE DRUGS

Any drug used in the anti-TB regiment is supposed to have an effective sterilizing activity that is capable of shortening the duration of treatment. Currently, a four-drug regiment is used consisting of INH, RIF, pyrazinamide (PZA) and ethambutol (EMB). Resistance to first line anti-TB drugs has been linked to mutations in at least 10 genes; *katG, inhA, ahpC, kasA* and *ndh* for INH resistance; *rpoB* for RIF resistance, *embB* for EMB resistance, *pncA* for PZA resistance and *rpsL* and *rrs* for STR resistance.

ISONIAZID

KatG

INH or isonicotinic acid hydrazide, was synthesized in the early 1900's but its anti-TB action was first detected in 1951 (Heym et al., 1999; Slayden and Barry, III, 2000; Rattan et al., 1998). INH enters the cell as a prodrug that is activated by a catalase peroxidase encoded by *katG*. The peroxidase activity of the enzyme is necessary to activate INH to a toxic substance in the bacterial cell (Zhang et al., 1992). This toxic substance subsequently affects intracellular targets such as mycolic acid biosynthesis which are an important component of the cell wall. A lack of mycolic acid synthesis eventually results in loss of cellular integrity and the bacteria die (Barry, III et al., 1998). Middlebrook et al. initially demonstrated that a loss of catalase activity can result in INH resistance (Middlebrook, 1954).

Subsequently genetic studies demonstrated that transformation of INH-resistant *Mycobacterium smegmatis* and *M. tuberculosis* strains with a functional *katG* gene restored INH susceptibility and that *katG* deletions give rise to INH resistance (Zhang et al., 1992; Zhang et al., 1993). However, mutations in this gene are more frequent than deletions in clinical isolates and these can lower the activity of the enzyme. Most mutations are found between codons 138 and 328 with the most commonly observed gene alteration being at codon 315 of the *katG* gene (Slayden and Barry, III, 2000). The Ser315Thr substitution is estimated to occur in 30-60% of INH resistant isolates (Ramaswamy and Musser, 1998; Musser et al., 1996; Slayden and Barry, III, 2000). The *katG* 463 (CGG-CTG) (Arg-Leu) amino acid substitution is the most common polymorphism found in the *katG* gene and is not associated with INH resistance.

ahpC

It has been observed that a loss of *katG* activity due to the S315T amino acid substitution is often accompanied by an increase in expression of an alkyl hydroperoxide reductase (*ahpC*) protein that is capable of detoxifying damaging organic peroxides (Sherman et al., 1996). Five different nucleotide alterations have been identified in the promoter region of the *ahpC* gene, which lead to over expression of *ahpC* and INH resistance (Ramaswamy and Musser, 1998). *AhpC* overexpression exerts a detoxifying effect on organic peroxides within the cell and protects the bacteria against oxidative damage but does not provide protection against INH. *KatG* expression can also be up regulated under conditions of oxidative stress. The correlation between polymorphic sites in the *ahpC* regulatory region with INH resistance in *M. tuberculosis* requires further examination.

inhA

One of the targets for activated INH is the protein encoded by the *inhA* locus. InhA is an enoyl–acyl carrier protein (ACP) reductase which is proposed to be the primary target for resistance to INH and ethionamide (ETH) (Banerjee et al., 1994). ETH, a second line drug, is a structural analog of INH that is also thought to inhibit mycolic acid biosynthesis and several studies have suggested that low-level INH resistance is correlated with resistance to ETH. Activated INH binds to the InhA-NADH complex to form a ternary complex that results in inhibition of mycolic acid biosynthesis. Six point mutations associated with INH resistance within the structural *inhA* gene have been identified (Ile16Thr, Ile21Thr, Ile21Val, Ile47Thr, Val78Ala and Ile95Pro) (Ramaswamy and Musser, 1998; Basso and Blanchard, 1998). A Ser94Ala substitution results in a decreased binding affinity of *inhA* for NADH, resulting in mycolic acid synthesis inhibition. Although these mutations in the structural *InhA* gene are associated with INH resistance, it is not frequently reported in clinical isolates. *InhA* promoter mutations are more frequently seen and are present at positions -24(G-T), -16(A-G), or -8(T-G/A) and -15(C-T). These promoter mutations result in over expression of *inhA* leading to low level INH resistance. To date approximately 70-80% of INH resistance in clinical isolates of *M. tuberculosis* can be attributed to mutations in the *katG* and *inhA* genes (Ramaswamy and Musser, 1998).

kasA

There seems to be considerable dispute within the literature as to the role of *kasA* as a possible target for INH resistance (Sherman et al., 1996). This gene encodes a β-ketoacyl-ACP synthase involved in the synthesis of mycolic acids. Mutations have been described in this gene that confer low levels of INH resistance. Genotypic analysis of the *kas*A gene reveals 4 different amino acid substitutions involving codon 66 (GAT-AAT), codon 269 (GGT-AGT), codon 312 (GGC-AGC) and codon 413 (TTC-TTA) (Ramaswamy and Musser, 1998; Mdluli et al., 1998). However, similar mutations were also found in INH susceptible isolates (Lee et al., 1999; Piatek et al., 2000). Nevertheless, the possibility of *kasA* constituting an additional resistance mechanism should not be completely excluded.

ndh

In 1998 another mechanism for INH resistance in *M. smegmatis* was described by Miesel et al. (1998). The *ndh* gene encodes NADH dehydrogenase that is bound to the active site of *inhA* to form the ternary complex with activated INH. Structural studies have shown that a reactive form of INH attacks the NAD(H) co-factor and generates a covalent INH-NAD adduct. Mutations in the *ndh* gene, encoding NADH dehydrogenase, cause defects in the enzymatic activity. Thus, defects in the oxidation of NADH to NAD result in NADH accumulation and NAD depletion (Lee et al., 2001). These high levels of NADH can then inhibit the binding of the INH-NAD adduct to the active site of the InhA enzyme (Rozwarski et al., 1998; Miesel et al., 1998). Prominent point mutations in the *ndh* gene at codons 110 and 268 (T110A and R268H) were detected in 9.5% of INH resistant samples. These similar mutations were not detected in the INH susceptible group (Lee et al., 2001).

RIFAMPINCIN

RIF was fist introduced in 1972 as an anti-TB drug and has excellent sterilizing activity (Rattan et al., 1998; Ramaswamy and Musser, 1998). The action of RIF in combination with PZA has allowed a shortening of routine TB treatment from 1 year to 6 months. RIF in combination with INH forms the backbone of short-course chemotherapy. It is interesting to note that mono resistance to INH is common but mono resistance to RIF is quite rare. It has thus been proposed that resistance to RIF can be used as a surrogate marker for MDR-TB as nearly 90% of RIF resistant strains are also INH resistant (Somoskovi et al., 2001). RIF interferes with transcription by the DNA-dependent RNA polymerase. RNA polymerase is composed of four different subunits (α, β, β' and σ) encoded by *rpoA, rpoB, rpoC* and *rpoD* genes respectively. RIF binds to the β-subunit hindering transcription and thereby killing the organism. Extensive studies on the *rpoB* gene in RIF resistant isolates of *M. tuberculosis* identified a variety of mutations and short deletions in the gene. A total of 69 single nucleotide changes; 3 insertions, 16 deletion and 38 multiple nucleotide changes have been reported (Herrera et al., 2003). More than 95% of all missense mutations are located in a 51bp core region (rifampin resistance determining region) of the *rpoB* gene between codons 507-533 with the most common changes in codons Ser531Leu, His526Tyr and Asp516Val. These changes occur in more than 70% of RIF resistant isolates (Rattan et al., 1998; Ramaswamy and Musser, 1998; Herrera et al., 2003). Furthermore, the minimal inhibitory concentration (MIC) showed that high level of RIF resistance is associated with mutations in codon 526 and 531, whereas alterations in codon 511, 516, 518 and 522 result in low level RIF resistance.

PYRAZINAMIDE

PZA, a nicotinamide analog, was first discovered to have anti-TB activity in 1952. PZA targets an enzyme involved in fatty-acid synthesis and is responsible for killing persistent tubercle bacilli in the initial intensive phase of chemotherapy (Somoskovi et al., 2001). However, during the first two days of treatment, PZA has no bactericidal activity against rapidly growing bacilli (Zhang and Mitchison, 2003). PZA on the other hand has effective sterilizing activity and shortens the chemotherapeutic regiment from 12 to 6 months. PZA is a prodrug which is converted to its active form, pyrazinoic acid (POA) by the pyrazinamidase (PZase) encoded by *pncA*. The activity of PZA is highly specific for *M. tuberculosis,* as it has no effect on other mycobacteria. *Mycobacterium bovis* is naturally resistant to PZA due to a unique C-G point mutation at position 169 of the *pncA* gene. PZA is only active against *M. tuberculosis* at acidic pH where POA accumulates in the cytoplasm due to an ineffective efflux pump. Accumulation of POA results in the lowering of intracellular pH to a level that inactivates a vital fatty acid synthase (Zimhony et al., 2004). Cloning and characterization of the *M. tuberculosis pncA* gene by Scorpio et al. (Scorpio and Zhang, 1996) showed that *pncA* mutations conferred PZA resistance. Various *pncA* mutations have been identified in more than 70% of PZA resistant clinical isolates scattered throughout the *pncA* gene but thus far no mutational hot spot has been identified (Scorpio and Zhang, 1996; Sreevatsan et al., 1997b; Scorpio et al., 1997).

ETHAMBUTOL

EMB, a first line drug, is used in combination with other drugs and is specific to the mycobacteria. EMB inhibits an arabinosyl transferase (*embB*) involved in cell wall biosynthesis (Takayama and Kilburn, 1989). Telenti et al. (1997) identified 3 genes, designated *emb*CAB, that encode homologous arabinosyl transferase enzymes involved in EMB resistance. Various studies have identified five mutations in codon 306 [(ATG-GTG), (ATG-CTG), (ATG-ATA), (ATG-ATC) and (ATG-ATT)] which result in three different amino acid substitutions (Val, Leu and Ile) in EMB-resistant isolates (Lee et al., 2002; Sreevatsan et al., 1997c; Mokrousov et al.,

2002b; Ramaswamy et al., 2000). These five mutations are associated with 70-90% of all EMB resistant isolates (Ramaswamy and Musser, 1998). Missense mutations were identified in three additional codons: Phe285leu, Phe330Val and Thr630Ile in EMB resistant isolates. MIC's were generally higher for strains with Met306Leu, Met306Val, Phe330Val and Thr630Ile substitutions than those organisms with Met306Ile substitutions. Mutations outside of codon 306 are present but quite rare. However a number of EMB phenotypic resistant isolates (about 30%) still lack an identified mutation in *embB*. There is therefore a need to fully understand the mechanism of EMB resistance to account for EMB resistant isolates.

STREPTOMYCIN

STR, an aminocyclitol glycoside, is an alternative first line anti-TB drug recommended by the WHO (Cooksey et al., 1996). STR is therefore used in the retreatment of TB cases together with the four drug regimen that includes INH, RIF, PZA and EMB (Brzostek et al., 2004). The effect of STR has been demonstrated to take place at the ribosomal level (Telenti et al., 1993). STR interacts with the 16S rRNA and S12 ribosomal protein (*rrs* and *rpsL)* (Escalante et al., 1998; Finken et al., 1993; Sreevatsan et al., 1996; Abbadi et al., 2001), inducing ribosomal changes, which cause misreading of the mRNA and inhibition of protein synthesis. Although STR is a recommended anti-TB drug, is it less effective against *M. tuberculosis* than INH and RIF. Point mutations in STR resistant isolates have been reported in *rrs* and *rpsL* genes in 65-67% of STR resistant isolates (Ramaswamy and Musser, 1998). In the *rrs* gene a C-T transition at positions 491, 512 and 516, and a A-C/T transversion at position 513 were observed in the highly conserved 530 loop. The 530 loop region is part of the aminoacyl–tRNA binding site and is involved in the decoding process (Carter et al., 2000). The C-T transition at codon 491 is not responsible for resistance to STR as it occurs in both STR resistant and susceptible isolates but is strongly associated with the global spread of *M. tuberculosis* with a Western Cape F11 genotype (van Rie et al., 2001; Victor et al., 2001). Other mutations in the 915 loop [903 (C-A/G) and 904 (A-G)] have also been reported to have an association with STR resistance (Carter et al., 2000). Mutations in the *rpsL* gene at codon 43 (AAG-AGG/ACG) (Lys-Arg/Thr) and codon 88 (AAG-AGG/CAG) (Lys-Arg/Gln) are associated with STR resistance. MIC analysis of STR resistant isolates indicate that amino acid replacements in the *rpsL* genes correlate with a high level of resistance, whereas mutations in the *rrs* gene correlate with an intermediate level of resistance (Cooksey et al., 1996) (Meier et al., 1996). In addition, it has been suggested that low levels of STR resistance are also associated with altered cell permeability or rare mutations which lie outside of the *rrs* and *rpsL* genes.

SECOND LINE DRUGS USED IN TB TREATMENT

According to the WHO the following drugs can be classified as second line drugs: aminoglycosides (kanamycin and amikacin) polypeptides (capreomycin, viomycin and enviomycin), fluoroquinolones (ofloxacin, ciprofloxacin, and gatifloxacin), D-cycloserine and thionamides (ethionamide and prothionamide) (World Health Organization, 2001). Unfortunately, second-line drugs are inherently more toxic and less effective than first-line drugs (World Health Organization, 2001). Second line drugs are mostly used in the treatment of MDR-TB and as a result prolong the total treatment time from 6 to 9 months (Cheng et al., 2004). The current understanding of molecular mechanisms associated with resistance to second line drugs are summarized in Table 3. The phenotypic methods to detect resistance to second line drugs are less well established and the molecular mechanisms of resistance are also less defined.

FLUOROQUINOLONES

Ciproflaxin (CIP) and ofloxacin (OFL) are the two fluoroquinolones (FQs) used as second-line drugs in MDR-TB treatment (World Health Organization, 2001). The quinolones target and inactivate DNA gyrase, a type II DNA topoisomerase (Cynamon and Sklaney, 2003; Ginsburg et al., 2003; Rattan et al., 1998). DNA gyrase is encoded by *gyrA* and *gyrB* (Rattan et al., 1998; Takiff et al., 1994) and introduces negative supercoils in closed circular DNA molecules (Rattan et al., 1998; Ramaswamy and Musser, 1998). The quinolone resistance-determining region (QRDR) is a conserved region in the *gyrA* (320bp) and *gyrB* (375bp) genes (Ginsburg et al., 2003) which is the point of interaction of FQ and gyrase (Ginsburg et al., 2003). Missense mutations in codon 90, 91, and 94 of *gyrA* are associated with resistance to FQs (Takiff et al., 1994; Xu et al., 1996). A 16-fold increase in resistance was observed for isolates with a Ala90Val substitution, a 30-fold increase for Asp94Asn or His94Tyr and a 60-fold increase for Asp94Gly (Xu et al., 1996). A polymorphism at *gyrA* codon 95 is not associated with FQ resistance, and is used, with the *katG*463 polymorphism, to classify *M. tuberculosis* into 3 phylogenetic groups (Sreevatsan et al., 1997a).

AMINOGLYCOSIDES

Kanamycin (KAN) and Amikacin (AMI) are aminoglycosides which inhibit protein synthesis and thus cannot be used against dormant *M. tuberculosis*. Aminoglycosides bind to bacterial ribosomes and disturb the elongation of the peptide chain in the bacteria. Mutations in the *rrs* gene encoding for 16s rRNA are associated with resistance to KAN and AMI. Nucleotide changes at positions 1400, 1401 and 1483 of the *rrs* gene have been found to be specifically associated with KAN resistance (Suzuki et al., 1998). An A→G change at codon 1400 in the *rrs* gene showed resistance to KAN of MICs more that 200 μg/ml (Taniguchi et al., 1997; Suzuki et al., 1998).

Table 3. Properties of Resistance to Various Second-line Anti-TB Drugs.

Second-line drug	Gene locus	Gene product	Known polymorphism	Most frequently mutated codons associated with resistance	MIC [a] (μg/ml)	Methods for genotypic detection of resistance	Reference
Fluoroquinolones Ofloxacin Cipromycin	*gyr*A	DNA gyrase	*gyrA* 95	*gyr* 90, 91, 94	OFL: 1.0-2.0 CIP: 0.5-4.0	Cloning and expression PCR-SSCP[b] DNA sequencing	(Takiff et al., 1994; Pletz et al., 2004; Rattan et al., 1998; Ginsburg et al., 2003; Cheng et al., 2004)
Aminoglycosides Kanamycin Amikacin	*rrs*	16 S rRNA		*rrs* position 1400	KAN: > 200 AMI: > 256	IS*6110*-RFLP PCR-RFLP[c]	(Ramaswamy and Musser, 1998; Takiff et al., 1994; Taniguchi et al., 1997; Suzuki et al., 1998; Ramaswamy et al., 2004; Vannelli et al., 2002)
Ethionamide	*inhA*	enoyl-ACP reductase		*inhA*21, 94, 44	≥ 25	DNA Sequencing	(Baulard et al., 2000; Morlock et al., 2003; Cynamon and Sklaney, 2003)
	ethA	Flavin monooxygenase			≥200		
	ethR	Transcriptional repressor					
D- cycloserine	*alr*	D-alanine racemase			≥ 300	Cloning and expression DNA Sequencing	(Caceres et al., 1997; Feng and Barletta, 2003)
	ddl	D-alanine: D-alanine ligase					
Viomycin	*rrs*	16S rRNA			20	DNA Sequencing	(Ramaswamy and Musser, 1998; Taniguchi et al., 1997; Suzuki et al., 1995)

a) Minimum inhibitory concentration. b) Polymerase chain reaction- single strand conformation polymorphism. c) Polymerase chain reaction- Restriction fragment length polymorphism.

ETHIONAMIDE

Ethionamide (ETH) is an important drug in the treatment of MDR-TB, and is mechanistically and structurally analogous to INH. Like INH, ETH is also thought to be a prodrug that is activated by bacterial metabolism. The activated drug then disrupts cell wall biosynthesis by inhibiting mycolic acid synthesis. Mutations in the promoter of the *inhA* gene are associated with resistance to INH and ETH (Morlock et al., 2003). *EthA* catalyses a two step activation of ETH and gene alterations leading to reduced EthA activity lead to ETH resistance (Engohang-Ndong et al., 2004; Morlock et al., 2003; Vannelli et al., 2002). The expression of *ethA* is under the control of the neighbouring *ethR* gene encoding a repressor. *EthR* negatively regulates the expression of *ethA*, by binding upstream of *ethA* to suppress *ethA* expression (Engohang-Ndong et al., 2004).

D-CYCLOSERINE

D-cycloserine (DCS) is a cyclic analog of D-alanine which is one of the central molecules of the cross linking step of peptidoglycan assembly (Ramaswamy and Musser, 1998; Feng and Barletta, 2003; David, 2001; Caceres et al., 1997). DCS inhibits cell wall synthesis by competing with D-Alanine for the enzymes D-alanyl-D-alanine synthetase (Ddl) and D-alanine racemase (Alr) and also inhibiting the synthesis of these proteins. Over expression of *alr* cause DCS resistance. A G→T transversion in the *alr* promoter may lead to the overexpression of *alr* (Feng and Barletta, 2003; Ramaswamy and Musser, 1998).

PEPTIDES

Viomycin (VIO) and capreomycin (CAP) are basic peptide antibiotics that inhibit prokaryotic protein synthesis and are used as second-line anti-TB drugs. Earlier studies have shown that resistance to VIO in *M. smegmatis* is caused by alterations in the 30S or 50S ribosomal subunits (Taniguchi et al., 1997). Mutations in the *rrs* gene that encodes the 16S rRNA is associated with resistance to VIO and CAP, specifically a G→A or G→T nucleotide change at position 1473 (Taniguchi et al., 1997).

MOLECULAR METHODS TO PREDICT DRUG RESISTANCE

M. tuberculosis is a very slow growing organism and the use of molecular methods for the identification of mutations in resistance-causing genes may offer a means to rapidly screen *M. tuberculosis* isolates for antibiotic resistance. Mutation screening methods are fast and include methods such as DNA sequencing, probe based hybridization methods, PCR-RFLP, single–strand conformation polymorphism (SSCP), heteroduplex analysis (HA), molecular beacons and ARMS-PCR (Victor et al., 2002). The end results for each of these methods are given as a combined photo in Fig. 2.

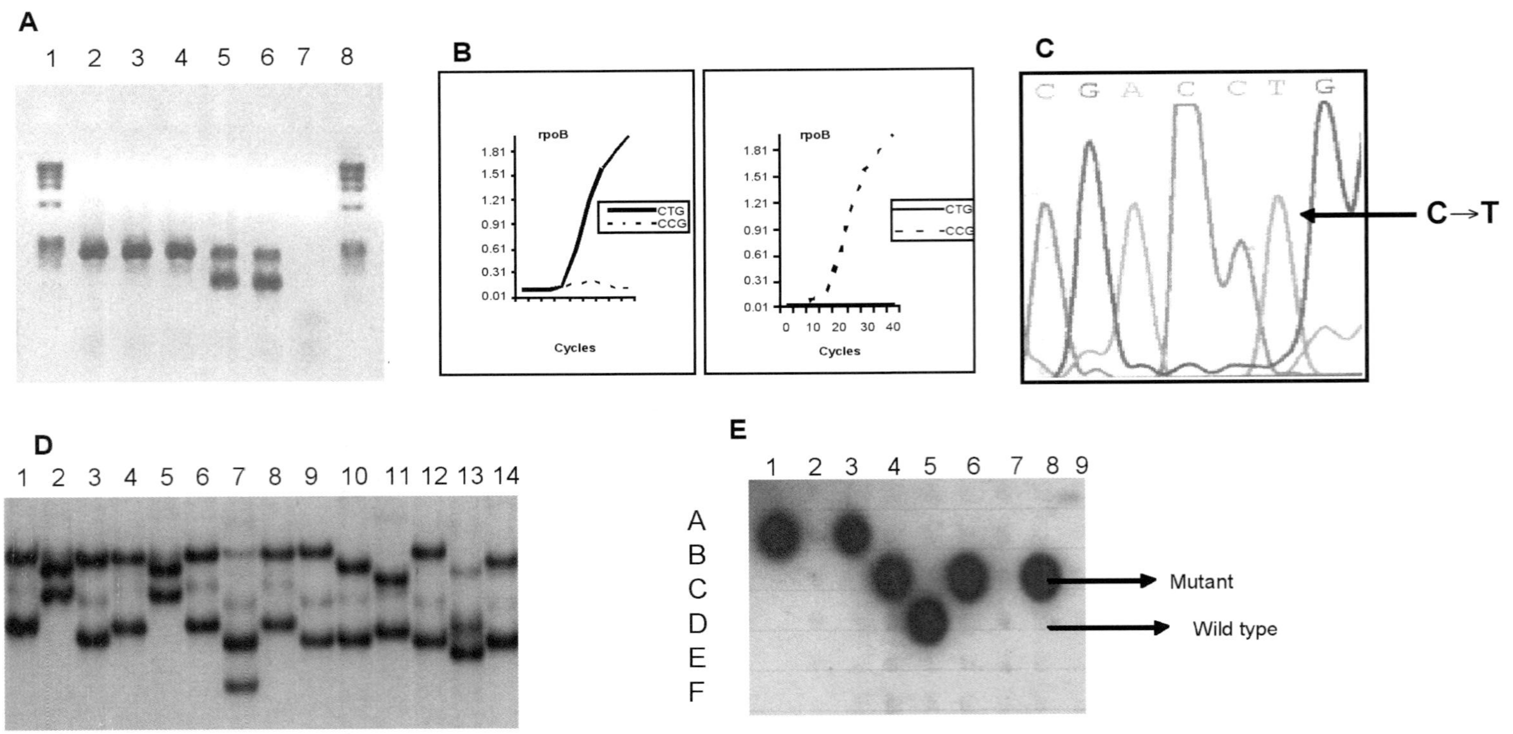

Fig. 2. Molecular methods for detecting gene mutations associated with resistance to anti-TB drugs. A-E are typical examples of final results obtained by PCR-based mutation screening methods. The DNA template can be pure DNA extracted from a culture, sputum or crude DNA templates prepared from culture or sputum. In all the examples PCR amplification of the DNA is followed by the different mutation detection methods (A-E). A: ARMS-PCR analysis of *embB* gene. Lane 1 and 8 = Molecular marker and Samples 5 and 6 are mutants. B: Molecular beacon genotyping of the *rpoB* gene showing a T→C mutation. C: Sequence analysis showed a C→T mutation at nt161 in the *pncA* gene D: SSCP analysis with samples having different mutations (visualized as different band mobility shifts) in the of *rpoB* gene. Heteroduplex analysis would give similar band mobility shifts on the gel. E: DOT-BLOT analysis with *rpsL 43* mutant probe showing wild type (lane D8) and mutant (lane C8) controls and mutant (STR resistant) clinical isolates in B1, B3, C4, D5, C6.

SEQUENCING

PCR amplification followed by DNA sequencing is the most widely used technique to identify mutations associated with drug resistance in TB (Victor et al., 2002). This technique is costly and require expertise, which make it unpractical for use in routine laboratories, especially in developing countries, where simple, cost effective drug susceptibility testing is needed (Victor et al., 2002).

PROBE-BASED HYBRIDIZATION METHODS

In these assays, amplified PCR products of genes known to confer drug resistance are hybridized to an allele-specific labeled probe that is complementary to the wild type or mutant sequence of the gene. This can then be visualized by autoradiography, enhanced chemiluminescence, alkaline phosphatase or other detection systems. These methods include the Dot-blot and Line blot essays and the commercially available INNO-LIPA RIF-TB test (Innogenetics, Belgium) (Victor et al., 1999; Mokrousov et al., 2004).

PCR-RESTRICTION FRAGMENT LENGTH POLYMORPHISM (PCR-RFLP)

Mutations associated with resistance can be identified by digestion of amplified PCR products with a restriction enzyme that cuts at the specific polymorphic DNA sequence followed by gel electrophoresis. Since not all mutations result in the gain or loss of a restriction site, general use of RFLP to screen for mutations associated with drug resistance is limited (Victor et al., 2002).

SINGLE STRANDED CONFORMATION POLYMORPHISM ANALYSIS (SSCP)

SSCP is a gel based method that can detect short stretches of DNA approximately 175-250bp in size. Small changes in a nucleotide sequence result in differences in secondary structures as well as measurable DNA mobility shifts that are detected on a non-denaturing polyacrylamide gel. To date various studies have applied PCR-SSCP to identify mutational changes associated with drug resistance in *M. tuberculosis* for frontline drugs like, RIF and INH (Kim et al., 2004; Cardoso et al., 2004; Fang et al., 1999; Heym et al., 1995; Pretorius et al., 1995). However, PCR-SSCP analysis has been found to be technically demanding and not sufficiently sensitive. Furthermore SSCP conditions must be carefully evaluated since not all mutations will be detected under the same conditions.

HETERODUPLEX ANALYSIS (HA)

HA depends on the conformation of duplex DNA when analysed in native gels. Heteroduplexes are formed when PCR amplification products from known wild type and unknown mutant sequences are heated and re-annealed. The DNA strand will form a mismatched heteroduplex if there is a sequence difference between the strands of the wild type and tested DNA. These heteroduplexes have an altered electrophoretic mobility when compared to homoduplexes, since the mismatches tend to retard the migration of DNA during electrophoresis. There are two types of heterodoplexes. The "bubble" type is formed between DNA fragments with single base differences and the bulge type is formed when there are deletions or insertions present within the two fragments. Recently, temperature mediated HA has been applied to the detection of mutations associated with mutations in *rpoB, katG, rpsL, embB* and *pncA* genes (Mohamed et al., 2004; Cooksey et al., 2002). Neither HA nor the SSCP analysis are 100% sensitive although Rosetti et al. found that HA detected more mutants (Nataraj et al., 1999). However, HA has certain disadvantages in that it has been found to be insensitive to G-C rich regions and is very time consuming (Nataraj et al., 1999).

MOLECULAR BEACONS

Molecular beacons are single-stranded oligonucleotide hybridization probes which can be used as amplicon detector probes in diagnostic assays. A beacon consists of a stem-loop structure in which the stem contains a fluorophore on one arm and a quencher on the other end of the arm. The loop contains the probe which is complementary to the target DNA. If the molecular beacon is free in a solution it will not fluoresce, because the stem places the fluorophore so close to the non-fluorescent quencher that they transiently share electrons, eliminating the ability of the fluorophore to fluoresce. However, in the presence of complementary target DNA the probe undergo a conformational change that enables them to fluoresce brightly. Different colored fluorophores (different primers) can be used simultaneously to detect multiple targets (each target will give a different color) in the same reaction. Molecular beacons are very specific and can discriminate between single nucleotide substitutions. Thus they are ideally suited for genotyping and have been used in the detection of drug resistance in *M. tuberculosis* (El Hajj et al., 2001; Piatek et al., 2000; Piatek et al., 1998).

AMPLIFICATION REFRACTORY MUTATION SYSTEM (ARMS)-PCR

ARMS also known as allelic specific PCR (ASPCR) or PCR amplification of specific alleles (PASA) is a well established technique used for the detection of any point mutation or small deletions (Newton et al., 1989). ARMS-PCR, is usually a multiplex reaction where three (or more) primers are used to amplify the same region simultaneously. One of the three primers is specific for the mutant allele and will work with a common primer during amplification. The mismatch is usually located at or near the 3' end of the primer. The third primer will work with the same common primer to generate an amplified fragment which is larger than the fragment from the mutant allele primer - this serves as an internal control for amplification. Amplification is detected by gel electrophoresis and the genotypic classification is determined by assessing which amplification products are present. An amplification product should always be present in the larger internal control amplified fragment; if this is the case then the absence or presence of the smaller product will indicate the presence or absence of a mutant allele. This technique has successfully been used for the detection of mutations associated with RIF resistance in *M. tuberculosis* (Fan et al., 2003). Fig. 2A indicates how the amplified products in the multiplex reaction are distinguished on a gel.

APPLICATIONS

One of the major advantages of PCR based methods is the speed by which the result can be obtained (Siddiqi et al., 1985; Snider, Jr. et al., 1981; Tarrand and Groschel, 1985). It is envisaged that molecular techniques may be important adjuncts to traditional culture based procedures to rapidly screen for drug resistance. Prospective analysis and intervention to prevent transmission may be particularly helpful in areas with ongoing transmission of drug resistant strains as reported previously (van Rie et al., 1999). In addition, molecular prediction may also be useful in drug surveillance studies to further improve the confidence limit of the data in these studies if this test is performed on a subset of the samples. Enhanced efforts are necessary to better understand the molecular mechanisms of resistance to second line anti-TB drugs in clinical isolates. However, implementation for both rapid diagnosis and surveillance requires proper quality control guidelines and controls, which is currently not in place yet for molecular prediction of drug resistance in TB. Although molecular methods are more rapid, and can be done directly from a clinical sample there are important limitations when compared to conventional phenotypic methods. These include a lack of sensitivity since not all molecular mechanisms leading to drug resistance are known, therefore not all resistant isolates will be detected. Molecular methods may also predict resistance genotypes that are expressed at levels that may not clinically be relevant (Victor et al., 2002).

TRANSMISSION AND EPIDEMIC DRUG RESISTANT STRAINS

There is much debate about the relative contribution of acquired and primary resistance to the burden of drug resistant TB in different communities. This controversy focuses on whether MDR strains are transmissible or whether the mutations that confer drug resistance also impair the reproductive function of the organism (fitness of the strain). Evidence that MDR strains do have the potential for transmission comes from a series of MDR-TB outbreaks that have been reported over the past decade. These have been identified in hospitals (Fischl et al., 1992; Edlin et al., 1992; Bifani et al., 1996; Cooksey et al., 1996), amongst health care workers (Beck-Sague et al., 1992; Pearson et al., 1992; Jereb et al., 1995) and in prisons (Valway et al., 1994) and have focused attention on MDR-TB as a major public health issue. Application of molecular epidemiological methods was central to the identification and description of all these outbreaks.

The most extensive MDR-TB outbreak reported to date occurred in 267 patients from New York, who were infected by Beijing/W genotype (Frieden et al., 1996). This cluster of cases included drug resistant isolates that were resistant to all first-line anti-TB drugs. The authors speculate that the delay in diagnosis and administering appropriate therapy resulted in prolonging infectiousness and placed healthcare workers and other hospital residents (or contacts) at risk of infection for nosocomial infection. This difficult-to-treat strain has subsequently disseminated to other US cities and Paris and the authors showed by using molecular methods, how this initially fully drug susceptible strain clonally expanded to result in a MDR phenotype by sequential acquisition of resistance conferring mutations in several genes (Bifani et al., 1996). Since then, the drug resistant Beijing/W genotype has been the focus of extensive investigations and Beijing drug resistant and susceptible genotypes have been found to be widely spread throughout the world (Glynn et al., 2002), including in South Africa (van Rie et al., 1999) and Russia (Mokrousov et al., 2002a). Beijing/W genotypes can be identified by their characteristic multi-banded IS*6110* restriction fragment-length polymorphism (RFLP) patterns, a specific spoligotype pattern characterized by the presence of spoligotype spacers 35-43 (Bifani et al., 2002) and resistance conferring gene mutations. Although these data led many to propose that Beijing/W strains behaved differently from other strains, more recent work suggests that MDR outbreaks are not limited to the Beijing/W genotype. Smaller outbreaks involving other MDR-TB genotypes have been reported in other settings such as the Czech Republic, Portugal and Norway (Kubin et al., 1999; Portugal et al., 1999). However, since much of the MDR burden falls in developing countries in which routine surveillance does not usually include molecular fingerprinting, little is known about the characteristics of circulating drug resistant strains in much of the world. It is therefore possible that there are other MDR strains, as widespread as Beijing/W, which have not been recognized and reported as such.

FUTURE

Enhanced efforts are necessary to better understand the molecular mechanisms of resistance in second line anti-TB drugs in clinical isolates. The next generation of molecular methods for the prediction of drug resistance in *M. tuberculosis* will possibly consists of matrix hybridization formats such as DNA oligonucleotide arrays on slides or silicon micron chips (Castellino, 1997; Vernet et al., 2004), particularly if these systems can be fully automated and re-used. This may be particularly useful for mutations in the *rpoB* gene, which can serve as a marker for MDR-TB (Watterson et al., 1998) and also for the multiple loci that are involved in INH resistance (Table 1). Selection of a limited number of target mutations which enable the detection of the majority of drug resistance (van Rie et al., 2001) would be useful in this strategy. It is essential that developments for new techniques must consider the fact that the majority of drug resistant cases occur in resource-poor countries (Raviglione et al., 1995) and therefore the methodologies must not only be cheap but also robust.

There are other rapid methods which do not depend on the detection of mutations to predict drug resistance. One promising method is phage amplification technology in which mycobacteriophages (bacteriophages specific for mycobacteria) are used as an indicator of the presence of viable *M. tuberculosis* in a clinical specimen (Albert et al., 2002; Eltringham et al., 1999; McNerney et al., 2000). The phage assay can also be adapted for the detection of drug resistance.

Application of rapid methods to break chains of ongoing transmission of drug resistant TB will increasingly become important as recent mathematical modeling indicate that the burden of MDR-TB cannot be contained in the absence of specific efforts to limit transmission (Cohen and Murray, 2004; Blower and Chou, 2004). This may include rapid detection of drug resistance by molecular methods.

ACKNOWLEDGEMENTS

The authors would like thank the South African National Research Foundation (GUN 2054278), IAEA (SAF6008), The Welcome Trust (Ref. 072402/Z/03/Z) and the NIH (R21 A155800-01) for support.

REFERENCES

Abbadi, S., Rashed, H.G., Morlock, G.P., Woodley, C.L., El Shanawy, O., and Cooksey, R.C. (2001). Characterization of IS6110 restriction fragment length polymorphism patterns and mechanisms of antimicrobial resistance for multidrug-resistant isolates of *Mycobacterium tuberculosis* from a major reference hospital in Assiut, Egypt. J. Clin. Microbiol. *39*, 2330-2334.

Albert, H., Heydenrych, A., Brookes, R., Mole, R.J., Harley, B., Subotsky, E., Henry, R., and Azevedo, V. (2002). Performance of a rapid phage-based test, FASTPlaqueTB, to diagnose pulmonary tuberculosis from sputum specimens in South Africa. Int. J. Tuberc. Lung Dis. *6*, 529-537.

Banerjee, A., Dubnau, E., Quemard, A., Balasubramanian, V., Um, K.S., Wilson, T., Collins, D., de Lisle, G., and Jacobs, W.R., Jr. (1994). *inhA*, a gene encoding a target for isoniazid and ethionamide in *Mycobacterium tuberculosis*. Science *263*, 227-230.

Barry, C.E., III, Lee, R.E., Mdluli, K., Sampson, A.E., Schroeder, B.G., Slayden, R.A., and Yuan, Y. (1998). Mycolic acids: structure, biosynthesis and physiological functions. Prog. Lipid Res. *37*, 143-179.

Basso, L.A. and Blanchard, J.S. (1998). Resistance to antitubercular drugs. Adv. Exp. Med. Biol. *456*, 115-144.

Bastian, I., Rigouts, L., Van Deun, A., and Portaels, F. (2000). Directly observed treatment, short-course strategy and multidrug-resistant tuberculosis: are any modifications required? Bull. World Health Organ *78*, 238-251.

Bastian, I., Stapledon, R., and Colebunders, R. (2003). Current thinking on the management of tuberculosis. Curr. Opin. Pulm. Med. *9*, 186-192.

Baulard, A.R., Betts, J.C., Engohang-Ndong, J., Quan, S., McAdam, R.A., Brennan, P.J., Locht, C., and Besra, G.S. (2000). Activation of the pro-drug ethionamide is regulated in mycobacteria. J. Biol. Chem. *275*, 28326-28331.

Beck-Sague, C., Dooley, S.W., Hutton, M.D., Otten, J., Breeden, A., Crawford, J.T., Pitchenik, A.E., Woodley, C., Cauthen, G., and Jarvis, W.R. (1992). Hospital outbreak of multidrug-resistant *Mycobacterium tuberculosis* infections. Factors in transmission to staff and HIV-infected patients. JAMA *268*, 1280-1286.

Bifani, P.J., Mathema, B., Kurepina, N.E., and Kreiswirth, B.N. (2002). Global dissemination of the *Mycobacterium tuberculosis* W-Beijing family strains. Trends Microbiol. *10*, 45-52.

Bifani, P.J., Plikaytis, B.B., Kapur, V., Stockbauer, K., Pan, X., Lutfey, M.L., Moghazeh, S.L., Eisner, W., Daniel, T.M., Kaplan, M.H., Crawford, J.T., Musser, J.M., and Kreiswirth, B.N. (1996). Origin and interstate spread of a New York City multidrug-resistant *Mycobacterium tuberculosis* clone family. JAMA *275*, 452-457.

Blower, S.M. and Chou, T. (2004). Modeling the emergence of the 'hot zones': tuberculosis and the amplification dynamics of drug resistance. Nat. Med. *10*, 1111-1116.

Blumberg, H.M., Burman, W.J., Chaisson, R.E., Daley, C.L., Etkind, S.C., Friedman, L.N., Fujiwara, P., Grzemska, M., Hopewell, P.C., Iseman, M.D., Jasmer, R.M., Koppaka, V., Menzies, R.I., O'Brien, R.J., Reves, R.R., Reichman, L.B., Simone, P.M., Starke, J.R., and Vernon, A.A. (2003). American Thoracic Society/Centers for Disease Control and Prevention/Infectious Diseases Society of America: treatment of tuberculosis. Am. J. Respir. Crit Care Med. *167*, 603-662.

Brzostek, A., Sajduda, A., Sliwinski, T., Augustynowicz-Kopec, E., Jaworski, A., Zwolska, Z., and Dziadek, J. (2004). Molecular characterisation of streptomycin-resistant *Mycobacterium tuberculosis* strains isolated in Poland. Int. J. Tuberc. Lung Dis. *8*, 1032-1035.

Caceres, N.E., Harris, N.B., Wellehan, J.F., Feng, Z., Kapur, V., and Barletta, R.G. (1997). Overexpression of the D-alanine racemase gene confers resistance to D-cycloserine in *Mycobacterium smegmatis*. J. Bacteriol. *179*, 5046-5055.

Cardoso, R.F., Cooksey, R.C., Morlock, G.P., Barco, P., Cecon, L., Forestiero, F., Leite, C.Q., Sato, D.N., Shikama Md, M.L., Mamizuka, E.M., Hirata, R.D., and Hirata, M.H. (2004). Screening and Characterization of Mutations in Isoniazid-Resistant *Mycobacterium tuberculosis* Isolates Obtained in Brazil. Antimicrob. Agents Chemother. *48*, 3373-3381.

Carter, A.P., Clemons, W.M., Brodersen, D.E., Morgan-Warren, R.J., Wimberly, B.T., and Ramakrishnan, V. (2000). Functional insights from the structure of the 30S ribosomal subunit and its interactions with antibiotics. Nature *407*, 340-348.

Castellino, A.M. (1997). When the chips are down. Genome Res. *7*, 943-946.

Cheng, A.F., Yew, W.W., Chan, E.W., Chin, M.L., Hui, M.M., and Chan, R.C. (2004). Multiplex PCR amplimer conformation analysis for rapid detection of *gyrA* mutations in fluoroquinolone-resistant *Mycobacterium tuberculosis* clinical isolates. Antimicrob. Agents Chemother. *48*, 596-601.

Cohen, T. and Murray, M. (2004). Modeling epidemics of multidrug-resistant *M. tuberculosis* of heterogeneous fitness. Nat. Med. *10*, 1117-1121.

Cohn, D.L., Bustreo, F., and Raviglione, M.C. (1997). Drug-resistant tuberculosis: review of the worldwide situation and the WHO/IUATLD Global Surveillance Project. International Union Against Tuberculosis and Lung Disease. Clin. Infect. Dis. *24 Suppl 1*, S121-S130.

Cooksey, R.C., Morlock, G.P., Holloway, B.P., Limor, J., and Hepburn, M. (2002). Temperature-mediated heteroduplex analysis performed by using denaturing high-performance liquid chromatography to identify sequence polymorphisms in *Mycobacterium tuberculosis* complex organisms. J. Clin. Microbiol. *40*, 1610-1616.

Cooksey, R.C., Morlock, G.P., McQueen, A., Glickman, S.E., and Crawford, J.T. (1996). Characterization of streptomycin resistance mechanisms among *Mycobacterium tuberculosis* isolates from patients in New York City. Antimicrob. Agents Chemother. *40*, 1186-1188.

Crofton, J. and Mitchison, D. (1948). Streptomycin resistance in pulmonary tuberculosis. Br. Med. J *2*, 1009-1015.

Cynamon, M.H. and Sklaney, M. (2003). Gatifloxacin and ethionamide as the foundation for therapy of tuberculosis. Antimicrob. Agents Chemother. *47*, 2442-2444.

David, S. (2001). Synergic activity of D-cycloserine and beta-chloro-D-alanine against *Mycobacterium tuberculosis*. J. Antimicrob. Chemother. *47*, 203-206.

Dooley, S.W. and Simone, P.M. (1994). The extent and management of drug-resistant tuberculosis: the American experience. Clinical tuberculosis. London: Chapman and Hall 171-189.

Edlin, B.R., Tokars, J.I., Grieco, M.H., Crawford, J.T., Williams, J., Sordillo, E.M., Ong, K.R., Kilburn, J.O., Dooley, S.W., Castro, K.G., and. (1992). An outbreak of multidrug-resistant tuberculosis among hospitalized patients with the acquired immunodeficiency syndrome. N. Engl. J. Med. *326*, 1514-1521.

El Hajj, H.H., Marras, S.A., Tyagi, S., Kramer, F.R., and Alland, D. (2001). Detection of rifampin resistance in *Mycobacterium tuberculosis* in a single tube with molecular beacons. J. Clin. Microbiol. *39*, 4131-4137.

Eltringham, I.J., Wilson, S.M., and Drobniewski, F.A. (1999). Evaluation of a bacteriophage-based assay (phage amplified biologically assay) as a rapid screen for resistance to isoniazid, ethambutol, streptomycin, pyrazinamide, and ciprofloxacin among clinical isolates of *Mycobacterium tuberculosis*. J. Clin. Microbiol. *37*, 3528-3532.

Engohang-Ndong, J., Baillat, D., Aumercier, M., Bellefontaine, F., Besra, G.S., Locht, C., and Baulard, A.R. (2004). *EthR*, a repressor of the *TetR/CamR* family implicated in ethionamide resistance in mycobacteria, octamerizes cooperatively on its operator. Mol. Microbiol. *51*, 175-188.

Escalante, P., Ramaswamy, S., Sanabria, H., Soini, H., Pan, X., Valiente-Castillo, O., and Musser, J.M. (1998). Genotypic characterization of drug-resistant *Mycobacterium tuberculosis* isolates from Peru. Tuber. Lung Dis. *79*, 111-118.

Espinal, M.A. (2003). The global situation of MDR-TB. Tuberculosis. (Edinb.) *83*, 44-51.

Fan, X.Y., Hu, Z.Y., Xu, F.H., Yan, Z.Q., Guo, S.Q., and Li, Z.M. (2003). Rapid detection of *rpoB* gene mutations in rifampin-resistant *Mycobacterium tuberculosis* isolates in shanghai by using the amplification refractory mutation system. J. Clin. Microbiol. *41*, 993-997.

Fang, Z., Doig, C., Rayner, A., Kenna, D.T., Watt, B., and Forbes, K.J. (1999). Molecular evidence for heterogeneity of the multiple-drug-resistant *Mycobacterium tuberculosis* population in Scotland (1990 to 1997). J. Clin. Microbiol. *37*, 998-1003.

Feng, Z. and Barletta, R.G. (2003). Roles of *Mycobacterium smegmatis* D-Alanine: D-Alanine Ligase and D-Alanine Racemase in the Mechanisms of Action of and Resistance to the Peptidoglycan Inhibitor D-Cycloserine. Antimicrob. Agents Chemother. *47*, 283-291.

Finken, M., Kirschner, P., Meier, A., Wrede, A., and Bottger, E.C. (1993). Molecular basis of streptomycin resistance in *Mycobacterium tuberculosis*: alterations of the ribosomal protein S12 gene and point mutations within a functional 16S ribosomal RNA pseudoknot. Mol. Microbiol. *9*, 1239-1246.

Fischl, M.A., Uttamchandani, R.B., Daikos, G.L., Poblete, R.B., Moreno, J.N., Reyes, R.R., Boota, A.M., Thompson, L.M., Cleary, T.J., and Lai, S. (1992). An outbreak of tuberculosis caused by multiple-drug-resistant tubercle bacilli among patients with HIV infection. Ann. Intern. Med. *117*, 177-183.

Frieden, T.R., Sherman, L.F., Maw, K.L., Fujiwara, P.I., Crawford, J.T., Nivin, B., Sharp, V., Hewlett, D., Jr., Brudney, K., Alland, D., and Kreisworth, B.N. (1996). A multi-institutional outbreak of highly drug-resistant tuberculosis: epidemiology and clinical outcomes. JAMA *276*, 1229-1235.

Ginsburg, A.S., Grosset, J.H., and Bishai, W.R. (2003). Fluoroquinolones, tuberculosis, and resistance. Lancet Infect. Dis. *3*, 432-442.

Glynn, J.R., Whiteley, J., Bifani, P.J., Kremer, K., and van Soolingen, D. (2002). Worldwide Occurrence of Beijing/W Strains of *Mycobacterium tuberculosis*: A Systematic Review. Emerg. Infect. Dis. *8*, 843-849.

Herrera, L., Jimenez, S., Valverde, A., Garci, Aranda, M.A., and Saez-Nieto, J.A. (2003). Molecular analysis of rifampicin-resistant *Mycobacterium tuberculosis* isolated in Spain (1996-2001). Description of new mutations in the *rpoB* gene and review of the literature. Int. J. Antimicrob. Agents *21*, 403-408.

Heym, B., Alzari, P.M., Honore, N., and Cole, S.T. (1995). Missense mutations in the catalase-peroxidase gene, *katG*, are associated with isoniazid resistance in *Mycobacterium tuberculosis*. Mol. Microbiol. *15*, 235-245.

Heym, B., Saint-Joanis, B., and Cole, S.T. (1999). The molecular basis of isoniazid resistance in *Mycobacterium tuberculosis*. Tuber. Lung Dis. *79*, 267-271.

Jereb, J.A., Klevens, R.M., Privett, T.D., Smith, P.J., Crawford, J.T., Sharp, V.L., Davis, B.J., Jarvis, W.R., and Dooley, S.W. (1995). Tuberculosis in health care workers at a hospital with an outbreak of multidrug-resistant *Mycobacterium tuberculosis*. Arch. Intern. Med. *155*, 854-859.

Kim, B.J., Lee, K.H., Yun, Y.J., Park, E.M., Park, Y.G., Bai, G.H., Cha, C.Y., and Kook, Y.H. (2004). Simultaneous identification of rifampin-resistant *Mycobacterium tuberculosis* and nontuberculous mycobacteria by polymerase chain reaction-single strand conformation polymorphism and sequence analysis of the RNA polymerase gene (*rpoB*). J. Microbiol. Methods *58*, 111-118.

Kubin, M., Havelkova, M., Hyncicova, I., Svecova, Z., Kaustova, J., Kremer, K., and van Soolingen, D. (1999). A multidrug-resistant tuberculosis microepidemic caused by genetically closely related *Mycobacterium tuberculosis* strains. J. Clin. Microbiol. *37*, 2715-2716.

Lee, A.S., Lim, I.H., Tang, L.L., Telenti, A., and Wong, S.Y. (1999). Contribution of *kasA* analysis to detection of isoniazid-resistant *Mycobacterium tuberculosis* in Singapore. Antimicrob. Agents Chemother. *43*, 2087-2089.

Lee, A.S., Teo, A.S., and Wong, S.Y. (2001). Novel mutations in ndh in isoniazid-resistant *Mycobacterium tuberculosis* isolates. Antimicrob. Agents Chemother. *45*, 2157-2159.

Lee, H.Y., Myoung, H.J., Bang, H.E., Bai, G.H., Kim, S.J., Kim, J.D., and Cho, S.N. (2002). Mutations in the *embB* locus among Korean clinical isolates of *Mycobacterium tuberculosis* resistant to ethambutol. Yonsei Med. J. *43*, 59-64.

Lienhardt, C. and Ogden, J.A. (2004). Tuberculosis control in resource-poor countries: have we reached the limits of the universal paradigm? Trop. Med. Int. Health *9*, 833-841.

Martin-Casabona, N., Xairo, M.D., Gonzalez, T., Rossello, J., and Arcalis, L. (1997). Rapid method for testing susceptibility of *Mycobacterium tuberculosis* by using DNA probes. J. Clin. Microbiol. *35*, 2521-2525.

McNerney, R., Kiepiela, P., Bishop, K.S., Nye, P.M., and Stoker, N.G. (2000). Rapid screening of *Mycobacterium tuberculosis* for susceptibility to rifampicin and streptomycin. Int. J. Tuberc. Lung Dis. *4*, 69-75.

Mdluli, K., Slayden, R.A., Zhu, Y., Ramaswamy, S., Pan, X., Mead, D., Crane, D.D., Musser, J.M., and Barry, C.E., III (1998). Inhibition of a *Mycobacterium tuberculosis* beta-ketoacyl ACP synthase by isoniazid. Science *280*, 1607-1610.

Meier, A., Sander, P., Schaper, K.J., Scholz, M., and Bottger, E.C. (1996). Correlation of molecular resistance mechanisms and phenotypic resistance levels in streptomycin-resistant *Mycobacterium tuberculosis*. Antimicrob. Agents Chemother. *40*, 2452-2454.

Middlebrook, G. (1954). Isoniazid-resistance and catalase activity of tubercle bacilli. Am. Rev. Tuberc. *69*, 471-472.

Miesel, L., Weisbrod, T.R., Marcinkeviciene, J.A., Bittman, R., and Jacobs, W.R., Jr. (1998). NADH dehydrogenase defects confer isoniazid resistance and conditional lethality in *Mycobacterium smegmatis*. J. Bacteriol. *180*, 2459-2467.

Mohamed, A.M., Bastola, D.R., Morlock, G.P., Cooksey, R.C., and Hinrichs, S.H. (2004). Temperature-mediated heteroduplex analysis for detection of *pncA* mutations associated with pyrazinamide resistance and differentiation between *Mycobacterium tuberculosis* and *Mycobacterium bovis* by denaturing high-performance liquid chromatography. J. Clin. Microbiol. *42*, 1016-1023.

Mokrousov, I., Bhanu, N.V., Suffys, P.N., Kadival, G.V., Yap, S.F., Cho, S.N., Jordaan, A.M., Narvskaya, O., Singh, U.B., Gomes, H.M., Lee, H., Kulkarni, S.P., Lim, K.C., Khan, B.K., van Soolingen, D., Victor, T.C., and Schouls, L.M. (2004). Multicenter evaluation of reverse line blot assay for detection of drug resistance in *Mycobacterium tuberculosis* clinical isolates. J. Microbiol. Methods *57*, 323-335.

Mokrousov, I., Filliol, I., Legrand, E., Sola, C., Otten, T., Vyshnevskaya, E., Limeschenko, E., Vyshnevskiy, B., Narvskaya, O., and Rastogi, N. (2002a). Molecular characterization of multiple-drug-resistant *Mycobacterium tuberculosis* isolates from northwestern Russia and analysis of rifampin resistance using RNA/RNA mismatch analysis as compared to the line probe assay and sequencing of the *rpoB* gene. Res. Microbiol. *153*, 213-219.

Mokrousov, I., Narvskaya, O., Limeschenko, E., Otten, T., and Vyshnevskiy, B. (2002b). Detection of ethambutol-resistant *Mycobacterium tuberculosis* strains by multiplex allele-specific PCR assay targeting *embB*306 mutations. J. Clin. Microbiol. *40*, 1617-1620.

Morlock, G.P., Metchock, B., Sikes, D., Crawford, J.T., and Cooksey, R.C. (2003). *ethA*, *inhA*, and *katG* loci of ethionamide-resistant clinical *Mycobacterium tuberculosis* isolates. Antimicrob. Agents Chemother. *47*, 3799-3805.

Musser, J.M., Kapur, V., Williams, D.L., Kreiswirth, B.N., van Soolingen, D., and van Embden, J.D. (1996). Characterization of the catalase-peroxidase gene (*katG*) and *inhA* locus in isoniazid-resistant and -susceptible strains of *Mycobacterium tuberculosis* by automated DNA sequencing: restricted array of mutations associated with drug resistance. J. Infect. Dis. *173*, 196-202.

Nataraj, A.J., Olivos-Glander, I., Kusukawa, N., and Highsmith, W.E., Jr. (1999). Single-strand conformation polymorphism and heteroduplex analysis for gel-based mutation detection. Electrophoresis *20*, 1177-1185.

Newton, C.R., Graham, A., Heptinstall, L.E., Powell, S.J., Summers, C., Kalsheker, N., Smith, J.C., and Markham, A.F. (1989). Analysis of any point mutation in DNA. The amplification refractory mutation system (ARMS). Nucleic Acids Res. *17*, 2503-2516.

Pablos-Mendez, A., Raviglione, M.C., Laszlo, A., Binkin, N., Rieder, H.L., Bustreo, F., Cohn, D.L., Lambregts-van Weezenbeek, C.S., Kim, S.J., Chaulet, P., and Nunn, P. (1998). Global surveillance for antituberculosis-drug resistance, 1994-1997. World Health Organization-International Union against Tuberculosis and Lung Disease Working Group on Anti-Tuberculosis Drug Resistance Surveillance[published erratum appears in N England J Med 1998 Jul 9;339(2):139]. N. Engl. J. Med. *338*, 1641-1649.

Pearson, M.L., Jereb, J.A., Frieden, T.R., Crawford, J.T., Davis, B.J., Dooley, S.W., and Jarvis, W.R. (1992). Nosocomial transmission of multidrug-resistant *Mycobacterium tuberculosis*. A risk to patients and health care workers. Ann. Intern. Med. *117*, 191-196.

Piatek, A.S., Telenti, A., Murray, M.R., el Hajj, H., Jacobs, W.R., Jr., Kramer, F.R., and Alland, D. (2000). Genotypic analysis of *Mycobacterium tuberculosis* in two distinct populations using molecular beacons: implications for rapid susceptibility testing. Antimicrob. Agents Chemother. *44*, 103-110.

Piatek, A.S., Tyagi, S., Pol, A.C., Telenti, A., Miller, L.P., Kramer, F.R., and Alland, D. (1998). Molecular beacon sequence analysis for detecting drug resistance in *Mycobacterium tuberculosis*. Nat. Biotechnol. *16*, 359-363.

Pletz, M.W., De Roux, A., Roth, A., Neumann, K.H., Mauch, H., and Lode, H. (2004). Early bactericidal activity of moxifloxacin in treatment of pulmonary tuberculosis: a prospective, randomized study. Antimicrob. Agents Chemother. *48*, 780-782.

Portugal, I., Covas, M.J., Brum, L., Viveiros, M., Ferrinho, P., Moniz-Pereira, J., and David, H. (1999). Outbreak of multiple drug-resistant tuberculosis in Lisbon: detection by restriction fragment length polymorphism analysis. Int. J. Tuberc. Lung Dis. *3*, 207-213.

Pretorius, G.S., van Helden, P.D., Sirgel, F., Eisenach, K.D., and Victor, T.C. (1995). Mutations in *katG* gene sequences in isoniazid-resistant clinical isolates of *Mycobacterium tuberculosis* are rare. Antimicrob. Agents Chemother. *39*, 2276-2281.

Ramaswamy, S. and Musser, J.M. (1998). Molecular genetic basis of antimicrobial agent resistance in *Mycobacterium tuberculosis*: 1998 update. Tuber. Lung Dis. *79*, 3-29.

Ramaswamy, S.V., Amin, A.G., Goksel, S., Stager, C.E., Dou, S.J., El Sahly, H., Moghazeh, S.L., Kreiswirth, B.N., and Musser, J.M. (2000). Molecular genetic analysis of nucleotide polymorphisms associated with ethambutol resistance in human isolates of *Mycobacterium tuberculosis*. Antimicrob. Agents Chemother. *44*, 326-336.

Ramaswamy, S.V., Dou, S.J., Rendon, A., Yang, Z., Cave, M.D., and Graviss, E.A. (2004). Genotypic analysis of multidrug-resistant *Mycobacterium tuberculosis* isolates from Monterrey, Mexico. J. Med. Microbiol. *53*, 107-113.

Rastogi, N., Goh, K.S., and David, H.L. (1989). Drug susceptibility testing in tuberculosis: a comparison of the proportion methods using Lowenstein-Jensen, Middlebrook 7H10 and 7H11 agar media and a radiometric method. Res. Microbiol. *140*, 405-417.

Rattan, A., Kalia, A., and Ahmad, N. (1998). Multidrug-resistant *Mycobacterium tuberculosis*: molecular perspectives. Emerg. Infect. Dis. *4*, 195-209.

Raviglione, M.C., Snider, D.E., Jr., and Kochi, A. (1995). Global epidemiology of tuberculosis. Morbidity and mortality of a worldwide epidemic. JAMA *273*, 220-226.

Rozwarski, D.A., Grant, G.A., Barton, D.H., Jacobs, W.R., Jr., and Sacchettini, J.C. (1998). Modification of the NADH of the isoniazid target (*InhA*) from *Mycobacterium tuberculosis*. Science *279*, 98-102.

Scorpio, A., Lindholm-Levy, P., Heifets, L., Gilman, R., Siddiqi, S., Cynamon, M., and Zhang, Y. (1997). Characterization of *pncA* mutations in pyrazinamide-resistant *Mycobacterium tuberculosis*. Antimicrob. Agents Chemother. *41*, 540-543.

Scorpio, A. and Zhang, Y. (1996). Mutations in *pncA*, a gene encoding pyrazinamidase/nicotinamidase, cause resistance to the antituberculous drug pyrazinamide in tubercle bacillus. Nat. Med. *2*, 662-667.

Sherman, D.R., Mdluli, K., Hickey, M.J., Arain, T.M., Morris, S.L., Barry, C.E., III, and Stover, C.K. (1996). Compensatory *ahpC* gene expression in isoniazid-resistant *Mycobacterium tuberculosis*. Science *272*, 1641-1643.

Siddiqi, S.H., Hawkins, J.E., and Laszlo, A. (1985). Interlaboratory drug susceptibility testing of *Mycobacterium tuberculosis* by a radiometric procedure and two conventional methods. J. Clin. Microbiol. *22*, 919-923.

Slayden, R.A. and Barry, C.E., III (2000). The genetics and biochemistry of isoniazid resistance in *Mycobacterium tuberculosis*. Microbes. Infect. *2*, 659-669.

Snider, D.E., Jr., Good, R.C., Kilburn, J.O., Laskowski, L.F., Jr., Lusk, R.H., Marr, J.J., Reggiardo, Z., and Middlebrook, G. (1981). Rapid drug-susceptibility testing of *Mycobacterium tuberculosis*. Am. Rev. Respir. Dis. *123*, 402-406.

Somoskovi, A., Parsons, L.M., and Salfinger, M. (2001). The molecular basis of resistance to isoniazid, rifampin, and pyrazinamide in *Mycobacterium tuberculosis*. Respir. Res. *2*, 164-168.

Sreevatsan, S., Escalante, P., Pan, X., Gillies, D.A., Siddiqui, S., Khalaf, C.N., Kreiswirth, B.N., Bifani, P., Adams, L.G., Ficht, T., Perumaalla, V.S., Cave, M.D., van Embden, J.D., and Musser, J.M. (1996). Identification of a polymorphic nucleotide in *oxyR* specific for *Mycobacterium bovis*. J. Clin. Microbiol. *34*, 2007-2010.

Sreevatsan, S., Pan, X., Stockbauer, K.E., Connell, N.D., Kreiswirth, B.N., Whittam, T.S., and Musser, J.M. (1997a). Restricted structural gene polymorphism in the *Mycobacterium tuberculosis* complex indicates evolutionarily recent global dissemination. Proc. Natl. Acad. Sci. U. S. A *94*, 9869-9874.

Sreevatsan, S., Pan, X., Zhang, Y., Kreiswirth, B.N., and Musser, J.M. (1997b). Mutations associated with pyrazinamide resistance in *pncA* of *Mycobacterium tuberculosis* complex organisms. Antimicrob. Agents Chemother. *41*, 636-640.

Sreevatsan, S., Stockbauer, K.E., Pan, X., Kreiswirth, B.N., Moghazeh, S.L., Jacobs, W.R., Jr., Telenti, A., and Musser, J.M. (1997c). Ethambutol resistance in *Mycobacterium tuberculosis*: critical role of *embB* mutations. Antimicrob. Agents Chemother. *41*, 1677-1681.

Sterling, T.R., Lehmann, H.P., and Frieden, T.R. (2003). Impact of DOTS compared with DOTS-plus on multidrug resistant tuberculosis and tuberculosis deaths: decision analysis. BMJ *326*, 574.

Suzuki, Y., Katsukawa, C., Inoue, K., Yin, Y., Tasaka, H., Ueba, N., and Makino, M. (1995). Mutations in *rpoB* gene of rifampicin resistant clinical isolates of *Mycobacterium tuberculosis* in Japan. Kansenshogaku Zasshi *69*, 413-419.

Suzuki, Y., Katsukawa, C., Tamaru, A., Abe, C., Makino, M., Mizuguchi, Y., and Taniguchi, H. (1998). Detection of kanamycin-resistant *Mycobacterium tuberculosis* by identifying mutations in the 16S rRNA gene. J. Clin. Microbiol. *36*, 1220-1225.

Takayama, K. and Kilburn, J.O. (1989). Inhibition of synthesis of arabinogalactan by ethambutol in *Mycobacterium smegmatis*. Antimicrob. Agents Chemother. *33*, 1493-1499.

Takiff, H.E., Salazar, L., Guerrero, C., Philipp, W., Huang, W.M., Kreiswirth, B., Cole, S.T., Jacobs, W.R., Jr., and Telenti, A. (1994). Cloning and nucleotide sequence of *Mycobacterium tuberculosis gyrA* and *gyrB* genes and detection of quinolone resistance mutations. Antimicrob. Agents Chemother. *38*, 773-780.

Taniguchi, H., Chang, B., Abe, C., Nikaido, Y., Mizuguchi, Y., and Yoshida, S.I. (1997). Molecular analysis of kanamycin and viomycin resistance in *Mycobacterium smegmatis* by use of the conjugation system. J. Bacteriol. *179*, 4795-4801.

Tarrand, J.J. and Groschel, D.H. (1985). Evaluation of the BACTEC radiometric method for detection of 1% resistant populations of *Mycobacterium tuberculosis*. J. Clin. Microbiol. *21*, 941-946.

Telenti, A., Imboden, P., Marchesi, F., Lowrie, D., Cole, S., Colston, M.J., Matter, L., Schopfer, K., and Bodmer, T. (1993). Detection of rifampicin-resistance mutations in *Mycobacterium tuberculosis*. Lancet *341*, 647-650.

Telenti, A., Philipp, W.J., Sreevatsan, S., Bernasconi, C., Stockbauer, K.E., Wieles, B., Musser, J.M., and Jacobs, W.R., Jr. (1997). The emb operon, a gene cluster of *Mycobacterium tuberculosis* involved in resistance to ethambutol. Nat. Med. *3*, 567-570.

Valway, S.E., Richards, S.B., Kovacovich, J., Greifinger, R.B., Crawford, J.T., and Dooley, S.W. (1994). Outbreak of multi-drug-resistant tuberculosis in a New York State prison, 1991. Am. J. Epidemiol. *140*, 113-122.

van Rie, A., Warren, R., Mshanga, I., Jordaan, A.M., van der Spuy, G.D., Richardson, M., Simpson, J., Gie, R.P., Enarson, D.A., Beyers, N., van Helden, P.D., and Victor, T.C. (2001). Analysis for a limited number of gene codons can predict drug resistance of *Mycobacterium tuberculosis* in a high-incidence community. J. Clin. Microbiol. *39*, 636-641.

van Rie, A., Warren, R.M., Beyers, N., Gie, R.P., Classen, C.N., Richardson, M., Sampson, S.L., Victor, T.C., and van Helden, P.D. (1999). Transmission of a multidrug-resistant *Mycobacterium tuberculosis* strain resembling "strain W" among noninstitutionalized, human immunodeficiency virus-seronegative patients. J. Infect. Dis. *180*, 1608-1615.

Vannelli, T.A., Dykman, A., and Ortiz de Montellano, P.R. (2002). The antituberculosis drug ethionamide is activated by a flavoprotein monooxygenase. J. Biol. Chem. *277*, 12824-12829.

Vernet, G., Jay, C., Rodrigue, M., and Troesch, A. (2004). Species differentiation and antibiotic susceptibility testing with DNA microarrays. J. Appl. Microbiol. *96*, 59-68.

Victor, T.C., Jordaan, A.M., van Rie, A., van der Spuy, G.D., Richardson, M., van Helden, P.D., and Warren, R. (1999). Detection of mutations in drug resistance genes of *Mycobacterium tuberculosis* by a dot-blot hybridization strategy. Tuber. Lung Dis. *79*, 343-348.

Victor, T.C., van Helden, P.D., and Warren, R. (2002). Prediction of drug resistance in *M. tuberculosis*: molecular mechanisms, tools, and applications. IUBMB. Life *53*, 231-237.

Victor, T.C., van Rie, A., Jordaan, A.M., Richardson, M., Der Spuy, G.D., Beyers, N., van Helden, P.D., and Warren, R. (2001). Sequence polymorphism in the rrs gene of *Mycobacterium tuberculosis* is deeply rooted within an evolutionary clade and is not associated with streptomycin resistance. J. Clin. Microbiol. *39*, 4184-4186.

Victor, T.C., Warren, R., Butt, J.L., Jordaan, A.M., Felix, J.V., Venter, A., Sirgel, F.A., Schaaf, H.S., Donald, P.R., Richardson, M., Cynamon, M.H., and van Helden, P.D. (1997). Genome and MIC stability in *Mycobacterium tuberculosis* and indications for continuation of use of isoniazid in multidrug-resistant tuberculosis. J. Med. Microbiol. *46*, 847-857.

Walley, J. (1997). DOTS for TB: it's not easy. Afr. Health *20*, 21-22.

Watterson, S.A., Wilson, S.M., Yates, M.D., and Drobniewski, F.A. (1998). Comparison of three molecular assays for rapid detection of rifampin resistance in *Mycobacterium tuberculosis*. J. Clin. Microbiol. *36*, 1969-1973.

World Health Organization (1997). Anti-tuberculosis drug resistance surveillance 1994 - 1997 (WHO/TB/97.229).

World Health Organization (2000). Anti-tuberculosis drug resistance in the world. report no. 2. Prevalence and trends.

World Health Organization (2001). Guidelines for drug susceptibility testing for second-line anti-tuberculosis drugs for DOTS-plus.

World Health Organization (2003). Global Tuberculosis Control.

Xu, C., Kreiswirth, B.N., Sreevatsan, S., Musser, J.M., and Drlica, K. (1996). Fluoroquinolone resistance associated with specific gyrase mutations in clinical isolates of multidrug-resistant *Mycobacterium tuberculosis* [published erratum appears in J Infect Dis 1997 Apr;175(4):1027]. J. Infect. Dis. *174*, 1127-1130.

Zhang, Y., Garbe, T., and Young, D. (1993). Transformation with *katG* restores isoniazid-sensitivity in *Mycobacterium tuberculosis* isolates resistant to a range of drug concentrations. Mol. Microbiol. *8*, 521-524.

Zhang, Y., Heym, B., Allen, B., Young, D., and Cole, S. (1992). The catalase-peroxidase gene and isoniazid resistance of *Mycobacterium tuberculosis*. Nature *358*, 591-593.

Zhang, Y. and Mitchison, D. (2003). The curious characteristics of pyrazinamide: a review. Int. J. Tuberc. Lung Dis. *7*, 6-21.

Zimhony, O., Vilcheze, C., and Jacobs, W.R., Jr. (2004). Characterization of *Mycobacterium smegmatis* expressing the *Mycobacterium tuberculosis* fatty acid synthase I *(fas1)* gene. J. Bacteriol. *186*, 4051-4055.

Chapter 6

Virulence Factors of Nontuberculosis Mycobacteria

P.L.C. Small

ABSTRACT

Nontuberculosis mycobacteria are environmental mycobacteria which are widespread in nature in soil, water, animals and invertebrates. Although almost one third of these species have been associated with human disease, fewer than 15 species are important human pathogens. Several factors have led to increased interest in environmental mycobacterial pathogens: 1) The spread of HIV has resulted in a large population of immunosuppressed people leading to the identification of several new mycobacterial pathogens (such as *Mycobacterium genovese*), as well as increasing the incidence of well known organisms such as *Mycobacterium avium* 2) The spread of *Mycobacterium ulcerans* which causes serious disease in immunocompetent hosts and contains unique virulence determinants 3) The growing awareness that some environmental pathogens such as *Mycobacterium marinum* can provide important insight into the pathogenesis of *Mycobacterium tuberculosis* and 4) The fact that genome sequences for *M. avium* (http://www.ebi.ac.uk/genomes/AE016958.html), *M. marinum* (www.sanger.ac.uk/Projects/M_marinum), and *M. ulcerans* (http://genopole.pasteur.fr/Mulc/BuruList.html) are available. Although virulence determinants have not been identified in the vast majority of environmental mycobacterial pathogens, there is a growing body of literature on *M. avium* subspecies *avium*, *Mycobacterium kansasii, M. marinum* and *M. ulcerans* that will be explored in this chapter.

INTRODUCTION

Over 95 species of mycobacteria have been identified, and this no doubt represents a minority of the species found in nature since new species are constantly being described. (Heckart et al., 2001; Falkingham, 2002; Koh et al., 2002; Daniel et al., 2004; Tortoli, 2004). Although a handful of these consistently cause disease in a human host, over 30 species of environmental mycobacteria (EM) cause occasional infections, particularly in severely immunocompromised hosts (Tortoli et al., 1994; Bloch et al., 1998; Floyd et al., 1996; Katoch, 2004; Okhusu et al., 2004). Since 1990 over ten new pathogenic environmental mycobacterial species (PEM) have been identified, primarily from HIV infected individuals (Tortoli, 2004). Despite the large number of PEM, an important and overlooked fact is that over 60% of environmental mycobacteria have never been isolated from infected humans. Along with the fact that the same group of EM are predominant PEM in widely separated parts of the world suggests that these bacteria do possess specific virulence determinants (Alcaide et al., 1997; Koh et al., 2002; Primm et al., 2004). Thus it is important to realize that even in severely immunocompromised patients who are exposed to a large number of different mycobacterial species, only a relatively small number of PEM are capable of causing disease.

Human-adapted pathogens such as *M. tuberculosis* have evolved strategies for survival within and spread between human hosts. In contrast PEM which cause disease in humans are rarely associated with person-to-person spread. Although many PEM species have been identified, a relatively small number of these account for the majority of cases (Table 1). In some cases, infection requires an immunocompromised host, but a significant number of PEM such as *M. kansasii*, *M. marinum*, and *M. ulcerans* cause persistent disease primarily in immunocompetent humans (Iredell et al., 1992; van der Werf et al., 1999; Bartralot et al., 2000; Kanathur et al., 2001). Mycobacteria are often thought of as soil bacteria, but a survey of PEM suggests that aquatic habitats may be a richer source of mycobacterial species than soil (Pedley et al., 2004; Falkinham, 2002; 2003). Thus *M. marinum, M. avium, Mycobacterium fortuitum, Mycobacterium cheloni*, *M. kansasii*, *Mycobacterium abscessans* and *M. ulcerans* are all associated with an aquatic habitat. The impact of PEM on human health is limited by two factors: 1) The inability of PEM to spread from person-to-person and 2) Human exposure to habitats in which PEM are found.

MYCOBACTERIA AS ENVIRONMENTAL PATHOGENS

The designation of a mycobacterial species as a soil or water bacterium leads to the assumption that organisms are distributed widely within the habitat. However, bacteria are never randomly distributed within an environment. In fact it is likely that most PEM occupy very specific niches within the environment. This becomes apparent when an attempt is made to determine where a particular PEM replicates in the environment. A cursory glance at a pond or river reveals the complexity of any specific habitat. For example an aquatic environment may include vegetation, fish, frogs, insects, snails, and a large number of protists as well as water, rocks, shells, leaves and branches. Terrestrial environments are equally complex. Thus is not surprising that little is known about precisely where in the environment most PEM grow. Nonetheless, a few basic factors probably play key roles in the environmental distribution of *Mycobacteria* spp. These are: 1) the hydrophobic cell envelope (Rostogi and Barrow, 1994), 2) mycobacterial physiology, particularly optimal growth temperature (Schulze-Robbecke and Buchholtz, 1992) and 3) the association between mycobacteria and specific non-human hosts. The hydrophobicity of the mycobacterial cell envelope precludes random distribution of mycobacteria in an aquatic environment, and most if not all species of mycobacteria readily form

Table 1. Pathogenic environmental pathogens.			
Species	Disease	Host factors	Source
M. abcessans	skin, soft tissue pulmonary	 cystic fibrosis, HIV	soil, water
M. avium	Pulmonary Disseminated lymphadenitis	underlying pulmonary disease in older women immunosuppression, HIV children	water, soil, birds, animals insects, amoeba
M. cheloni	pulmonary, lymphadenitis, skin disseminated	 HIV	tap water, salmonid fish, turtles,manatees
M fortuituim	skin, synovitis keratitis pulmonary disseminated	 pre-existing mycobacterial infection HIV	water, fish, dogs, cats, cattle
M. genavense	disseminated	HIV	birds, ferrets, cats, tap water
M. gordonae	skin disseminated	 HIV	water, widespread tap water
M. kansasii	pulmonary tenosynovitis skin cellulitis	urban	tap water, dogs
M. malmoense	lymphadenitis, pulmonary	children underlying disease	
M. marinum	skin, tenosynovitis disseminated	 HIV	fish, frogs, snakes, fresh water
M. scrofulaceum	cervical lymphadenitis pulmonary disease	children adults	water
M. smegmatis	skin	accidental or surgical trauma	water, fish, dogs, cats
M. terrae	cutaneous,(hand) tenosynovitis pulmonary disseminated	 HIV HIV	soil
M. ulcerans	cutaneous, osteomyelitis	exposure to slow moving water	slow moving water, aquatic insects
X. xenopi	pulmonary, tenosynovitis, osteomyelitis disseminated	immunosupression, HIV	water, hot water

biofilms (Recht, and Kolter, 2001; Marsollier et al., 2004). With a few exceptions such as *M. haemophilum*, the nutritional requirements of mycobacterial species are simple. However, oxygen availability as well as temperature play important roles in determining where mycobacteria are found in the environment. PEM are unlikely to be encountered in an anaerobic environment and many PEM grow very poorly or at all above 34°C (MacCallum and Tolhurst, 1948; Linell and Norden, 1954) which is likely to explain the fact that *M. leprae, M. marinum* and *M. ulcerans* grow only in cooler parts of the human body. Finally the relationship between PEM and non-human hosts also restricts distribution of PEM in humans. Although the relationship between PEM such as *M. marinum* and *M. fortuitum* with frogs (Trott et al., 2004) and fish (Ucko et al., 2002; Decostere et al., 2004) is well known, what is less appreciated is that a large number of mycobacterial species have also been isolated from manatees (Sato et al., 2003), turtles (Oros et al., 2003), insects (Pai et al, 2003) , and protists (Greub, and Raoult, 2004). Many PEM have enjoyed a long evolutionary relationship with a non-human host species which has resulted in the acquisition of genetic factors which also enable them to successfully infect human hosts.

VIRULENCE DETERMINANTS OF ENVIRONMENTAL PATHOGENS

Although the virulence of PEM is much less well understood than that of *M. tuberculosis* complex organisms, it is striking that all PEM, with the exception of *M. ulcerans* are facultative intracellular pathogens capable of survival and replication within naïve macrophages. This virulence strategy is conserved by mycobacteria as they replicate in a wide range of hosts including snails, frogs, fish, seals, dogs, cats and humans. Although all of these PEM replicate in a restricted phagosomal compartment in the absence of lysosomal fusion, there are nonetheless some species-specific differences in the bacterial-host cell interactions that will be discussed later. Vertebrate hosts uniformly respond to infections with most PEM by mounting a granulomatous response (Ashford et al, 2001; Pressler et al., 2002; Sato et al., 2003). However for PEM, in contrast to *M. tuberculosis*, the human host is a dead end since PEM are incapable of person-to-person spread.

In the past 20 years considerable progress has been made towards understanding the basis of virulence in *M. tuberculosis*. This work has flourished with the development of robust tools for transposon mutagenesis, genetic complementation, and clever strategies such as signature tagged mutagenesis for identifying attenuating mutations. With the notable exception of *M. marinum*, however, the application of these tools to the study of PEM pathogenesis has lagged considerably behind that of *M. tuberculosis*. Although definitive identification of specific virulence determinants requires the genetic tools of mutational analysis and complementation, information gleaned from bioinformatics, comparative cell biology and animal studies has provided potential insight into the pathogenesis of PEM. The availability of the genome sequence for *M. avium* subsp. paratuberculosis, *M. marinum* and *M. ulcerans* has provided much useful information for studying the virulence of these organisms.

Out of the 30 or more PEM identified, evidence for the role of specific virulence determinants in these organisms is limited to a rather small number of species: These include *M. avium, M. cheloni/abcessus, M. fortuitum, M. kansasii, M. marinum* and *M. ulcerans* (Bartralot et al., 2000; Katoch, 2004, Primm et al., 2004). It is not surprising, considering the shared pathogenic life style of *M. tuberculosis* and PEM that many of the virulence determinants identified in *M. tuberculosis* are also present in PEM (Table 2). For example, PEM contain genes for leucine, purine or isocitrate lyase biosynthesis as well as those for ESAT6 or CFP10 with a specific role in virulence (Gey Van Pittius et al., 2001).

In addition to shared virulence genes, there are a number of virulence determinants specific to a particular PEM species. Most, though not all species-specific virulence determinants encode cell envelope structures. In PEM, as in *M. tuberculosis*, changes in colonial morphology have led to the identification of attenuating mutations (Calmette et al., 1926; Linell and Norden, 1954; Belisle et al.,

Table 2. DNA and Amino Acid (AA) similarity between *M. tuberculosis* virulence-associated genes and *M. avium* subsp. *paratuberculosis* and *M. marinum* orthologues.
M. tuberculosis genes: http://genolist.pasterur.fr/Tuberculist
M. marinum genes: http://www.sanger.ac.uk/Projects/M_marinum
M. avium subsp. *paratuberculosis*: http://pathogenomics.ahc.edu

M. tuberculosis	*M. marinum* homologs		*M. avium paratuberculosis* homologs	
Gene name	DNA Identity	Similar	DNA Identity	AA Identity/Similar
esxa (ESAT6)	83%	90%	None	None
esxb (CFP10)	90%	84%	None	None
pks15	80%	91%	73%	71% / 80%
pks1	86%	84%	60%	49% / 61%
phoP	86%	90%	84%	87% / 89%
secA2	85%	95%	70%	87% / 90%
fadD28	81%	87%	70%	64% / 79%
pcaA	82%	87%	82%	78% / 85%
icl	87%	94%	89%	92% / 98%
plcA	81%	93%	54%	28% / 44%
plcB	80%	88%	53%	36% / 47%
plcC	79%	94%	55%	23% / 41%
plcD	75%	77%	57%	27% / 42%
sigH	86%	94%	85%	87% / 90%
hbhA	84%	77%	82%	71% / 76%
mtrA	90%	83%	90%	86% / 87%
mtrB	87%	89%	82%	84% / 89%
katG	73%	78%	70%	66% / 77%
mas	85%	93%	62%	58% / 69%
tylA	80%	88%	80%	71% / 71%
leuC	84%	90%	85%	88% / 92%
trpD	80%	74%	81%	70% / 75%

1989; Cangelosi et al., 2001). Thus the presence of strain specific glycopeptidolipids, glycolipids and polyketides in the cell envelope of *M. avium* (Chatterjee et al., 1989, Minnikin et al., 1989, and polyketides in *M. ulcerans* (George et al., 1999) play major roles in the virulence of these organisms. Finally, gene duplication, genomic deletions and rearrangements, insertion elements and horizontal transfer have undoubtedly also played important roles in shaping the differential virulence of PEM (Krzywubska et al., 2004; Stinear et al., 2004).

It is intriguing that a large number of species of PEM that are of relatively low virulence such as *M. terrae*, can nonetheless cause severe, persistent infections in humans (Bartralot et al., 2000; Torteli, 2004). Though the list of PEM is long, the number of patients infected by most PEM is quite small. However, the high antibiotic resistance of many PEM along with the growing populations of immunocompromised patients has led to an increasing recognition of the importance of these pathogens. The lack of genetic tools for investigating the virulence of many PEM has made it impossible to assign a particular gene a role in virulence. However, genomics, animal studies, cell biology and epidemiological tools have been used to glean information about the virulence of a handful of PEM (including *M. avium, M. marinum* and *M. ulcerans*) and preliminary studies have provided insight into the pathogenesis of *M. fortuitum, M. cheloni/abcessus* and *M. kansasii*. The virulence of several of these species will be discussed in more detail below.

M. AVIUM SUBSP. *AVIUM*

ECOLOGY AND EPIDEMIOLOGY

The *M. avium* complex was historically treated as part of a group of slow-growing mycobacteria designated the *M. avium-intracellulare* complex. However, molecular tools clearly separate *M. avium* and *M. intracellulare* into separate species (Primm et al., 2004). Despite the exclusion of *M. intracellulare* strains, *M. avium* species remains a heterogeneous group of strains, often referred to as the *M. avium* complex (MAC) which is divided into three subspecies: *M. avium* subsp. *avium* (MAA), which includes many strains isolated from human and animal hosts and which is a wide-spread pathogen of immunocompromised hosts; M. *avium* subsp. *paratuberculosis* (MAP), the causative agent of Johne's disease in cattle; and *M. avium* subsp. *sylvaticum* a pathogen of birds and other animals. MAC are among the most widely distributed mycobacteria in nature. They have been isolated from water, soil, air, dust, plant material, insects, cattle, pigs, birds, amoebae and humans (Mijs et al., 2002; Falkinham, 2003; Greub, and Raoult, 2004; Pedley et al., 2004). In fact almost any search for mycobacteria in the environment will yield MAC. Considering the large number of host species and environmental niches occupied by these organisms, it is not surprising that MAC comprises a very heterogeneous group of strains (Mazurek et al., 1993). Although as in other mycobacterial species, the cell wall contains a repertoire of mycolic acids, a unique feature of the *M.*

avium cell wall is the presence of a highly heterogeneous and immunogenic class of glycopeptidolipids (GPLs) some of which are strain-specific (ss) (Eckstein et al., 2000; Smole et al. 2002, Semret et al., 2003). The *M. avium* ssGPLs are encoded in a roughly 50 kb cluster of genes (*ser2* locus) which varies in length due to differences in the number and types of IS elements present (Laurent et al., 2003). Biochemically GPLs consist of a core characterized by a tetrapeptide structure linked to 6-deoxy-L-talose (Chatterjee and Khoo, 2001; Eckstein et al., 2003). This core structure is extensively modified by the addition of oligosaccharide side chains and these form the basis for serovar designation. Further heterogeneity in GPLs is generated by the presence of IS elements within and between genes in the GPL cluster (Mazurek et al., 1993; Semret et al., 2004). Despite some correlation between specific *M. avium* serovars and differential virulence, the significant amount of heterogeneity found outside of the ssGPL cluster in different strains makes these differences difficult to interpret (Torrelles et al., 2000; Smole, et al, 2002; Ohkusu et al., 2004).

PATHOGENESIS AND VIRULENCE TRAITS

MAA gained prominence as the cause of the most severe of all mycobacterial infections (except for *M. tuberculosis*) threatening the lives of severely immunocompromised patients with HIV (Ohkusu et al., 2004). MAA infection causes a wide spectrum of disease which depends primarily on host immunity (Table 1). In HIV-infected patients MAA causes a severe systemic, intracellular infection. In patients with chronic lung disease, MAA causes chronic progressive pulmonary disease and in children it has replaced *M. scrofulaceum* as the most frequent cause of regional lymphadenitis (Katoch, 2004; Primm et al., 2004). Like other mycobacterial pathogens, MAA is a facultative intracellular parasite that replicates within macrophages and it is assumed that this ability is responsible for its virulence. The fact that MAA is a Biolevel 2 pathogen has made it a convenient model for use in studying mycobacterial-macrophage interactions (Russell, 2001; Daigle et al., 2002; Danelishvili et al., 2004; McGarvey et al., 2004; Pietersen et al., 2004). A number of investigators have shown that the intracellular trafficking of MAA and *M. tuberculosis* are extremely similar, if not identical. Both species replicate within an endosome that is delayed in maturation and as a result fails to fuse with the lysosome (Ullrich et al., 2000; Russell, 2001). A large number of metabolic genes have been identified in *M. tuberculosis* which play a role in intracellular replication. Very similar genes are also present in MAA and these would be expected to play the same role in intracellular replication in MAA (Table 2). Several investigators have studied the expression of *M. avium* genes within macrophages (Daigle et al., 2002; Hou et al., 2002). The genes identified in these studies included a large number of genes involved in metabolism (*icl*), lipid biosynthesis (*pks*), iron acquisition (iron and mycobactin biosynthesis), as well as a number of other genes such as *mce* and PPE homologues which are predicted to play a role in the virulence of *M. tuberculosis*. Using a *M. avium*-GFP promoter library, Danelishvili et al. (2004) identified a number of promoters up-regulated within macrophages and confirmed this data with mouse studies for a sub-set of genes. Although an analysis of the expression of genes within macrophages provides insight into how *M. avium* responds to the intracellular environment, a great deal of work will need to be done to determine the importance of these genes to virulence. Also annotation of the *M. avium* genome is incomplete and so ortholog assignment could change. The host response to *M. avium* has also been studied *in vitro* using macrophages. The lists of macrophage genes up-regulated by intracellular *M. tuberculosis* and *M. avium* contain significant over-lap suggesting that the macrophage reacts in a similar, though not identical way to infection with these two (McGarvey et al., 2004).

Despite the similarities between MAA and *M. tuberculosis* in their intracellular environment, there is one important bacterial-host cell interaction in which the MAA and *M. tuberculosis* show striking differences. Whereas *M. tuberculosis* exhibits cytotoxicity for host cells, MAA does not (Sano et al., 2002; Hsu et al., 2003). In vitro MAA can be maintained for weeks within macrophages whereas *M. tuberculosis* -infected macrophages are killed within a few days. Cell-contact cytotoxicity has recently been associated with two gene products of the RD1 region, ESAT6 and CFP10 in *M. tuberculosis* and *M. marinum* (Hsu et al., 2003; Gao et al., 2004). In both of these species, deletion of genes encoding esat6 and cfp10 ablates cytotoxicity. The Esat6 gene locus has undergone considerable rearrangement and duplication in *M. tuberculosis*. *M. tuberculosis* contains 5 copies of the ESAT 6 cluster of genes (Gey Van Pittius; et al., 2001; Marmiesse et al., 2004). Although genomic analysis suggests that homologous copies of genes within these regions are functional, it is clear that the functions of individual gene copies are not identical since deletion of a single gene is sufficient for loss of cytotoxicity. The ESAT6/CFP10 proteins in *M. tuberculosis* are encoded by the gene cluster in RD1 (designated Region 1) (Gey van Pittius, 2001; Brodin et al., 2004). A comparison of the five ESAT-6 gene clusters in *M. tuberculosis* with those in *M. avium* 104 (the genome sequence strain) shows that *M. avium* has the RD1 deletion and so lacks Rv3861-3883c which includes the genes encoding ESAT-6 (Rv3874) and CFP10 (Rv3875). The loss of these genes is the major attenuating event that produced BCG from *M. bovis*. Whether the absence of these genes in *M. avium* is associated with the lack of cytotoxicity and is associated with reduced virulence in *M. avium* awaits genetic proof.

It was noted early on, that colony morphology in *M. avium* is not a stable trait and a number of colonial morphotypes have been described (Reddy et al., 1996). Differences in colony morphology in mycobacterial species, as in many bacterial species reflect changes in cell wall constituents and these in turn are often related to virulence. Three types of variable colony morphotypes or transitions have been described in *M. avium* both among natural populations of organisms as well as in laboratory passaged strains. These are: 1) an irreversible smooth to rough transition;

2) a reversible smooth- transparent to smooth- opaque transition; and 3) a reversible red-white switch expressed as the differential ability of bacteria to bind Congo red dye (Pedley et al., 2004). The genetic changes underlying these morphotypes are not fully understood. It is clear though that changes in colony morphology are due to changes in cell wall constituents. However, the genetic pathways involved are complex since a specific morphotype can result from lesions in seemingly unrelated genes.

The biochemical basis of the smooth to rough transition has been clearly associated with deletions in the GPLs which are likely due to homologous recombination between IS elements within the GPL cluster (Ekstein et al., 2000). The location of IS elements in *M. avium* is a major source of heterogeneity between strains (Semret et al., 2004). In the case of the genome sequence strain, *M. avium* 104, it is clear that a 10 kb deletion in the oligosaccharide side chain of an ssGPL cluster results in a rough colony (Belisle et al., 1993). Although there is some evidence from experiments using isogenic strains that the smooth morphotype may be more virulent than the rough, there are also some rough morphotypes that appear equally virulent as the smooth morphotypes (Pedrosa et al., 1994). The situation is likely to be similar to that found with LPS deletions in *Escherichia coli* where the differential virulence of smooth versus rough colonies depends on the type and extent to which LPS has been shortened through deletion. Rough *M. avium* mutants which have undergone a deletion of the entire ssGPL cluster are avirulent as are "deep" rough LPS mutants in *E. coli*. This is not surprising since such mutations produce drastic changes in the cell wall resulting in pleomorphic phenotypes.

A second source of colonial variation is characterized by a reversible transition between smooth translucent and smooth opaque colonies (Pedrosa et al., 1994; Pedley et al., 2004). There is evidence to strongly support the greater virulence of the smooth transparent over the smooth opaque morphotype in that: 1) patient and animal isolates consist almost entirely of smooth transparent colonies 2) when organisms from a smooth opaque colony are introduced into an animal, they are re-isolated almost entirely as smooth transparent colonies (Pedrosa et al., 1994), and 3) organisms from smooth transparent colonies are better able to replicate within macrophages than are organisms from smooth opaque colonies (Ohkusu et al., 2004) . The genetic basis of this transition is unknown.

A third morphotype variation, the white-red switch, is expressed in terms of the ability to bind the dye Congo Red (Cangelosi et al., 1999). Congo Red binding has been used as a basis for detecting pathogenicity in a large number of genetically unrelated organisms, including *Bordetella pertussis, Shigella flexneri, Yersinia species, Pseudomonas* plant pathogens, *Neisseria* species and *Aeromonas* fish pathogens as well as for detecting amyloid fibrils in human patients. Although the specific mechanism of Congo Red binding is unknown, evidence suggests that Congo Red has the ability to intercalate between protein molecules (Khurana et al., 2001). Although Congo Red binding may be a useful marker for virulence, it lacks specificity even within a single MAA isolate (Laurent et al., 2003) and the genetic basis of Congo Red binding in different strains is likely to differ.

In *M.avium* several lines of evidence suggest that white transparent organisms exhibit heightened virulence compared to red transparent organisms. They are also more drug resistant. White transparent colonies are highly over-represented in patient isolates, and white transparent isolates are also more capable of replication within macrophages and mice (Mukherjee et al., 2001). Further, following introduction of red transparent organisms into animals, *in vivo* switching results in the re-isolation of populations highly enriched for white transparent bacteria. One of the determinants underlying Congo Red binding was recently identified in a screen of drug resistant *M. avium* isolated through transposon mutagenesis. One of the drug resistant mutants showed both reduced Congo Red binding and reduced ability to survive in THP-1 human macrophages (Laurent et al., 2003). Bioinformatics reveals that Maa2520 encodes a hypothetical protein ortholog of *M. tuberculosis* Rv1697. However, it was not possible to obtain further evidence to support the role of this gene in virulence since the study was performed in an avirulent smooth opaque strain. Transposon mutagenesis has not yet been successfully performed in the virulent white translucent morphotype. It was also clear from this study that the Congo Red phenotype is is not specific for a single virulence trait since some of the Congo Red negative mutants obtained were fully capable of intracellular replication (Philalay et al., 2004). In a separate study, transposon mutagenesis conducted in stable red and white opaque variants of clinical isolate of Maa 104 resulted in the isolation of 8 white colonies out of 2,500 mutants. Two insertions occurred in a putative acyl transferase, *crs*, located upstream of the ssGPL cluster whereas other mutants contained single insertions in a diverse collection of genes (Laurent et al., 2003).

In summary, investigations into the virulence of *M. avium* are complicated by the genetic diversity between strains, the genetic instability of strains, and the inability so far to achieve transposon mutagenesis in the most virulent phenotype of *M. avium* (the white transparent morphotype). However, both transposon mutagenesis and allelic exchange have recently been reported for *M. avium* in avirulent isolates (Otero et al., 2003; Irani et al., 2004) and it is likely that methods for use with virulent isolates will follow shortly. However, the fact that *M. avium* strains are not clonal, and that patient isolates represent a large repertoire of isolates suggests that virulence may be an extremely complex strain-specific genotype in *M. avium*.

MYCOBACTERIUM KANSASII

EPIDEMIOLOGY, TAXONOMY AND ECOLOGY

M. kansasii is a slow growing photochromogenic mycobacterium capable of producing serious pulmonary disease in immunocompetent humans (Kanathur et al., 2001; Koh et al., 2002). Of all the PEM that cause pulmonary disease, *M. kansasii* is considered the most virulent. Nontheless, as with *M. avium* infections, *M. kansasii* infections have increased dramatically as a result of HIV (Smith et al., 2003). In a study conducted in San Francisco, 12% of *M. kansasii* infected patients lacked signs of immunosuppression or underlying disease (Bloch et al., 1998). Compared with *M. avium*, *M. kansasii* infections are more sensitive to antibiotics and for this reason considerably easier to treat. This fact has contributed to making *M. kansasii* a rather neglected mycobacterial species despite its virulence.

A number of molecular tools including restriction fragment length polymorphism, pulsed-field gel electrophoresis, amplified fragment length polymorphism analysis and PCR restriction analysis have been used to group *M. kansasii* isolates. The probes used in these studies have been derived from the major polymorphic tandem repeat, IS*1652*, the *hsp65* gene, and the 16S-23S intergenic region. Each method of analysis separates *M. kansasii* strains into 5 major subspecies which are found world wide (Alcaide et al., 1997; 2004 Picardeau et al., 1997; Santin et al., 2004; Zhang et al., 2004). *M. kansasii* Type I is highly associated with human disease in immunocompetent as well as immunocompromised patients and this association is particularly striking in the USA, Europe, Japan and Australia (Ross et al., 1992; Alcaide et al., 1997; 2004 Picardeau et al., 1997; Santin et al., 2004; Zhang et al., 2004). Evidence from an Italian study shows a strong relationship between Type II isolates (Tortoli et al., 1994) and HIV infection whereas biotypes III, IV and V have only been isolated from the environment (Tortoli, 2003). Thus epidemiological evidence strongly suggests that Type I *M. kansasii* isolates carry specific virulence determinants which enable them to cause human disease.

As with *M. avium*, acquisition of *M. kansasii* is assumed to be through contact with the environment. Although the organism is frequently isolated from tap water and thus is likely to be an urban infection, it is rarely isolated from natural sources such as rivers, soil or dust (Alcaide et al., 1997; Taillard et al., 2003; Zhang et al., 2004). However, *M. kansasii* has been isolated from a large number of animal species including cattle, swine, cockroaches, dogs, manatees and turtles (Oros et al., 2003; Pai et al., 2003; Sato et al., 2004; Southern, 2004).

PATHOGENESIS AND VIRULENCE TRAITS

M. kansasii infection presents primarily as a pulmonary infection in immunocompetent and immunocompromised patients. The pathogenesis and histopathology of the disease is essentially similar to that of *M. tuberculosis*. Disseminated infection with *M. kansasii* even in HIV-infected individuals is uncommon (Smith et al., 2003). Although *M. kansasii* generally responds well to antibiotic treatment, the prognosis of the infection depends primarily on whether the infected person is immunocompromised. Occasional extrapulmonary infections have been reported of which the vast majority are cutaneous infections in immunocompromised persons (Bartralot et al., 2000; Kotb et al., 2001; Tzen et al., 2001).

Remarkably little is known about the virulence traits of *M. kansasii*. Early descriptions of *M. kansasii* reported the occurrence of rough and smooth colonial morphotypes (Fregnan et al., 1961). Evidence from mouse studies suggested that rough, more strongly catalase positive isolates were more virulent (Wayne et al., 1962; Collins et al., 1981). The basis for the difference in colonial morphology in *M. kansasii* is associated with biochemical changes in two cell wall lipids which were long recognized as species-specific antigens. Analysis of cell envelope lipids extracted from stable rough and smooth strains allowed identification and structural analysis of the *M. kansasii* species-specific glycolipids. Whereas the cell envelope of both smooth and rough morphotypes contains a phenolic glycolipid, initially described as Mycoside A, smooth isolates contain an additional unique glycolipid family of eight lipids called the trehalose-containing lipooligosaccharides (Belisle et al., 1989). The association of loss of a cell wall glycolipid with a gain in virulence does not lead to a straightforward explanation. It has been proposed that the loss of the trehalose-containing lipooligosaccharides leads to greater exposure of lipoarabaninmannan (LAM) and the phenolic glycolipid, both of which are highly immunomodulatory molecules (Belisle et al., 1989). However, isogenic strains were not used in these studies and the genetic basis for glycolipid production in *M. kansasii* has not been further elucidated.

There is also evidence that the mycobacterial cell wall lipids (LAM) and the precursor molecules (LM) may play a role in the virulence of *M. kansasii* as they do in *M. tuberculosis* despite the fact that *M. kansasii* has a unique mannan core (Guerardel et al., 2003). In a recent study these lipids were extracted from a virulent clinical isolate of *M. kansasii* and assayed for their ability to induce pro-inflammatory cytokines from differentiated THP-1 human macrophages. *M. kansasii* LM elicited a strong pro-inflammatory response characterized by induction of IL-8 and TNF-alpha (Vignal et al., 2003). This response was dependent on TLR2/CD14 receptors and required the presence of an acute phase serum protein, LPB, which is involved in the transfer of glycolipids to CD14 (Vignal et al., 2003). *M. kansasii* LAM did not elicit an inflammatory response. A second lipid-mediated phenotype described for numerous pathogenic mycobacteria is the ability to prevent or delay the apoptosis of infected cells (Fratazzi et al., 1999). Infection of cells with *M. tuberculosis* or *M. kansasii* results in extensive cell death by 5 days. However, *M. kansasii*-mediated cell death was characterized by extensive apoptosis, whereas cell death by virulent *M. tuberculosis* complex organisms showed little apoptosis (Keane et al., 2002). More recently, purified *M. kansasii* LM was shown to have potent apoptosis-inducing activities (Guerardel et al., 2003). Unfortunately different strains of *M. kansasii* were used in these two studies. The study using whole bacteria was conducted with an ATCC strain referred to as an "avirulent" bacterium, whereas studies using purified LM and LAM used material derived from a virulent clinical isolate of *M. kansasii*. However, both studies suggest that a difference between the virulence of *M. kansasii* and *M. tuberculosis* could reside in the differential ability to produce apoptosis.

The RD1 region has been little explored in *M. kansasii*. However, it has been shown that serum from *M. kansasii* patients reacts to ESAT6 antigen suggesting that the locus is intact in this organism (Arend et al., 2002).

In summary, the pathogenesis of *M. tuberculosis* and *M. kansasii* appear to be very similar although a striking difference is the inability of *M. kansasii* to be transmitted by person-to-person contact. Secondly, although HIV infection has resulted in increased incidence of *M. kansasii*, disseminated disease is much less common with *M. kansasii* than with *M. tuberculosis* or *M. avium*. This may be because *M. kansasii* has particular tropism for pulmonary tissue or alternatively because it is unable to grow in less oxygen rich areas in the body. Although evidence suggests a role for cell wall lipids in the virulence of *M. kansasii*, there is no genetic proof for this. The fact that human disease is associated world- wide with a particular biovar of *M. kansasii*, Type I, points to the likely presence of specific virulence determinants in this serovar. There are a number of molecular methods such as subtractive hybridization which could be used to identify specific virulence determinants uniquely present in Type I *M. kansasii*. The relative host specificity of *M. kansasii* compared with that of *M. avium* is also of interest. Finally, investigations into the tissue tropism of *M. kansasii* could lead to insights into the pathogenesis of pulmonary tuberculosis.

MYCOBACTERIUM MARINUM

Although pulmonary disease is the most important manifestation of mycobacterial disease, a large number of mycobacterial pathogens are primarily associated with cutaneous or relatively superficial infections. These include *M. leprae, M. cheloni, M. fortuitum, M. gordonae, M. terrae M. marinum* and *M. ulcerans*. In the case of *M. leprae, M. marinum*, and *M. ulcerans*, the inability of the organisms to grow at 37° C appears to be a major determinant of tissue tropism (Clark and Shepard, 1963). *M. cheloni*, *M. fortuitum* and *M. marinum* are the primary causes of fish mycobacteriosis (Decostere et al., 2003). Thus exposure to aquatic environments, both natural and man-made, is the primary risk factor for infection (Iredell et al., 1992; Gluckman, 1995). Taxonomically *M. fortuitum* and *M. cheloni* are classified as fast growing mycobacteria, whereas *M. marinum* is identified as a slow grower. Of all the aquatic mycobacteria capable of causing human infection, *M. marinum* has received the most attention by the research community although clinically it is less frequently encountered than *M. cheloni* or *M. fortuitum*. *M. marinum* can cause persistent cutaneous infections in immunocompetent hosts, and rare lethal disseminated disease in immunocompromised individuals (Tchornobay et al., Parent et al., 1995). However, research on *M. marinum* has grown exponentially in the past decade for the following reasons: 1) the organism is closely related to *M. tuberculosis* and mutations in many *M. tuberculosis* genes required for intracellular replication can be complemented by the *M. marinum* orthologue, 2) the organism has a doubling time of 4 hrs which greatly facilitates genetic manipulation, 3) the pathogenesis of *M. marinum* can be explored in its natural aquatic hosts, 4) the *M. marinum* genome sequence is available and 5) the evolutionary relationship between *M. tuberculosis, M. marinum* and *M. ulcerans*, the causative agent of Buruli ulcer, is likely to provide insight into the evolution of virulence in mycobacterial pathogens. However, there are also important ways in which *M. marinum* differs from *M. tuberculosis*. It does not cause pulmonary tuberculosis even in immunocompromised persons or even where systemic infection occurs, nor is *M. marinum* capable of person-to-person spread.

EPIDEMIOLOGY, TAXONOMY AND ECOLOGY

M. marinum is widely distributed within vertebrate hosts in marine and fresh water environments (Decostere et al., 2004; Iredell et al., 1992). Within these environments the organism has been isolated from a number of poikilothermic aquatic species including snakes, frogs, turtles, snails and crustaceans. *M. marinum* is by far the most important mycobacterial pathogen of fish. In aquarium- raised fish *M. marinum* is transmitted by feeding and is a serious cause of morbidity to aquarium maintained or commercially farmed fish (Decostere et al., 2003). In contrast to other water-associated pathogens such as *M. avium* and *M. kansasii, M. marinum* has not been found in tap water, presumably due to the organisms' sensitivity to chlorination. This sensitivity to chlorination may also explain the changing epidemiology. Fifty years ago a frequent source of human *M. marinum* infections was swimming pools (Linell and Norden, 1954). However, with the introduction of better chlorination, swimming pool associated *M. marinum* infections have become rare. At present, those at highest risk for *M. marinum* infections are people who raise fish in confined environments either in an aquarium at home or commercially.

DNA-DNA hybridization studies conducted between *M. marinum* isolates suggest that there is considerable strain heterogeneity (Tonjum et al., 1998). A molecular analysis of 21 *M. marinum* strains using 16S rRNA and *hsp65* genes and RFLP analysis readily distinguished between fish isolates from marine and fresh water environments in Israel (Ucko et al.,2002). Scrutiny of a more diverse collection of *M. marinum* isolates using multilocus sequence typing (MLST) (Stinear et al., 2000) and amplified fragment length polymorphism (AFLP) shows that *M. marinum* can be separated into two clusters, one of which includes isolates from fish, and the other which includes isolates from human disease (Stinear et al., 2000; van der Sar et al., 2004). Although all strains are pathogenic for fish, the following evidence suggests that Cluster I strains may contain specific virulence determinants lacking in Cluster II strains: 1) Cluster I strains cause far more serious disease in fish 2) Cluster I strains are over-represented among human isolates and 3) Cluster I strains are more closely related to *M. ulcerans*, a serious human pathogen. Thus isolation of *M. marinum* from humans may select for more virulent strains. Like other PEM, person-to-person spread for *M. marinum* does not occur and all available evidence suggests that human infections arise from contact with aquatic reservoirs. The strength of epidemiological data is dependent on sampling and the fact that strains from diseased humans are much more likely to be cultured than are strains from diseased fish or frogs is a confounding factor in epidemiological studies. Whereas strain collections from human patients are likely to represent a significant proportion of human infections, culture collections obtained from fish are likely to represent a small proportion of strains present in fish.

PATHOGENESIS AND VIRULENCE

M. marinum disease in poikilothermic animals and humans differs in at least one important way; whereas *M. marinum* causes disseminated disease in poikilothermic animals (Ramakrishnan et al., 1997; van der Sar, et el., 2004, Decostere et al., 2004; Ruley et al., 2004), with rare exceptions, it causes localized, cutaneous infections in humans (Iredell et al., 1992; Gluckman, 1995; Zenone et al., 1999; Aubrey et al., 2002). Human infections in immunocompetent individuals are readily treated with antibiotics and even untreated lesions may resolve within many months. In some cases, extension of a cutaneous infection can lead to involvement of underlying tissue resulting in tenosynovitis, osteomyelitis, arthritis and bursitis. Disseminated *M. marinum* infection is a rare event and only occurs in the presence of severe immunosuppression. In these patients the organism can be isolated from blood, bone marrow and internal organs and the infection is often fatal (Tchornobay et al., 1992; Parent et al., 1995; Holmes et al., 1999). The histopathology of human and animal lesions is similar in that *M. marinum* forms characteristic granulomas in humans, fish and frogs, and the majority of organisms are found intracellularly within macrophages.

BACTERIAL PHYSIOLOGY

The most important aspect of *M. marinum* physiology from the standpoint of pathogenesis is the restricted growth temperature. Optimal growth occurs at 30-32°C and most isolates grow poorly if at all at 37°C. It is likely that this temperature restricted growth severely limits the ability of the organism to cause systemic disease. In the initial description of *M. marinum* (formerly described as *M. balnei*), it was reported that the ability of the bacteria to grow at 37°C was both strain and media dependent (Linell and Norden, 1954). Three strains, one from a human infection, one from a strand of human hair at an outlet pipe from a swimming pool, and one from the cement side of the swimming pool were investigated for temperature dependent survival. All strains were capable of survival at 45°C but not at 56°C. However, growth of *M. marinum* was media dependent. Whereas no growth occurred from any strain at 37°C on Lowenstein-Jensen, abundant growth occurred at 37°C on glycerol agar. Although systemic infection with *M. marinum* is rare, when it does occur organisms are found within liver, spleen, and bone marrow showing that temperature restricted growth is not due to an intrinsic inability of the bacterium to grow at body temperature. Further proof for this comes from studies on *M. marinum* isolates obtained from systemic disease (Tchornobay et al., 1992; Kishihara et al., 1993). Although histopathology suggested that enormous bacterial growth had occurred within infected organs, on laboratory media these isolates produced very scant growth at 37°C (Small, unpublished results).

Like many mycobacterial species, *M. marinum* undergoes an irreversible smooth to rough transition (Linell and Norden, 1954; Smith and Jiji, 1975). Early experiments in mice showed that both morphotypes were equally virulent and both morphotypes have been isolated from human patients. It is likely that rough and smooth colonial morphology can be produced by a number of genetic changes since it has been recently shown that an avirulent smooth mutant can be obtained from a virulent rough parental strain by an alteration of mycolic acid biosynthesis (Gao et al., 2003). *M. marinum* and *M. tuberculosis* contain similar classes of lipids including phenolic glycolipids, phthiocerol diesters and similar classes of mycolic acids. Genetic studies described below show that at least in some cases *M. marinum* and *M. tuberculosis* orthologs are functionally equivalent.

With respect to growth rate *M. marinum* represents an enigma. Taxonomically *M. marinum* falls clearly within the family of slow-growing mycobacteria and is closely related to *M. tuberculosis* and *M. ulcerans*. However, whereas the doubling time of *M. tuberculosis* and *M. ulcerans* is in excess of 20 hrs, *M. marinum* has a doubling time of 4 hrs. The difference in growth rate between *M. marinum* and *M. ulcerans* is especially surprising considering their close relatedness. Although early studies suggested that *M. marinum* might have two copies of the rRNA operon, recent work provides definitive evidence that *M. marinum* has a single rRNA operon (Helguera-Repetto et al., 2004). Helguera-Repetto et al. suggest that the slower growth rate of *M. tuberculosis* with respect to *M. marinum* is due to the presence of the larger number of IS elements in the slower growing mycobacteria. The same argument could be used to explain the extremely long doubling time (36-80 hrs) of *M. ulcerans* compared with *M. marinum*.

GENETIC ANALYSIS OF *M. MARINUM*

In the past 10 years, the genetic toolbox for use with *M. tuberculosis* has expanded enormously. Transposon mutagenesis, signature-tagged mutagenesis (STM), allelic exchange, *sacB* counter selection, DNA micro- array analysis, and a number of expression technologies have all been used to identify new virulence determinants in *M. tuberculosis*. The same tools can be used with *M. marinum*. The obvious advantage in working with *M. marinum* is that colonies appear on plates within 2-3 weeks rather than the 3-4 weeks required for *M. tuberculosis*. Genetic analysis of virulence in *M. marinum* has been used to identify promoters up-regulated inside macrophages, or *in vivo* within granulomas (Barker et al., 1998; Chan et al., 2002). STM has led to the identification of a number of genes required for survival of *M. marinum* in a goldfish infection model (Talaat et al., 1998; Ruley et al., 2004). These studies show a strong similarity between the host response to *M. tuberculosis* and *M. marinum* and reaffirm the likely importance of cell associated lipids, the RD1 region, and the PE/PPE family genes for virulence. Mutations in many *M. marinum* virulence genes have been successfully complemented with *M. tuberculosis* orthologues (Gao et al., 2004). Two *M. marinum* PE/PE-PGRS genes with close *M. tuberculosis* homologues were required for survival within the granuloma in a frog model of infection. Although these large gene families had long been hypothesized to play a role in the virulence of *M. tuberculosis*, this work with *M. marinum* was the first to prove the association (Ramakrishnan, 2000). The close taxonomic relationship between *M. marinum* and *M. tuberculosis* is demonstrated by the fact that DNA relatedness of *M. marinum* and *M. tuberculosis* orthologues is generally greater than 80% (Table 2).

M. MARINUM-HOST CELL INTERACTIONS

The hallmark of mycobacterial virulence is the ability of mycobacteria to evade phagosomal-lysosomal fusion and by doing so to replicate within naïve macrophages. *M. marinum*, like *M. tuberculosis* inhibits phagosomal-lysosomal fusion and replicates in an intracellular compartment which does not undergo acidification (Ramakrishnan and Falkow, 1994; Barker et al., 1998). Thus, *M. marinum* causes an arrest in phagosomal maturation (Vieira et al., 2004). A second cellular phenotype shared by *M. marinum* and *M. tuberculosis* is that both are cytotoxic for host cells (Hsu et al., 2003; Gao et al., 2004). However, the intracellular biology of *M. marinum* and *M. tuberculosis* differs in one potentially important aspect. In cell assays, *M. tuberculosis* replicates primarily within a membrane-bound compartment, although there is some evidence from electron microscopy that a small proportion of *M. tuberculosis* escape to the cytoplasm (McDonough et al., 1993). In contrast to this, *M. marinum* escapes from the phagosome where it forms actin tails and replicates in the cytoplasm, much like *Shigella* species (Stamm et al., 2003).

Several lines of evidence suggest that the ability of *M. marinum* to replicate within a delayed phagosome requires the presence of specific cell -associated lipids. A number of gene expression studies have shown that genes involved in the synthesis of these lipids are expressed by *M. marinum* within macrophages (Barker et al., 1998; Ramakarishnan et al., 2000; Chan et al., 2003; Ruley et al., 2004). Solid evidence for the requirement of specific cell- associated lipids in intracellular growth has been provided by studies of mutants lacking genes involved in lipid biosynthesis (Gao et al., 2003). Using transposon mutagenesis, Gao et al. isolated a mutant in *kasB* which is involved in mycolic acid biosynthesis in *M. tuberculosis* and *M. marinum*. Biochemical analysis of mycolic acids in the *kasB* mutant showed that the proximal portion of the meromycolate chain in this mutant was shortened by 2-4 carbons. This seemingly modest change in mycolic acid was accompanied by a change from a rough to smooth colonial morphotype, decreased cording, increased cell wall permeability and a defect in the ability to replicate within macrophages. KasB mutants exhibited heightened drug sensitivity presumably due to possession of a more permeable cell membrane. The *kasB* defect in *M. marinum* could be complemented by the *kasB* from *M. tuberculosis* demonstrating the shared functionality of these genes.

A second class of cell wall lipids, the phosphotidyinositol mannosides (PIMs) have also been recently shown to have a potential role in virulence in *M. marinum*. In *M. tuberculosis* PIMs are involved in the attachment of the bacteria to phagocytic cells, although they do not appear to play a role in intracellular replication. A transposon library of *M. marinum* mutants was constructed in a virulent rough strain and screened for alterations in colony morphology in an attempt to identify genes involved in the synthesis of cell wall lipids. One of the mutants obtained in this screen produced a translucent border around the cell. Biochemical evidence showed that the mutant contained an insertion in *pimF*, a gene encoding a mannosyltransferase required for the biosynthesis of PIMs. The cell envelope of this mutant had a decrease in LAM and LM supporting the hypothesis that PIMs in *M. marinum* as in *M. tuberculosis* are late precursors in the biosynthesis of mycolic acids (Alexander et al., 2004). The defect in PIM biosynthesis could be complemented in the *M. marinum* mutant by *pimF* from *M. tuberculosis* or *M. marinum*. Uptake of the *M. marinum pimF* mutant by J774 murine macrophages was decreased considerably compared with WT although intracellular replication was not affected. This phenotype is in agreement with results from similar studies conducted with *M. tuberculosis pimF* mutants and is thought to be secondary to a defect in LAM biosynthesis.

The RD1 locus in *M. marinum* as in *M. tuberculosis* plays a multifunctional role in cell biology and pathogenesis. The *M. tuberculosis* RD1 locus is proposed to encode a novel secretory pathway required for the secretion of the small protein antigens ESAT6 and CFP10. Insertion in any loci within this region ablates the secretion of these proteins in *M. tuberculosis*. The role of the RD1 locus in the pathogenesis of *M. tuberculosis* is well established (Hsu et al., 2003; Brodin et al., 2004; Guinn et al., 2004) although the particular mechanism underlying attenuation of these mutants is not well understood. *M. tuberculosis* mutants lacking ESAT 6 replicate within macrophages, but are not cytolytic and therefore cannot escape from a cell and spread to adjacent cells in a monolayer. *M. tuberculosis* lacking ESAT6 shows reduced replication and spread within the mouse lung as well as a defect in the ability to spread to other organs. *M. marinum* contains a full repertoire of ESAT-6 family genes as well as an intact RD1 region. The ESAT6 and CFP10 genes of *M. tuberculosis* and *M. marinum* are highly conserved (Table 2) and are functionally equivalent in *M. marinum*. When a transposon library of *M. marinum* mutants was screened for inability to lyse sheep red blood cells, eight mutants were obtained all of which had insertions in the *M. marinum* RD1 (RV-3871-3881c) region (Gao et al., 2004). Mutations in the *M. marinum* homologues of Rv3866-3868, Rv3876, Rv 3878, Rv 3879c and Rv3881c result in a plethora of virulence-associated phenotypes: 1) loss of haemolytic activity 2) loss of the ability to lyse macrophages and epithelial cells following uptake 3) loss of cell-contact cytolysis, 4) inhibition of the ability to spread within a tissue culture monolayer, and 5) decreased virulence to zebra fish. As with insertions in homologous *M. tuberculosis* genes, secretion of ESAT-6 and CFP10 was lacking in *M. marinum* RD1 mutants. The RD1 associated phenotypes in *M. marinum* mutants were complemented by introduction of the homologous genes from either *M. tuberculosis* or *M. marinum* providing further validation of the *M. marinum* model for studying the pathogenesis of *M. tuberculosis*.

VIRULENCE IN ANIMALS

Exciting work *in vivo* with *M. marinum* using animal models has contributed important information to our understanding of persistence and recurrence of mycobacterial infection. A number of infection models have been used to explore the interaction of *M. marinum* with the host. These include *Caenorhabditis elegans*, (Couillault et al., 2002), *Dictyostelium discoideum*, (Solomon et al., 2003) and *Drosophila* (Dionne et al., 2003;). However, the unique advantage offered by *M. marinum* is the ability to investigate the virulence of a mycobacterial pathogen in frogs and fish, its natural hosts. The signature pathology of mycobacterial infections is the formation of the granuloma along with the ability of mycobacteria to remain dormant within the granuloma for many years. Because of the difficulty of obtaining access to mycobacteria within human granulomas, there are few studies on specific gene expression and identification of genes required for intra-granuloma survival. Seminal studies with *M. marinum* have led the way for understanding the biology and physiology of the granuloma. Using leopard frogs (*Rana pipiens*) and more recently zebra fish (*Danio rerio*) it has been possible to look at the expression of genes within the granuloma (Ramakrishnan et al., 2000; Bouley et al., 2001; Chan et al., 2002). This work has led to the identification of mycobacterial genes required for survival within the granuloma, as well as provided insight into the nature of the granuloma. Proteins from the glycine rich PE-PGRS family in *M. tuberculosis* had long been promoted as potential virulence genes. Work from the Ramakrishnan laboratory using the frog model showed that specific PE-PGRS genes were expressed within the granuloma, and that loss of these genes results in loss of the ability to replicate within macrophages (Ramakrishnan et al., 2000). The expression of *M. marinum* promoters was studied at eight and 77 weeks post infection as well as in culture medium and macrophages (Chan et al., 2002). Three patterns of promoter expression were identified in these studies: 1) promoters expressed in broth, macrophages and granulomas 2) promoters expressed in macrophages and granulomas and 3) promoters expressed only in the granuloma. A complex set of promoters expressed within granulomas included those for PE, PPE family genes, genes involved in lipid biosynthesis, genes involved in stress or starvation pathways (*recC, fprA, dcd* and *ald*) and regulatory genes (*pknB*). Attempts to produce a granuloma -like expression pattern *in vitro* by manipulation of the growth environment was only partially successful. Whereas expression of three granuloma expressed promoters was induced at low pH *in vitro*, and one was expressed by low magnesium, none of the granuloma expressed promoters were induced *in vitro* during stationary phase or hypoxic conditions (Chan et al., 2002). Several important results issued from these studies. As predicted, a specific set of genes are expressed differentially in the granuloma and this set only marginally overlaps with that of macrophage expressed genes. An unexpected finding which runs contrary to accepted dogma showed that most granuloma expressed promoters are also expressed during logarithmic stage suggesting that the majority of bacteria in a granuloma are not in a dormant state.

The zebra fish model has been especially valuable for dissection of mycobacterial virulence, partially because the transparency of the fish allows visual observation of infection, but also because the zebra fish genome has been sequenced (Davis et al., 2002; Cosma et al., 2004; Gao et al., 2004). The toll-like receptors TLR1 and TLR2 are present in zebra fish (Meijer et al., 2004) and these are known to interact with mycobacterial pathogens in humans. The *M. marinum*-zebra fish model has provided the first dynamic studies of the mycobacterial granuloma showing that the granuloma, at least in fish, is not the isolated, static compartment it was once held to be. In re-infection studies using differentially labeled fluorescent *M. marinum*, it has been demonstrated that mycobacteria are trafficked to pre-existing granulomas via host cells, and that the cellular constituents of the granuloma represent a dynamic compartment both with respect to host cells and mycobacteria (Cosma et al., 2004). Interestingly, trafficking of bacteria to the granuloma shows species specificity since *M. marinum*, but not an enteric bacterium *Salmonella arizonae,* were trafficked to previously established *M. marinum* granulomas. This may have relevance to human infections since pulmonary disease with PEM has been shown to occur often in the context of prior pulmonary *M. tuberculosis* where old tuberculosis scars are infected with PEM.

Mature granulomas in fish, as in humans, are often characterized by the presence of a necrotic, caseous center. The caseous center of these granulomas has been thought be an isolated impenetrable compartment. However, studies show that *M. marinum* can penetrate the caseum of previously infected granulomas demonstrating that even this compartment is not isolated. These findings have stimulated a great deal of speculation concerning the differential advantage of the granuloma to the bacterium versus the host (Flynn, 2004).

In summary, *M. marinum* has emerged as a powerful model for exploring the pathogenesis of *M. tuberculosis*. *M. tuberculosis* and *M. marinum* appear to reside in identical compartments at least within the macrophage. *M. marinum* contains orthologs of the major virulence determinants in *M. tuberculosis*, and many, if not all features of the granulomatous response to *M. tuberculosis* and *M. marinum* are shared. It is likely investigations into the virulence of *M. marinum* will continue to yield important new insight into the virulence of *M. tuberculosis*. Nonetheless, there are limitations to any model system. The genome of *M. marinum* is at least 2 Mbp larger than that of *M. tuberculosis* and it is likely that some of this additional genetic material plays a role in *M. marinum* pathogenesis. Although a strength of the *M.*

marinum system is the ability to study a mycobacterial pathogen in its natural host, there are many immunological differences between fish and humans. *M. tuberculosis* can infect many sites in the body but the vast majority of disease is pulmonary and without the propensity to colonize the lung *M. tuberculosis* would be a much less significant pathogen. *M. marinum*, even in the severely immunocompromised host does not cause pulmonary disease. Although restricted growth temperature almost certainly explains the tissue tropism of *M. marinum*, it is unlikely to be the whole story. Even in situations where *M. marinum* is able to replicate in deep organs, the organism shows no particular predilection for replication in the lungs. Although a great deal of effort is being expended to show the equivalency of "fish tuberculosis" with human tuberculosis, investigations into differences between *M. marinum* and *M. tuberculosis* and the diseases they cause could also lead to better understanding of the determinants that make *M. tuberculosis* and its disease unique.

MYCOBACTERIUM ULCERANS

M. ulcerans is the causative agent of Buruli ulcer, an emerging infectious disease which is important in rural areas primarily in the tropics. Although the organism has been found globally, the majority of Buruli ulcer cases occur in West Africa and Australia (Hayman, 1985; van der Werf et al., 1999; Bafende et al., 2004). Following *M. tuberculosis* and *M. leprae*, *M. ulcerans* is the most prevalent and most virulent mycobacterial pathogen. In contrast to most PEM, *M. ulcerans* is almost exclusively associated with disease in the immunocompetent host. *M. ulcerans* represents a further anomaly among mycobacterial pathogens in the following respects: 1) It is an extracellular pathogen 2) it produces a family of immunosuppressive, toxic polyketide- derived macrolides required for virulence 3) virulence in *M. ulcerans* is plasmid-encoded and 4) *M. ulcerans* cannot be isolated from natural water sources in endemic regions despite the fact that exposure to slow moving water represents the only documented risk factor for disease.

TAXONOMY, EPIDEMIOLOGY AND ECOLOGY

M. ulcerans is a slow growing mycobacterial pathogen which is extremely closely related to *M. marinum* (Tonjum et al., 1998). Taxonomic studies based on 16s rRNA show that *M. ulcerans* and *M. marinum* differ by only a single base-pair. However, a number of molecular methods used to investigate the relationship between *M. marinum* and *M. ulcerans* including MLST, and lipid analysis (Portaels et al., 1996; Tonjum et al., 1998; Stinear et al., 2000; Chemlal et al., 2001) clearly distinguish the two species. The two species can be readily distinguished by two molecular criteria: *M. ulcerans* contains in total over 300 copies of the IS elements IS2404 and IS2606 (Stinear et al., 1999; Stinear et al., 2000; Chemlal et al., 2001) which are absent from *M. marinum*; *M. ulcerans* contains a large plasmid which encodes a toxic, immunosuppressive macrolide, mycolactone (George, et al; 1999; Stinear et al., 2004). The growing body of taxonomic data on these strains, along with the discovery of another related aquatic pathogen present in West African Clawed frogs (Trott et al., 2004) suggests that there is an *M. marinum* complex from which *M. ulcerans* diverged fairly recently (Stinear et al., 2000). The genome sequences of *M. ulcerans* (http://genopole.pasteur.fr/Mulc/BuruList) and *M. marinum* (http://www.sanger.ac.uk/Projects/M_marinum/) are available although genome assembly is incomplete. Although *M. marinum* and *M. ulcerans* genes involved in basic metabolism generally show 100% DNA identity (Stinear et al., 2000), the genome sizes of these species are estimated to be quite different. The current estimate of genome size for *M. marinum* are 6.7 Mbp whereas the *M. ulcerans* genome size is estimated as 4.6 Mbp. Evidence from DNA hybridization reveals much less heterogeneity among *M. ulcerans* species than among *M. marinum* species. Despite the fact that *M. ulcerans* strains show relatively little intra-species heterogeneity and the fact that all *M. ulcerans* cause similar human disease, there is geographical heterogeneity between strains. Analysis of the numbers and locations of IS*2404* and IS*2606* elements, variability within the upstream region of the 16s rRNA gene, and analysis of mycolactone toxins are all criteria which can be used to identify the source of an *M. ulcerans* isolate as African and Papua New Guinean, Asian, South American, Mexican, or Australian (Chemlal et al., 2001; Mve-Obiang et al., 2003).

The epidemiology of Buruli ulcer has been a subject of intense speculation since the first descriptions of the disease (Conner and Lunn, 1965). Although the first cases described were from Uganda, (Conner and Lunn, 1966) the organism was first isolated from a human ulcer in Australia (MacCallum and Tolhurst, 1948). Since then numerous studies have shown that the only strong risk factor for infection is proximity to slow moving water (Hayman, 1985; Marston et al., 1995; van der Terf et al., 1999; Stienstra et al., 2001; Guarner et al., 2003; Duker et al., 2004). Despite intense efforts both in Africa and Australia to culture organisms from water in endemic areas, not a single isolate of *M. ulcerans* has been obtained. The fact that Schistosomiasis and Buruli ulcer are both associated with exposure to slow moving water led to several studies to determine whether there was a correlation between infection with *M. ulcerans* and schistosomiasis. The results of these studies failed to identify an association between the two diseases (Stienstra et al., 2004). *M. ulcerans* has also been isolated from infected animals in Australia (Mitchell et al., 1987).

Preliminary evidence that IS*2404* was specific to *M. ulcerans* led to a number of studies in which IS*2404* was used in PCR analysis of environmental isolates to detect *M. ulcerans* (Stinear et al., 2000; Chemlal et al., 2001; Stienstra et al., 2003). These studies have produced hundreds of IS*2404* positive samples from such diverse aquatic organisms as water insects (particularly *Belostomatidae* and *Naucoridae*), small fish, tadpoles, snails, and other aquatic sources in Africa (Portaels, et al; 1999; Portales et al., 2001; Marsollier et al., 2002; Eddyani et al., 2004; Kotlowski et al., 2004). The yield of IS*2404* positive environmental samples is much higher in sites where Buruli ulcer is present than similar environments where the disease is absent. Within these environments, predatory aquatic insects, and insect-feeding fish have yielded the greatest number of IS*2404* positive samples. Extensive efforts to identify *M. ulcerans* in similar samples from Buruli ulcer endemic regions in Australia have met with very limited success and no positive insects have been found. Despite the large number of IS*2404* PCR positive samples from West Africa, only one *M. ulcerans* culture has been reported from the hundreds of IS*2404* positive samples analyzed. This isolate was obtained from a water insect (Naucoridae) in Cote d'Ivoire (Marsollier et al., 2002). In a laboratory infection model using *Naucoridae cimicoides* insects collected in France, *M. ulcerans* was easily cultured suggesting that the difficulty in culturing *M. ulcerans* from environmental samples is not due to the fact organisms are present in insects in a viable but non-culturable state. Recent evidence shows that the presence of IS*2404* is not as specific for *M. ulcerans* as previously thought (Chemlal et al., 2002; Trott et al., 2004) which may explain the inability to culture *M. ulcerans* from IS*2404* positive samples. The presence of IS*2404* in the newly discovered frog pathogen, *Mycobacterium liflandii* is particularly interesting since since *M. liflandii* does not grow in primary isolation on the mycobacterial media used to culture *M. ulcerans* (Trott et al., 2004). Further, the natural host of *M. liflandii* is *Xenopus tropicalis*, a species of West African Clawed frog that is wide-spread in Buruli ulcer endemic regions.

Nonetheless, there is provocative laboratory evidence that aquatic insects may be reservoirs for *M. ulcerans* infection. Naucoridae can be experimentally infected with *M. ulcerans* (Marsollier et al., 2002). In these studies the bacterium are found primarily in the salivary gland. Further, the only environmental isolate of *M. ulcerans* was obtained from similar insect species in West Africa (Marsoiller et al.,2002). Despite these intriguing findings, there is no epidemiological connection between insect bites and the occurrence of Buruli ulcer. This could be because of the long incubation of the disease, or because insect bites are not memorable in the tropics.

M. ulcerans like other mycobacteria are capable of forming biofilms, and it has recently been shown in the laboratory that they are capable of forming biofilms on aquatic vegetation. (Recht and Kolter, 2001; Marsoiller et al., 2004). However, the organisms have rarely been detected by PCR from plant materials, and they have never been cultured from this source despite efforts to do so . Though evidence strongly points to an environmental reservoir for *M. ulcerans* the precise location of *M. ulcerans* in the environment as well as the method of transmission remain elusive.

BACTERIAL PHYSIOLOGY

M. ulcerans shares a low optimal growth temperature of 30-34°C with *M. marinum*. However, *M. ulcerans* is much less able to survive above 35°C (Linell and Norden, 1954; Mve-Obiang et al., 1999) and *in vitro M. ulcerans* is killed by 48 h incubation at 37°C. Consistent with this, *M. ulcerans* has never been detected or isolated from blood or deep organs of infected people even in the presence of severe immunosuppression (Delaporte et al., 1994; Johnson et al., 2002; Ouattara et al., 2002;) Some isolates of *M. ulcerans* grow preferentially *in vitro* under microaerophilic conditions (Mve-Obiang et al.,1999). *M. ulcerans* doubling time is 36-80 hrs *in vitro*, compared with a doubling time of 4 hours for *M. marinum*. Although the lipid repertoire of *M. ulcerans* and *M. marinum* differs in some important respects, there is no evidence that the slower growth rate of *M. ulcerans* is related to differences in cell wall lipids. It is possible, however, that the slow replication of *M. ulcerans* may be due to the very large number of insertions sequences present.

PATHOGENESIS AND VIRULENCE

M. ulcerans causes a severe, persistent skin ulcer (Connor and Lunn, 1966; Hayman, 1985; van der Werf et al., 1999). In the majority of cases a single ulcer is present,

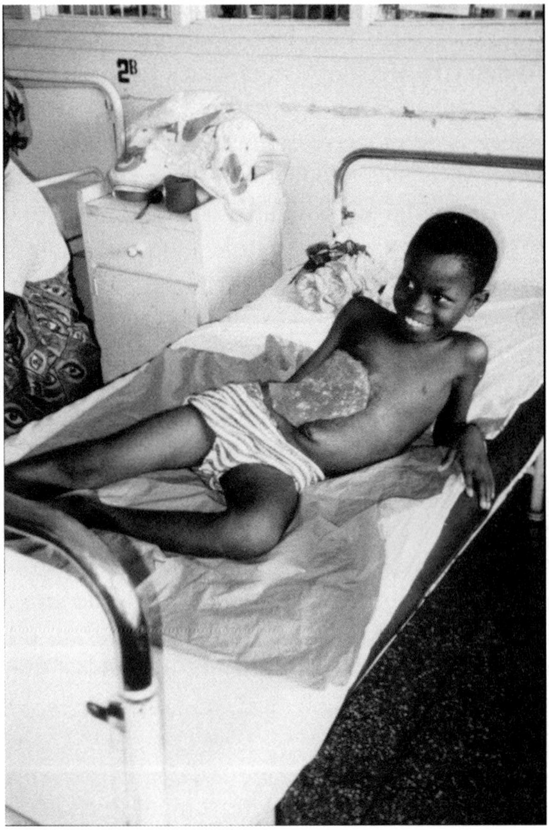

Fig. 1. Lesion caused by *M. ulcerans*.

though multiple lesions are not infrequent (Fig. 1). Single ulcers can be extremely large and may extend to 30% of a person's total skin surface (Debacker et al., 2004). Although the incubation time for Buruli ulcer is unknown, large lesions are assumed to be of many months or even years duration. Conclusive evidence for systemic disease is lacking and positive *M. ulcerans* cultures have never been obtained from blood or deep tissue of infected persons despite the fact the organism is readily cultured from infected skin.

M. ulcerans infection can present in several different forms: as a nodule, papule, or plaque (Hayman, 1985; van der Werf et al., 1999; Weir, 2002). As tissue is destroyed, the lesion progresses into a large ulcer, or a large area of edema under which lies necrotic tissue. In some cases local dissemination can be extensive resulting in osteomyelitis (Abalos et al., 2000; Debacker et al., 2004). Over 65% of Buruli ulcers occur on the arms or legs. As a result autoamputation is not infrequent (Asiedu et al.,1998; Teelken et al., 2003). Although *M. ulcerans* shares with *M. leprae* a tropism for skin, the distribution of the skin lesions is quite different. Whereas *M. leprae* lesions occur preferentially in cooler areas of skin such as the nose, fingers, or toes, these are extremely uncommon sites for Buruli ulcer. Fatalities due to Buruli ulcer either do not occur or are extremely rare although morbidity with the disease is considerable (Asiedu, 1998; van der Werf et al., 1999).

A remarkable aspect of Buruli ulcer is that, despite the considerable depth and extent of many Buruli ulcers, the lesions are painless. If the size of the ulcer is remarkable, the histopathology is even more so. Within Buruli ulcer lie discrete infection foci containing huge numbers of extracellular bacteria surrounded by an area of acellular necrosis (Fig. 2). Coagulation necrosis, resulting from destruction of adipose tissue is a prominent feature of these lesions (MacCallum and Tolhurst, 1948; Connor and Lunn, 1966; Hayman, 1985). The presence of necrosis extending far out from the foci of acid fast bacilli led to early suggestions that Buruli ulcer was a toxin-mediated disease (Connor and Lunn, 1965). Although there are areas in lesions, especially where healing is evident, where inflammatory cells are found, these areas are devoid of *M. ulcerans* organisms (Guarner et al., 2003). Instead, organisms are located focally within lesions where they appear to provide a barrier to the inflammatory response (Hayman, 1985; van der Werf et al., 1999). Scrutiny of thousands of sections from *M. ulcerans* infected tissue have failed to reveal a single acid-fast bacilli within an intact inflammatory cell. Not only is there local immunosuppression in Buruli ulcer, there is also evidence of systemic immunosuppression (Gooding et al., 2001; Gooding et al., 2002; Gooding et al., 2003; Okenu et al., 2004; Prevet et al., 2004). T-cells from Buruli ulcer patients fail to secrete IFN-gamma after stimulation with *M. ulcerans*, and have a specific anergy to other mycobacterial antigens such as PPD from *M. tuberculosis* (Gooding et al., 2001). However, there is no evidence that Buruli ulcer renders patients more susceptible to other infectious diseases.

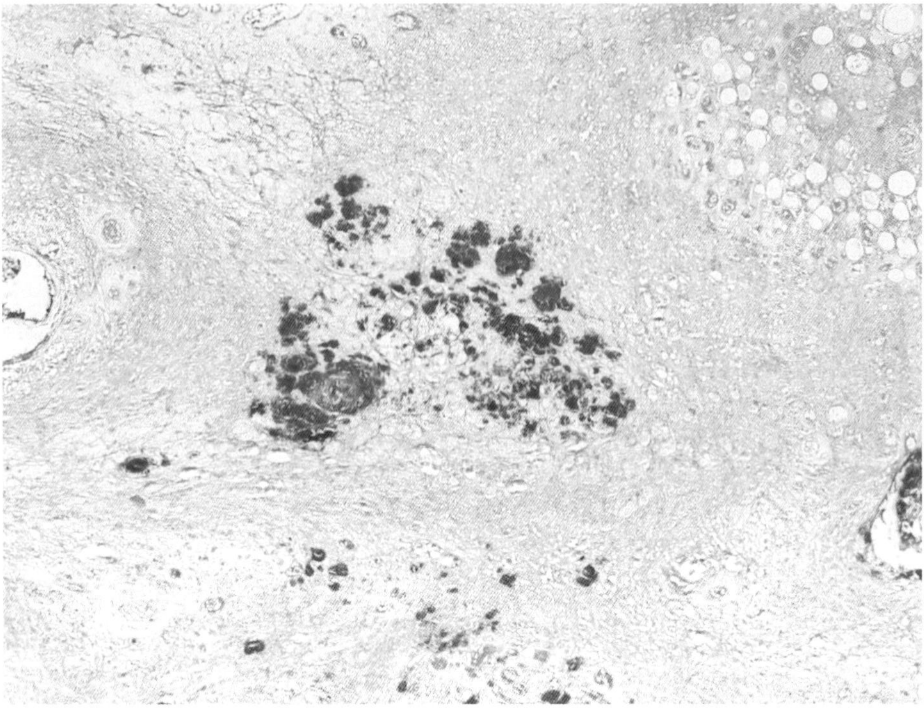

Fig. 2. Acellular necrosis in Buruli ulcer.

The unusual pathology of *M. ulcerans* infection is due to a toxic macrolide, mycolactone which, when injected into guinea pig skin, produces local immunosuppression, necrosis and apoptosis along with a characteristic Buruli ulcer (Kreig et al., 1974; George et al., 1999; van der Werf et al., 2003; Cope et al., 2004). Elucidation of the structure and complete organic synthesis of mycolactone A and B, the major macrolide species produced by African *M. ulcerans* isolates (Gunawardana et al., 1999; Fidanze et al., 2001; Song et al., 2002) show that mycolactone is a hybrid molecule composed of two polyketide chains. One chain cyclizes to form a 12-membered lactone ring (Fig. 3) and the second chain comprises a fatty acid side chain. Many macrolides, such as erythromycin, FK506, and rapamycin have immunosuppressive activities (Katz and Donodio, 1995). However, until the discovery of mycolactone in *M. ulcerans*, the biosynthesis of macrolides by prokaryotes had been restricted to a relatively few genera of non-pathogenic *Actinomycetales* such as *Streptomyces*, and *Saccharopolyspora* species.

There is considerable heterogeneity among mycolactone molecules. First, each strain of *M. ulcerans* produces several mycolactone variants or congeners due to imprecise processing by the polyketide synthase enzymes and secondly, *M. ulcerans* strains from different geographic regions produce geographically distinct mycolactones (Cadapan et al., 2001; Hong et al., 2003, Mve-Obiang et al., 2003;

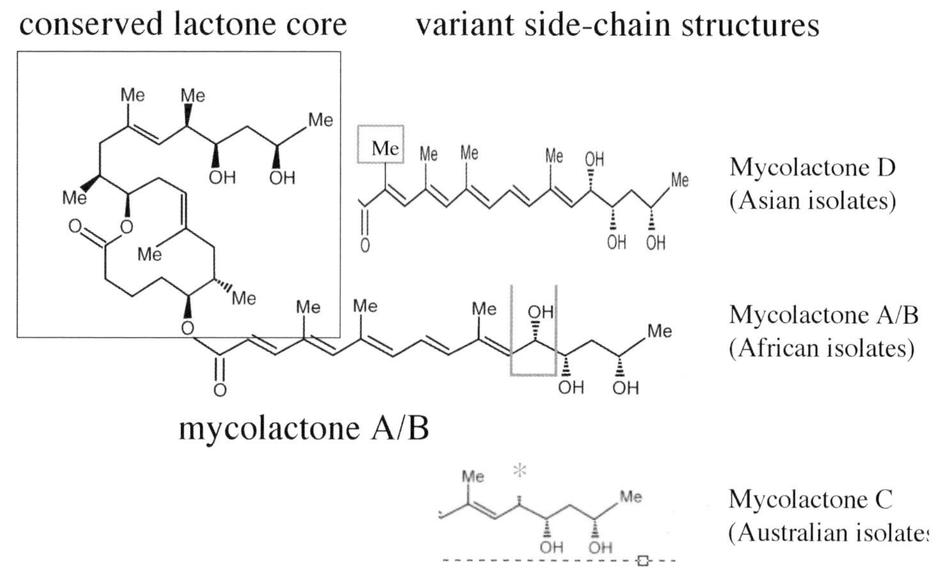

Fig. 3. Geographical variation in mycolactones.

Judd et al., 2004; Hong et al., 2004) although the lactone core structure is conserved in all mycolactones (Fig. 3). Considerable heterogeneity resides in the fatty acid side chain most of which is due to the number of hydroxyls present at C 12 on the fatty acid side chain.

Macrophages, fibroblasts, endothelial cells and lymphocytes are sensitive to mycolactone whereas epithelial cells are relatively resistant (George et al., 1999; Pahlevan et al., 1999, Dobos et al., 2001; Snyder and Small, 2003). In human tissue, the lack of toxicity for epithelial cells is also evident. Undermined edges and epithelial hyperplasia is a characteristic feature of Buruli ulcer (Hayman, (1996). Mycolactone produces cell rounding (6-10 hr), cell cycle arrest (36 hr) and apoptosis (48-72 hr) in cultured fibroblasts and macrophage cell lines at picogram amounts (George et al., 1999; Dobos et al., 2001; Snyder and Small et al., 2003).

As with other lipid immunosuppressants, mycolactone enters the cell by diffusion and is sequestered in the cytosol. An increase in caspases 2,3 and 6 can be detected by three days in mycolactone treated fibroblasts whereas little activity against substrates for caspases 4,8,9, granzyme and calpain can be detected (Snyder and Small, 2003). Although mycolactone produces a calcium flux following cellular uptake, chelation of intracellular calcium does not prevent this flux or inhibit apoptosis. These results suggest that the mechanism of mycolactone activity is different than that of other macrolide immunosuppressants such as FK506 or rapamycin. Recent studies show that the mycolactone core which is present as a precursor on the *M. ulcerans* cell surface causes necrosis as well as apoptosis

(Small, unpublished results). Many attempts have been made to use model systems to identify the cellular target for mycolactone. Whereas *C. elegans, Drosophila* and *Dictyostelium discoideum* have been useful for studying *M. marinum* virulence, *M. ulcerans* is not virulent in these systems (Small, unpublished data). Further, mycolactone lacks toxicity for *Dictyostelium*, *Drosophilia*, Gram negative or Gram positive bacteria, *Candida albicans*, *Sacchromyces cerevisiae*, and *Cryptococcus neoformans* (Small, unpublished data).

While the cell biology of mycolactone activity has been elusive, exciting progress has been made in deciphering the mycolactone biosynthetic pathway. Transposon mutagenesis along with DNA sequencing identified a 110 kb cluster of mycolactone biosynthesis genes on a large (150-180 kb) plasmid in *M. ulcerans* (Stinear et al., 2004). Plasmids are conspicuously absent from *M. tuberculosis, M. bovis* and *M. leprae*. Although the origin of the mycolactone plasmid is unknown, the plasmid *repA* gene is homologous (68% amino acid similarity) to *repA* in a cryptic plasmid, pJAZ38, in *M. fortuitum*. The discovery of the *M. ulcerans* plasmid is the first example of plasmid-mediated virulence in a mycobacterial species. The mycolactone gene cluster encodes three Type I modular polyketide synthases, two of which *(mlsA1, mlsA2)* encode the lactone core molecule, and a third *(mlsB)* which encodes the fatty acid side chain. The *mlsA1* gene is the largest single bacterial gene so far identified. Three accessory or tailoring genes are also present. These include genes encoding a P450 mono-oxygenase (*mlgC*), a FabH like ketosynthase, thought to be involved in joining the two polyketide chains, and a small thioesterase gene hypothesized to play a role in release of the growing polyketide chain.

Mycolactone is a yellow lipid that is present on the surface of the cell as well as being secreted. Initial identification of the mycolactone genes was identified by screening transposon insertion mutants for loss of yellow pigment. Transposon insertions in *mlsA1* ablated mycolactone production, where as insertions in *mlsB* resulted in production of the lactone core only. The extensive duplication between modules encoding the polyketide synthase genes is unprecedented and shows that these genes have evolved through multiple gene duplication and recombination events. The remaining portion of the mycolactone plasmid is comprised of large numbers of IS elements including many copies of IS*2606* and IS*2404*.

Analysis of mycolactone clusters from geographically diverse isolates reveals the genetic basis of some of the mycolactone heterogeneity. Australian isolates lack both the P450 hydroxylase and a hydroxyl at C12 on the fatty acid side chain of mycolactone C (Mve-Obiang et al., 2003; Judd et al., 2004). Despite the remarkably high redundancy within the mycolactone gene cluster which provides much opportunity for recombination, the mycolactone gene cluster appears to be remarkably stable within a particular geographic area. Strains collected over a 40 years period in Africa produce identical mycolactone congeners (Mve-Obiang et al., 2003). The fact that human infection is clearly incidental to the life of *M. ulcerans* suggests that there is not only strong natural selection for mycolactone production, but also that there may selection for production of a particular variant of mycolactone within a geographic area.

Aside from mycolactone, little is known about virulence determinants in *M. ulcerans*. *M. ulcerans* contains genes for phospholipase C and D which share DNA homology with the corresponding *M. tuberculosis* and *M. marinum* genes (Johanson et al., 1996; Gomez et al., 2000). Phospholipase activity has also been demonstrated in *M. ulcerans*. It is unlikely, however, that these genes play a role in cytotoxicity because a single transposon insertion in the mycolactone gene cluster totally ablates cytotoxicity in *M. ulcerans* (Stinear et al., 2004).

Whereas the virulence of *M. ulcerans* has clearly been shaped by horizontal transfer of plasmid DNA, it is likely that virulence has also been shaped by deletion. The *M. ulcerans* genome is about 2 Mb smaller than that of *M. marinum*. It is plausible that this reduction in genome size occurred as *M. ulcerans* adapted to life in a specific reservoir species. Of particular interest is the fact that the most virulent isolates of *M. ulcerans*, those from Africa and Australia, contain a partial deletion in RD1 and lack genes for ESAT6 and CFP10. Strains of *M. ulcerans* which contain an intact RD1 lesion provoke a greater immune response and produce less serious disease than do strains in which they are missing. The acquisition of plasmid-mediated cytotoxicity by an *M. marinum* like ancestor may have obviated the requirement for ESAT6-mediated cytotoxicity resulting in a unique extracellular virulence profile for *M. ulcerans*. Recent investigations on conjugation in mycobacteria show that ESAT6 and CFP10 may also play a role in mycobacterial conjugation (Flint et al., 2004). These results indicate that the deletion of ESAT6 and CFP10 genes from *M. ulcerans* may affect the plasmid content of these strains.

Though research on *M. ulcerans* disease has produced a number of fascinating findings regarding mycobacterial virulence, there are a number of intriguing questions remaining. Is the mycolactone plasmid transmissible and if so under what conditions? What is the cellular target of mycolactone? What is the mechanism of mycolactone-mediated immunosuppression? What is the role of mycolactone in the natural biology of *M. ulcerans*? Where is *M. ulcerans* maintained in the environment and how is it transmitted to humans? This last seemingly simple question is actually the most difficult to answer. Nonetheless, research in this area is likely to have the highest impact on patients suffering from Buruli ulcer.

CONCLUSIONS

Of the nearly one hundred species of mycobacteria identified, the vast majority of these are environmental mycobacteria which lack virulence for humans. Of the 30 or more environmental mycobacteria isolated as human pathogens, less than half of these are frequent causes of disease in immunocompetent hosts. The most important pathogens are *M. abscessans* (cystic fibrosis patients), *M. avium* (particular in immunosuppressed HIV patients), *M. cheloni, M. fortuitum, M. kansasii* and *M. ulcerans*.

M. *kansasii* and *M. ulcerans* are the most common PEM infections in immunocompetent hosts. Although *M. marinum* is not as widespread or serious pathogen as those listed above, its close taxonomic relationship with *M. tuberculosis* as well as its rapid doubling time have made it an important model for studying mycobacterial pathogenesis.

What can be learned from the study of pathogenic environmental mycobacteria?

It is likely that the ability of *M. tuberculosis* to infect and replicate within phagocytic cells evolved through the interaction of environmental mycobacteria with environmental hosts. An understanding of these relationships could provide insight into the pathogenesis of *M. tuberculosis*. Most PEM also contain species -specific virulence traits which can be useful for dissecting the requirements for intracellular survival. The rapid growth rate and availability of natural hosts for investigating the virulence of *M. marinum* has already provided exciting information about the nature of the granuloma which is likely to be relevant for the pathogenesis of *M. tuberculosis*. Further investigations into the virulence of *M. marinum* are likely to continue to provide insight into the virulence of *M. tuberculosis* even if the frog and human granulomas differ in some respects. Some PEM, such as *M. ulcerans*, are important pathogens in their own right. In addition to this, the unique immunosuppressive effects of the mycolactones may lead to insight into the virulence of polyketides in *M. tuberculosis* as well as provide information regarding the role of RD1 genes in virulence. The study of *M. ulcerans* may lead to better understanding of the roles and limitations of plasmids in mycobacterial virulence. Finally the study of species-specific mycobacterial virulence determinants is a fascinating endeavor in itself which may contribute broadly to an understanding of microbial pathogenesis.

WEB RESOURCES

http://www.sanger.ac.uk/Projects/M_marinum/
http://genopole.pasteur.fr/Mulc/BuruList
http://pathogenomics.ahc.umn.edu/map_index.htm
http://genolist.pasteur.fr/TubercuList/
http://www.who.int/gtb-buruli/

REFERENCES

Abalos, F.M., Aguiar, J. Sr, Guedenon, A., Portaels, F., and Meyers, W.M. (2000). *Mycobacterium ulcerans* infection (Buruli ulcer): a case report of the disseminated nonulcerative form. Ann. Diagn. Pathol. *4*, 386-90.

Alcaide, F., Richter, I., Bernasconi, C., Springer, B., Hagenau, C., Schulze-Robbecke, R., Tortoli, E., Martin, R., Bottger, E.C., and Telenti, A. (1997). Heterogeneity and clonality among isolates of *Mycobacterium kansasii*: implications for epidemiological and pathogenicity studies. J. Clin. Microbiol. *35*, 1959-1964.

Alexander, D.C., Jones, J.R., Tan, T., Chen, J.M., and Liu, J. (2004). PimF, a mannosyltransferase of mycobacteria, is involved in the biosynthesis of phosphatidylinositol mannosides and lipoarabinomannan. J. Biol. Chem. *279*, 18824-18833.

Arend, S.M., van Meijgaarden, K.E., de Boer, K., de Palou, E.C., van Soolingen, D., Ottenhoff, T.H., and van Dissel, J.T. (2002). Tuberculin skin testing and *in vitro* T cell responses to ESAT-6 and culture filtrate protein 10 after infection with *Mycobacterium marinum* or *M. kansasii*. Infect. Dis. *186*, 1797-1807.

Asiedu, K., and Etuaful, S. (1998). Socioeconomic implications of Buruli ulcer in Ghana: a three-year review. Am. J. Trop. Med. Hyg. *59*, 1015-22

Ashford, D.A., Whitney, E., Raghunathan, P., Cosivi, O. (2001) Epidemiology of selected mycobacteria tht infect humans and other animals. Rev. Sci. Tech. 20, 325-37.

Aubry, A., Chosidow, O., Caumes, E., Robert, J., and Cambau, E. (2002). Sixty-three cases of *Mycobacterium marinum* infection: clinical features, treatment, and antibiotic susceptibility of causative isolates. Arch. Intern. Med. *162*, 1746-1752.

Bafende, A.E., Phanzu, M.D., and Imposo, B.B. (2004). Buruli ulcer in the Democratic Republic of Congo: epidemiology, presentation and outcome. Trop. Doct. *34*, 82-84.

Barker, L.P., Brooks, D.M., and Small, P.L. (1998). The identification of *Mycobacterium marinum* genes differentially expressed in macrophage phagosomes using promoter fusions to green fluorescent protein. Mol. Microbiol. *29*, 1167-1177.

Bartralot, R., Pujol, R.M., Garcia-Patos, V., Sitjas, D., Martin-Casabona, N., Coll, P., Alomar, A., and Castells, A. (2000). Cutaneous infections due to nontuberculous mycobacteria: histopathological review of 28 cases. Comparative study between lesions observed in immunosuppressed patients and normal hosts. J. Cutan. Pathol. *27*, 124-129.

Belisle, J.T., Brennan, P.J. (1989). Chemical basis of rough and smooth variation in mycobacteria. J. Bacteriol. *171*, 3465–3470.

Belisle, J.T., Klaczkiewicz , K., Brennan, P.J., Jacobs, W.R. Jr., and Inamine, J.M. (1993). Rough morphological variants of *Mycobacterium avium*. Characterization of genomic deletions resulting in the loss of glycopeptidolipid expression. J. Biol. Chem. *268*, 10517-23.

Bloch, K.C., Zwerling, L., Pletcher, M.J., Hahn, J.A., Gerberding, J.L., Ostroff, S.M., Vugia, D.J., and Reingold, A.L. (1998). Incidence and clinical implications of isolation of *Mycobacterium kansasii*: results of a 5-year, population-based study. Ann. Intern. Med. *129*, 698-704.

Bouley, D.M., Ghori, N., Mercer, K.L., Falkow, S., and Ramakrishnan, L. (2001). Dynamic nature of host-pathogen interactions in *Mycobacterium marinum* granulomas. Infect. Immun. *69*, 7820-7831.

Brodin, P., Rosenkrands, I., Andersen, P., Cole, S.T., and Brosch, R. (2004). ESAT-6 proteins: protective antigens and virulence factors? Trends Microbiol. *12*, 500-508.

Cadapan, L.D., Arslanian, R.L., Carney, J.R., Zavala, S.M., Small, P.L., and Licari, P. (2001). Suspension cultivation of *Mycobacterium ulcerans* for the production of mycolactones. FEMS Microbiol. Lett. *205*, 385-389.

Calmette, A., Guerin, L., Negre, and Boquet, A. (1926). Premunition nes nouveaux des contra la tuberculose part le vaccin BCG, 1921-1926. Ann. Inst. Pasteur. (Paris) 89-133.

Cangelosi, G.A., Palermo, C.O., Laurent, J.P., Hamlin, A.M., Brabant, W.H. (1999).

Colony morphotypes on Congo red agar segregate along species and drug susceptibility lines in the *Mycobacterium avium*-intracellulare complex. Microbiology. *145*, 1317-24.

Cangelosi, G.A., Palermo, C.O., Bermudez, L.E. (2001). Phenotypic consequences of red-white colony type variation in *Mycobacterium avium*. Microbiology. *147*, 527-533.

Chan, K., Knaak, T., Satkamp, L., Humbert, O., Falkow, S., and Ramakrishnan, L. (2002). Complex pattern of *Mycobacterium marinum* gene expression during long-term granulomatous infection. Proc. Natl. Acad. Sci. U S A. *99*, 3920-3925.

Chatterjee, D. and Khoo, K.H. (2001). The surface glycopeptidolipids of mycobacteria: Structures and biological properties. Cell. Mol. Life Sci. 54. 3854-62.

Chatterjee, D., Hunter, S.W., McNeil, M., Jardine, I. And Brennan, P.J. (1989). Structure and function of mycobacterial glycolipids and glycopeptidolipids. Acta Leprol. 7 Suppl 1, 81-4.

Chemlal, K., De Ridder, K., Fonteyne, P.A., Meyers, W.M., Swings, J., and Portaels, F. (2001). The use of IS2404 restriction fragment length polymorphisms suggests the diversity of *Mycobacterium ulcerans* from different geographical areas. Am. J. Trop. Med. Hyg. *64*, 270-273.

Chemlal, K., Huys, G., Fonteyne, P.A., Vincent, V., Lopez, A.G., Rigouts, L., Swings, J., Meyers, W.M., and Portaels, F. (2001). Evaluation of PCR-restriction profile analysis and IS2404 restriction fragment length polymorphism and amplified fragment length polymorphism fingerprinting for identification and typing of *Mycobacterium ulcerans* and *M. marinum*. J. Clin. Microbiol. *39*, 3272-3278.

Chemlal, K., Huys, G., Laval, F., Vincent, V., Savage, C., Gutierrez, C., Laneelle, M.A., Swings, J., Meyers, W.M., Daffe, M., and Portaels, F. (2002). Characterization of an unusual *Mycobacterium*: a possible missing link between *Mycobacterium marinum* and *Mycobacterium ulcerans*. J. Clin. Microbiol. *40*, 2370-2380.

Clark, H.F., and Shepard, C.C. (1963). Effect of environmental temperatures on infection with *Mycobacterium marinum* (Balnei) of mice and a number of poikilothermic species. J. Bacteriol. *86*, 1057–1069.

Collins, F.M., and Cunningham, D.S. (1981). Systemic *Mycobacterium kansasii* infection and regulation of the alloantigenic response. Infect. Immun. *32*, 614-624.

Connor, D.H., and Lunn, H.F. (1965). Mycobacterium ulcerans infection (with comments on pathogenesis). Int. J. Lepr. *33,* 698-709

Connor, D.H., and Lunn, H.F. (1966). Buruli ulceration, a clinicopathologic study of 38 Ugandans with *Mycobacterium ulcerans* infection. Arch. Path. *81,* 183-99.

Cope, R.B., Stang, B., Valentine, B.A., and Bermudez, L.E. (2004). Topical exposure to exogenous ultraviolet-irradiated urocanic acid enhances *Mycobacterium ulcerans* infection in a Crl:IAF (HA)-hrBR hairless guinea-pig model of Buruli ulcer disease. Photodermatol. Photoimmunol. Photomed. *20*, 14-20.

Cosma, C.L., Humbert, O., and Ramakrishnan, L. (2004). Superinfecting mycobacteria home to established tuberculous granulomas. Nat. Immunol. *5*, 828-835.

Cosma, C.L., Sherman, D.R., and Ramakrishnan, L. (2003). The secret lives of the pathogenic mycobacteria. Annu. Rev. Microbiol. *57*, 641-676.

Couillault, C., Ewbank, J.J. (2002). Diverse bacteria are pathogens of *Caenorhabditis elegans*. Infect. Immun. *70*, 4705-4707.

Daigle, F., Hou, J.Y., and Clark-Curtiss, J.E. (2002). Microbial gene expression elucidated by selective capture of transcribed sequences (SCOTS). Methods Enzymol. 358, 108-22.

Daniel, A.K., Lee, R.E., Portaels, F., and Small, P.L. (2004). Analysis of Mycobacterium species for the presence of a macrolide toxin, mycolactone. Infect. Immun. *72*, 123-132.

Danelishvili, L., Poort, M.J., and Bermudez, L.E. (2004). Identification of *Mycobacterium* avium genes up-regulated in cultured macrophages and in mice. FEMS Microbiol. Lett. 239, 41-9.

Davis, J.M., Clay, H., Lewis, J.L., Ghori, N., Herbomel, P., and Ramakrishnan, L. (2002). Real-time visualization of mycobacterium-macrophage interactions leading to initiation of granuloma formation in zebrafish embryos. Immunity. *17*, 693-702.

Debacker, M., Aguiar, J., Steunou, C., Zinsou, C., Meyers, W.M., Guedenon, A., Scott, J.T., Dramaix, M., and Portaels, F (2004). *Mycobacterium ulcerans* disease (Buruli ulcer) in rural hospital, Southern Benin, 1997-2001. Emerg Infect Dis.*10*, 1391-8

Decostere, A., Hermans, K., and Haesebrouck, F. (2004). Piscine mycobacteriosis: a literature review covering the agent and the disease it causes in fish and humans. Vet. Microbiol. *99*, 159-166.

Delaporte, E., Savage, C., Alfandari, S., Piette, F., Leclerc, H., and Bergoend, H. (1994). Buruli ulcer in a Zairian woman with HIV infection. Ann. Dermatol. Venereol. *121*, 557-560.

Dionne, M.S., Ghori, N., and Schneider, D.S. (2003). *Drosophila melanogaster* is a genetically tractable model host for *Mycobacterium marinum*. Infect. Immun. *71*, 3540-3550.

Dobos, K.M., Small, P.L., Deslauriers, M., Quinn, F.D., and King, C.H. (2001). *Mycobacterium ulcerans* cytotoxicity in an adipose cell model. Infect. Immun. *69*, 7182-7186.

Duker, A.A., Carranza, E.J., and Hale, M. (2004). Spatial dependency of Buruli ulcer prevalence on arsenic-enriched domains in Amansie West District, Ghana: implications for arsenic mediation in *Mycobacterium ulcerans* infection. Int. J. Health Geogr. *15*, 19.

Eckstein, T. B., Inamine, J.M., Lambert, M.L. and Belisle J. (2000). A genetic mechanism for deletion of the *ser2* gene cluster and formation of rough morphological variants of *Mycobacterium avium*. J. Bacteriol. 182, 6177-6182

Eckstein, T.M., Belisle, J.T., and Inamine, J.M. (2003). Proposed pathway for the biosynthesis of serovar-specific glycopeptidolipids in *Mycobacterium avium* serovar 2. Microbiology. *149*, 2797-2807.

Eddyani, M., Ofori-Adjei, D., Teugels, G., De Weirdt, D., Boakye, D., Meyers, W.M., and Portaels. F. (2004). Potential role for fish in transmission of *Mycobacterium ulcerans* disease (Buruli ulcer): an environmental study. Appl. Environ. Microbiol. *70*, 5679-5681.

Falkinham, J.O. 3rd. 2003. Factors influencing the chlorine susceptibility of *Mycobacterium avium*, *Mycobacterium intracellulare*, and *Mycobacterium scrofulaceum*. Appl. Environ. Microbiol. *69*, 5685–5689.

Falkinham, J.O., 3rd. Nontuberculosis mycobacteria in the environment. 2002. Clin. Chest Med. 23, 529-51.

Fidanze, S., Song, F., Szlosek-Pinaud, M., Small, P.L., and Kishi, Y. (2001). Complete structure of the mycolactones. J. Am. Chem. Soc. *123*, 10117-10118.

Flint, J.L., Kowalski, J.C., Karnati, P.K., and Derbyshire, K.M. (2004).The RD1 virulence locus of *Mycobacterium tuberculosis* regulates DNA transfer in *Mycobacterium smegmatis*. Proc Natl Acad Sci U S A. 24;101, 12598-603.

Floyd, M.M., Guthertz, L.S., Silcox V.A, Duffey, P.S., Jang Y., Desmond, E.P., Crawford, J.T., and Butler, W.R. (1996). Characterization of an SAV organism and proposal of *Mycobacterium triplex* sp. J. Clin. Microbiol. 34, 2963-7.

Flynn, J.L. (2004). Mutual attraction: does it benefit the host or the bug? Nat Immunol. *5*, 778-9.

Fratazzi, C., Arbeit, R.D., Carini, C., Balcewicz-Sablinska, M.K., Keane, J., Kornfeld, H., and Remold, H.G. (1999). Macrophage apoptosis in mycobacterial infections. J. Leukoc. Biol. *66*, 763-764.

Fregnan, G.B., Smith, D.W., and Randall. (1961). Biological and chemical studies on mycobacteria. J. Bacteriol. *82,* 1961)

Gao, L.Y., Groger, R., Cox, J.S., Beverley, S.M., Lawson, E.H., and Brown, E.J. (2003). Transposon mutagenesis of *Mycobacterium marinum* identifies a locus linking pigmentation and intracellular survival. Infect. Immun. *71*, 922-929.

Gao, L.Y., Laval, F., Lawson, E.H., Groger, R.K., Woodruff, A., Morisaki, J.H., Cox, J.S., Daffe, M., and Brown, E.J. (2003). Requirement for *kasB* in Mycobacterium mycolic acid biosynthesis, cell wall impermeability and intracellular survival: implications for therapy. Mol. Microbiol. *49*, 1547-1563.

Gao, L.Y., Guo, S., McLaughlin, B., Morisaki, H., Engel, J.N., and Brown, E.J. (2004). A mycobacterial virulence gene cluster extending RD1 is required for cytolysis, bacterial spreading and ESAT-6 secretion. Mol. Microbiol. *53*, 1677-1693.

George, K.M., Chatterjee, D., Gunawardana, G., Welty, D., Hayman, J., Lee, R., and Small, P.L. (1999). Mycolactone: a polyketide toxin from *Mycobacterium ulcerans* required for virulence. Science. *283*, 854-857.

George, K.M., Pascopella, L., Welty, D.M., and Small, P.L. (2000). A *Mycobacterium ulcerans* toxin, mycolactone, causes apoptosis in guinea pig ulcers and tissue culture cells. Infect. Immun. 68, 877-883.

Gey Van Pittius, N.C. et al. (2001). The ESAT-6 gene cluster of *Mycobacterium tuberculosis* and other high G + C Gram positive bacteria. Genome Biol. *2*, 1-18.

Gluckman, S.J. (1995). *Mycobacterium marinum*. Clin. Dermatol. *13*, 273-276.

Gomez, A., Mve-Obiang, A., Vray, B., Remacle, J., Chemlal, K., Meyers, W.M., Portaels, F., and Fonteyne, P.A. (2000). Biochemical and genetic evidence for phospholipase C activity in *Mycobacterium ulcerans*. Infect Immun. 68, 2995-2997.

Gooding, T.M., Johnson, P.D., Campbell, D.E., Hayman, J.A., Hartland, E.L., Kemp, A.S., and Robins-Browne, R.M. (2001). Immune response to infection with *Mycobacterium ulcerans*. Infect. Immun. *69*, 1704-1707.

Gooding, T.M., Johnson, P.D., Smith, M., Kemp, A.S., and Robins-Browne, R.M. (2002). Cytokine profiles of patients infected with *Mycobacterium ulcerans* and unaffected household contacts. Infect. Immun. *70*, 5562-5567.

Gooding, T.M., Kemp, A.S., Robins-Browne, R.M., Smith, M., and Johnson, P.D. (2003). Acquired T-helper 1 lymphocyte anergy following infection with *Mycobacterium ulcerans*. Clin. Infect. Dis. *36*, 1076-1077.

Greub, G., and Raoult, D. (2004). Microorganisms resistant to free-living amoebae. Clin. Microbiol. Rev. *17*, 413-433.

Guarner, J., Bartlett, J., Whitney, E.A., Raghunathan, P.L., Stienstra, Y., Asamoa, K., Etuaful, S., Klutse, E., Quarshie, E., van der Werf, T.S., van der Graaf, W.T., King, C.H., and Ashford, D.A. (2003). Histopathologic features of *Mycobacterium ulcerans* infection. Emerg. Infect. Dis. *9*, 651-656.

Guerardel, Y., Maes, E., Briken, V., Chirat, F., Leroy, Y., Locht, C., Strecker, G., and Kremer, L. (2003). Lipomannan and lipoarabinomannan from a clinical isolate of *Mycobacterium kansasii*: novel structural features and apoptosis-inducing properties. J. Biol. Chem. 278, 36637-51.

Guinn, K.M., Hickey, M.J., Mathur, S.K., Zakel, K.L, Grotzke, J.E., Lewinsohn, D.M., Smith, S., and Sherman, D.R. (2004). Individual RD1-region genes are required for export of ESAT-6/CFP-10 and for virulence of *Mycobacterium tuberculosis*. Mol Microbiol. 2004 *51*:359-70.

Gunawardana, G., Chatterjee, D., George, K.M., Brennan, P., Whittern, D., and Small, P.L.C. (1999). Characterization of novel macrolide toxins, mycolactone A and B, from a human pathogen, *Mycobacterium ulcerans*. J. Am. Chem. Soc. *121*, 6092-6093.

Hayman J. (1985). Clinical features of Mycobacterium ulcerans infection. Australas J. Dermatol. *26*, 67-73

Hayman, J.A., Smith, I.M., and Flood, P. (1996). Pseudoepitheliomatous hyperplasia in *Mycobacterium ulcerans* infection. Pathology. *28*, 131-4.

Heckert RA, Elankumaran S, Milani A, and Baya A. (2001). Detection of a new *Mycobacterium* species in wild striped bass in the Chesapeake Bay. J. Clin. Microbiol. *39*, 710-715.

Helguera-Repetto, C., Cox, R.A., Munoz-Sanchez, J.L., and Gonzalez-y-Merchand, J.A. (2004). The pathogen *Mycobacterium marinum*, a faster growing close relative of *Mycobacterium tuberculosis*, has a single rRNA operon per genome. FEMS Microbiol. Lett. *235*, 281-288.

Holmes, G.F., Harrington, S.M., Romagnoli, M.J., and Merz, W.G. (1999). Recurrent, disseminated *Mycobacterium marinum* infection caused by the same genotypically defined strain in an immunocompromised patient. J. Clin. Microbiol. *37*, 3059-3061.

Hong, H., Gates, P.J., Staunton, J., Stinear, T., Cole, S.T., Leadlay, P.F., and Spencer, J.B. (2003). Identification using LC-MSn of co-metabolites in the biosynthesis of the polyketide toxin mycolactone by a clinical isolate of *Mycobacterium ulcerans*. Chem. Commun. (Camb). *22*, 2822-2823.

Hong, H., Spencer, J.B., Porter, J.L., Leadley, P.F., and Stinear, T. (2005). A novel mycolactone from a clinical isolate of *Mycobacterium ulcerans* provides additional toxin heterogeneity as a result of specific changes in the modular polyketide synthase. ChemBioChem. *6*, 1-5.

Hou, J.Y., Graham, J.E., and Clark-Curtiss, J.E. (2002). *Mycobacterium avium* genes expressed during growth in human macrophages detected by selective capture of transcribed sequences (SCOTS). Infect Immun. 70, 3714-26

Hsu, T., Hingley-Wilson, S.M., Chen, B., Chen, M., Dai, A.Z., Morin, P.M., Marks, C.B., Padiyar, J., Goulding, C., Gingery, M., Eisenberg, D., Russell, R.G., Derrick, S.C., Collins, F.M., Morris, S.L., King, C.H., and Jacobs, W.R. Jr. (2003). The primary mechanism of attenuation of bacillus Calmette-Guerin is a loss of secreted lytic function required for invasion of lung interstitial tissue. Proc. Natl. Acad. Sci. U S A. *100*, 12420-12425.

Irani, V.R., Lee, S.H., Eckstein, T.M., Inamine, J.M., Belisle, J.T., and Maslow, J.N. (2004). Utilization of a *ts-sacB* selection system for the generation of a *Mycobacterium avium* serovar-8 specific glycopeptidolipid allelic exchange mutant. Ann. Clin. Microbiol. Antimicrob. *3*, 18.

Iredell, J., Whitby, M., and Blacklock, Z. (1992). *Mycobacterium marinum* infection: epidemiology and presentation in Queensland 1971-1990. Med. J. of Aust. *157*, 596-598.

Johansen, K.A., Gill, R.E., Vasil, M.L. (1996). Biochemical and molecular analysis of phospholipase C and phospholipase D activity in mycobacteria.Infect Immun. *64*, 3259-66.

Johnson, R.C., Ifebe, D., Hans-Moevi, A., Kestens, L., Houessou, R., Guedenon, A., Meyers, W.M., and Portaels, F. (2002). Disseminated *Mycobacterium ulcerans* disease in an HIV-positive patient: a case study. AIDS. *16*, 1704-1705.

Judd, T. C.; Bischoff, A.; Kishi, Y.; Adusumilli, S.; Small, P. L. C. (2004). Structure determination of mycolactone C via total synthesis. Org. Lett.

Katz, L., and Donadio, S. (1995). Macrolides. Biotechnology. *28*, 385-420

Keane, J., Remold, H.G., and Kornfeld, H. (2002). Virulent *Mycobacterium tuberculosis* strains evade apoptosis of infected alveolar macrophages. J. Immunol. 164, 2016-20.

Kanathur, N., Shantaveerapa, H.N., Byrd, R.P. Jr, Mehta, J.B., and Roy, T.M. (2001). Nontubercular mycobacterial pulmonary infection in immunocompetent men. South Med. J. *94*, 719-723.

Katoch, V.M. (2004). Infections due to non-tuberculous mycobacteria (NTM). Ind. J. Med. Res. 120, 290-304.

Khurana, R., Uversky, V.N., Nielsen, L., and Fink, A.L. (2001). Is Congo red an amyloid-specific dye? J. Biol. Chem. *276*, 22715-22721.

Kishihara, Y., Nakashima, K., Nukina, H., Hayashi, J., and Kashiwagi, S. (1993). Two cases of acquired immunodeficiency syndrome with disseminated non-tuberculous mycobacterial infection. Kansenshogaku Zasshi. *67*, 1223-1227.

Koh, W. J., Kwon, O.J., Lee, K.S. Nontuberculous mycobacterial pulmonary diseases in immunocompetent patients. (2002). Kor. J. Radiol. 3, 145-157.

Kotb, R., Dhote, R., Garcia-Ricart, F., Permal, S., Carlotti, A., Arfi, C., and Christoforov, B. (2001). Cutaneous and mediastinal lymphadenitis due to *Mycobacterium kansasii*. J. Infect. *42*, 277-278.

Kotlowski, R., Martin, A., Ablordey, A., Chemlal, K., Fonteyne, P.A., and Portaels, F. (2004). One-tube cell lysis and DNA extraction procedure for PCR-based detection of *Mycobacterium ulcerans* in aquatic insects, molluscs and fish. J. Med. Microbiol. *53*, 927-933.

Krieg, R.E., Hockmeyer, W.T., and Meyers, W. (1974). Toxin of *Mycobacterium ulcerans*. Production and effects in guinea pig skin, Arch. Dermatol. *110*, 783-8.

Krzywinska, E., Krzywinski, J., and Schorey, J.S. (2004). Naturally occurring horizontal gene transfer and homologous recombination in *Mycobacterium*. Microbiology. *150*, 1707-1712.

Laurent, J.P., Haugem, K., Burnside, K., and Cangelosi, G. (2003). Mutational analysis of cell wall biosynthesis in *Mycobacterium avium*. J. Bacteriol. *185*, 5003-5006.

Linell, F., and Norden, A. (1954). *Mycobacterium balnei*, a new acid-fast bacillus occurring in swimming pools and capable of producing skin lesions in humans. Acta Tuberc.Scand.Suppl. *33*, 1-84.

MacCallum, P., and Tolhurst, et al. (1948). A new mycobacterial infection in man. J. Path Bacteriol. *60,* 93-122.

Marmiesse, M., Brodin, P., Buchrieser, C., Gutierrez, C., Simoes, N., Vincent, V., Glaser, P., Cole, S.T., and Brosch, R. (2004). Macro-array and bioinformatic analyses reveal mycobacterial 'core' genes, variation in the ESAT-6 gene family and new phylogenetic markers for the *Mycobacterium tuberculosis* complex. Microbiology. *150*, 483-496.

Marsollier, L., Robert, R., Aubry, J., Saint Andre, J.P., Kouakou, H., Legras, P., Manceau, A.L., Mahaza, C., and Carbonnelle, B. (2002). Aquatic insects as a vector for *Mycobacterium ulcerans*. Appl. Environ. Microbiol. *68*, 4623-4628.

Marsollier, L., Stinear, T., Aubry, J., Saint Andre, J.P., Robert, R., Legras, P., Manceau, A.L., Audrain, C., Bourdon, S., Kouakou, H., and Carbonnelle, B. (2004). Aquatic plants stimulate the growth of and biofilm formation by *Mycobacterium ulcerans* in axenic culture and harbor these bacteria in the environment. Appl. Environ. Microbiol. *70*, 1097-1103.

Marston, B.J., Diallo, M.O., Horsburgh, C.R Jr., Diomande, I., Saki, M.Z., Kanga, J.M., Patrice, G., Lipman, H.B., Ostroff, S.M., and Good, R.C. (1995). Emergence of Buruli ulcer disease in the Daloa region of Cote d'Ivoire. Am. J. Trop. Med. Hyg. *52*, 219-24.

Mazurek, G.H., Hartman, S., Zhang, Y., Brown, B.A., Hector, J.S., Murphy, D., and Wallace, R.J. Jr. (1993). Large DNA restriction fragment polymorphism in the *Mycobacterium avium-M. intracellulare* complex: a potential epidemiologic tool. J. Clin. Microbiol. *31*, 390-394.

McDonough, K.A., Kress, Y., Bloom, B.R. (1993). Pathogenesis of tuberculosis: interaction of *Mycobacterium tuberculosis* with macrophages. Infect. Immun. *61*, 2763-2773.

McGarvey, J.A., Wagner, D., and Bermudez, L.E. (2004).Differential gene expression in mononuclear phagocytes infected with pathogenic and non-pathogenic mycobacteria. Clin. Exp. Immunol. 136, 490-500.

Meijer, A.H., Gabby Krens, S.F., Medina Rodriguez, I.A., He, S., Bitter, W., Ewa Snaar-Jagalska, B., and Spaink, H.P. (2004). Expression analysis of the Toll-like receptor and TIR domain adaptor families of zebrafish. Mol. Immunol. *40*, 773-783.

Mijs, W., de Haas, P., Rossau, R., Van der Laan, T., Rigouts, L., Portaels, F., and van Soolingen, D. (2002). Molecular evidence to support a proposal to reserve the designation *Mycobacterium avium* subsp. avium for bird-type isolates and '*M. avium* subsp. hominissuis' for the human/porcine type of M. avium. Int. J. Syst. Evol. Microbiol. *52*, 1505-1518.

Minnikin, D.E., Ridell, M., Wallerstrom, G., Besra, G.S., Parlett, J.H., Bolton, R.C., and Magnusson, M. (1989). Comparative studies of antigenic glycolipids of mycobacteria related to the leprosy bacillus. Acta. Leprol. *7*, 51-54.

Mitchell, P.J., McOrist, S., and Bilney, R. (1987). Epidemiology of *Mycobacterium ulcerans* infection in koalas (*Phascolarctos cinereus*) on Raymond Island, southeastern Australia. J. Wildl Dis. *23*, 386-90.

Mukherjee, S., Petrofsky, M., Yaraei, K., Bermudez, L.E., and Cangelosi, G.A. (2001). The white morphotype of *Mycobacterium avium*-intracellulare is common in infected humans and virulent in infection models. J. Infect. Dis. *184*, 1480-1484.

Mve-Obiang, A., Lee, R.E., Portaels, F., and Small, P.L. (2003). Heterogeneity of mycolactones produced by clinical isolates of *Mycobacterium ulcerans*: implications for virulence. Infect. Immun. *71*, 774-783.

Mve-Obiang, A., Remacle, J., Palomino, J.C., Houbion, A., and Portaels, F. (1999). Growth and cytotoxic activity by *Mycobacterium ulcerans* in protein-free media. FEMS Microbiol. Lett. *181*, 153-157.

Ohkusu, K., Bermudez, L.E., Nash, K.A., MacGregor, R.R., and Inderlied, C.B. (2004). Differential virulence of *Mycobacterium avium* strains isolated from HIV-infected patients with disseminated *M. avium* complex disease. J. Infect. Dis. *190*, 1347-1354.

Okenu, D.M., Ofielu, L.O., Easley, K.A., Guarner, J., Spotts Whitney, E.A., Raghunathan, P.L., Stienstra, Y., Asamoa, K., van der Werf, T.S., van der Graaf, W.T., Tappero, J.W., Ashford, D.A., and King, C.H. (2004). Immunoglobulin M antibody responses to *Mycobacterium ulcerans* allow discrimination between cases of active Buruli ulcer disease and matched family controls in areas where the disease is endemic. Clin. Diagn. Lab. Immunol. *11*, 387-391.

Oros, J., Acosta, B., Gaskin, J.M., Deniz, S., and Jensen, H.E. (2003). *Mycobacterium kansasii* infection in a Chinese soft shell turtle (*Pelodiscus sinensis*). Vet. Rec. *152*, 474-476.

Otero, J., Jacobs, W.R. Jr, and Glickman, M.S. (2003). Efficient allelic exchange and transposon mutagenesis in *Mycobacterium avium* by specialized transduction. Appl. Environ. Microbiol. *69*, 5039-5044.

Ouattara, D., Meningaud, J.P., and Saliba, F. (2002). Multifocal forms of Buruli ulcer: clinical aspects and management difficulties in 11 cases. Bull. Soc. Pathol. Exot. *95*, 287-291.

Pahlevan, A.A., Wright, D.J., Andrews, C., George, K.M., Small, P.L., Foxwell, B.M. (1999). The inhibitory action of *Mycobacterium* ulcerans soluble factor on monocyte/T cell cytokine production and NF-kappa B function. J. Immunol. *163*, 3928-3935.

Pai, H.H., Chen, W.C., and Peng, C.F. (2003). Isolation of non-tuberculous mycobacteria from hospital cockroaches (*Periplaneta americana*). J. Hosp. Infect. *53*, 224-228.

Parent, L.J., Salam, M.M., Appelbaum, P.C., and Dossett, J.H. (1995). Disseminated *Mycobacterium marinum* infection and bacteremia in a child with severe combined immunodeficiency. Clin. Infect. Dis. *21*, 1325-1327.

Pedley, S., Bartram, J., Rees, G., Dufour, A., and Cotruvo, J.A. (2004). Pathogenic mycobacteria in water. (Geneva: World Health Organization).

Pedrosa, J., Florido, M., Kunze, Z.M., Castro, A.G., Portaels, F., McFadden, J., Silva, M.T., and Appelberg, R. (1994). Characterization of the virulence of *Mycobacterium avium* complex (MAC) isolates in mice. Clin. Exp. Immunol. *98*, 210-216

Philalay, J.S., Palermo, C.O., Hauge, K.A., Rustad, T.R., and Cangelosi, G. (2004). Genes required for intrinsic multidrug resistance in *Mycobacterium avium*. Antimicrob. Agents Chemother. *48*, 3412-3418.

Picardeau M, Prod'Hom G, Raskine L, LePennec MP, and Vincent V. (1997). Genotypic characterization of five subspecies of *Mycobacterium kansasii*. J. Clin. Microbiol. *35*, 25-32.

Pietersen R, Thilo L, and de Chastellier C. (2004). *Mycobacterium tuberculosis* and *Mycobacterium avium* modify the composition of the phagosomal membrane in infected macrophages by selective depletion of cell surface-derived glycoconjugates. Eur. J. Cell Biol. *83*, 153-8.

Portaels, F., Fonteyene, P.A., de Beenhouwer, H., de Rijk, P., Guedenon, A., Hayman, J., and Meyers, M.W. (1996). Variability in 3' end of 16S rRNA sequence of *Mycobacterium ulcerans* is related to geographic origin of isolates. J. Clin. Microbiol. 34 (4):962-5

Portaels, F., Chemlal, K., Elsen, P., Johnson, P.D., Hayman, J.A., Hibble, J., Kirkwood, R., and Meyers, W.M. (2001). *Mycobacterium ulcerans* in wild animals. Rev. Sci. Tech. *20*, 252-264.

Portaels, F., Elsen, P., Guimaraes-Peres, A., Fonteyne, P.A., and Meyers, W.M. (1999). Insects in the transmission of *Mycobacterium ulcerans* infection. Lancet. *353*, 986.

Pressler, B.M., Hardie, E.M., Pitulle, C., Hopwood, R.M., Sontakke, S., and Breitschwerdt, E.B. (2002). Isolation and identification of *Mycobacterium kansasii* from pleural fluid of a dog with persistent pleural effusion. J. Am. Vet Med Assoc. *220*, 1336-1340, 1313-1314.

Prevot, G., Bourreau, E., Pascalis, H., Pradinaud, R., Tanghe, A., Huygen, K., and Launois, P. (2004). Differential production of systemic and intralesional gamma interferon and interleukin-10 in nodular and ulcerative forms of Buruli disease. Infect. Immun. *72*, 958-965.

Primm, T.P., Lucero, C.A., and Falkingham, J.O. (2004). Health impacts of environmental mycobacteria. Clin. Microbiol. Rev. 17, 98-106.

Ramakrishnan, L., and Falkow, S. (1994). *Mycobacterium marinum* persists in cultured mammalian cells in a temperature-restricted fashion. Infect. Immun. *62*, 3222-3229.

Ramakrishnan, L., Valdivia, R.H., McKerrow, J.H., and Falkow, S. (1997). *Mycobacterium marinum* causes both long-term subclinical infection and acute disease in the leopard frog (Rana pipiens). Infect. Immun. *65*, 767-773.

Ramakrishnan, L., Federspiel, N.A., and Falkow, S. (2000). Granuloma-specific expression of *Mycobacterium* virulence proteins from the glycine-rich PE-PGRS family. Science. *288*, 1436-1439.

Recht, J., Kolter, R. (2001). Glycopeptidolipid acetylation affects sliding motility and biofilm formation in *Mycobacterium smegmatis*. J. Bacteriol. 183, 5718-24.

Reddy, V.M, , Luna-Herrera, J., and Gangadharam, P.R. (1996). Pathobiological significance of colony morphology in *Mycobacterium avium* complex. Microb Pathog. 21, 97-109.

Ross, B.C., Jackson, K., Yang, M., Sievers, A., and Dwyer, B. (1992). Identification of a genetically distinct subspecies of *Mycobacterium kansasii*. J. Clin. Microbiol. *30*, 2930-2933.

Rostogi, N. and Barrow, W.W. (1994). Cell envelope constituents and the multifaceted nature of Mycobacterium pathogenicity and drug resistance. Res. Microbiol. 145, 243-52.

Ruley, K.M., Ansede, J.H., Pritchett, C.L., Talaat, A.M., Reimschuessel, R., and Trucksis, M. (2004). Identification of *Mycobacterium marinum* virulence genes using signature-tagged mutagenesis and the goldfish model of mycobacterial pathogenesis. FEMS Microbiol. Lett. *232*, 75-81.

Russell, D.G. (2001). *Mycobacterium tuberculosis*: here today, and here tomorrow. Nat. Rev. Mol. Cell Biol. *2*, 569-577.

Sano K., Sato, K., Sano. C., Shimizu, T., and Tomioka, H. (2002). Comparative profiles of intramacrophage behavior of *Mycobacterium tuberculosis* and *Mycobacterium avium* complex with different levels of virulence. Microbiol. Immunol. 46, 483-6.

Santin, M., Alcaide, F., Benitez, M.A., Salazar, A., Ardanuy, C., Podzamczer, D., Rufi, G., Dorca, J., Martin, R., and Gudiol, F. (2004). Incidence and molecular typing of *Mycobacterium kansasii* in a defined geographical area in Catalonia, Spain. Epidemiol. Infect. *132*, 425-432.

Sato, T., Shibuya, H., Ohba, S., Nojiri, T., and Shirai, W. (2003). Mycobacteriosis in two captive Florida manatees (*Trichechus manatus latirostris*). J. Zoo Wildl. Med. *34*, 184-188.

Schulze-Robbecke, R., and Buchholtz, K. (1992). Heat susceptibility of aquatic mycobacteria. Appl. Environ. Microbiol. *58*, 1869-1873.

Semret, M., Zhai, G., Mostowy, S., Cleto, C., Alexander, D., Cangelosi, G., Cousins, D., Collins, D.M., van Soolingen, D., and Behr, M.A. (2004). Extensive genomic polymorphism within *Mycobacterium avium*. J. Bacteriol. *186*, 6332-6334.

Smith, A.G., and Jiji, R.M. (1975). Cutaneous infection due to a rough variant of *Mycobacterium marinum*. Am. J. Clin. Pathol. *64*, 263-70.

Smith, M.B., Molina, C.P., Schnadig, V.J., Boyars, M.C., and Aronson, J.F. (2003). Pathologic features of *Mycobacterium kansasii* infection in patients with acquired immunodeficiency syndrome. Arch. Pathol. Lab. Med. *127*, 554-560.

Smole, S.C. McAleese F., Ngampasutadol, J., Von Reyn, C.F., and Arbeit, R.D. (2002). Clinical and epidemiological correlates of genotypes within the *Mycobacterium avium* complex defined by restriction and sequence analysis of hsp65. J. Clin. Microbiol. 40, 3374-80.

Snyder, D.S., and Small, P.L. (2003). Uptake and cellular actions of mycolactone, a virulence determinant for *Mycobacterium ulcerans*. Microb. Pathog. 34, 91-101.

Solomon, J.M., Leung, G.S., and Isberg, R.R. (2003). Intracellular replication of *Mycobacterium marinum* within *Dictyostelium discoideum*: efficient replication in the absence of host coronin. Infect. Immun. *71*, 3578-3586.

Song, F., Fidanze, S., Benowitz, A.B., and Kishi, Y. (2002). Total synthesis of the mycolactones. Org. Lett. *4*, 647-650.

Southern, P.M. Jr. MD. (2004). Tenosynovitis caused by *Mycobacterium kansasii* associated with a dog bite. Am. J. Med. Sci. *327*, 258-261.

Stamm, L.M., Morisaki, J.H., Gao, L.Y., Jeng, R.L., McDonald, K.L., Roth, R., Takeshita, S., Heuser, J., Welch, M.D., and Brown, E.J. (2003). *Mycobacterium marinum* escapes from phagosomes and is propelled by actin-based motility. J. Exp. Med. *198*, 1361-1368.

Stienstra, Y., van der Graaf, W.T., Meerman, G.J., The, T.H., de Leij, L.F., van der and Werf, T.S. (2001). Susceptibility to development of *Mycobacterium ulcerans* disease: review of possible risk factors. Trop. Med. Int. Health. *6*, 554-562.

Stienstra, Y., van der Werf, T.S., Guarner, J., Raghunathan, P.L., Spotts Whitney, E.A., van der Graaf, W.T., Asamoa, K., Tappero, J.W., Ashford, D.A., and King, C.H. (2003). Analysis of an IS2404-based nested PCR for diagnosis of Buruli ulcer disease in regions of Ghana where the disease is endemic. J. Clin. Microbiol. *41*, 794-797.

Stienstra, Y., van der Werf, T.S., van der Graaf, W.T., Secor, W.E., Kihlstrom, S.L., Dobos, K.M., Asamoa, K., Quarshi, E., Etuaful, S.N., Klutse, E.Y., and King, C.H. (2004). Buruli ulcer and schistosomiasis: no association found. Am. J. Trop. Med. Hyg. *71*, 318-321.

Stinear, T., Davies, J.K., Jenkin, G.A., Hayman, J.A., Oppedisano, F., and Johnson, P.D. (2000). Identification of *Mycobacterium ulcerans* in the environment from regions in Southeast Australia in which it is endemic with sequence capture-PCR. Appl. Environ. Microbiol. *66*, 3206-3213.

Stinear, T., Davies, J.K., Jenkin, G.A., Portaels, F., Ross, B.C., Oppedisano, F., Purcell, M., Hayman, J.A., and Johnson, P.D. (2000). A simple PCR method for rapid genotype analysis of *Mycobacterium ulcerans*. J. Clin. Microbiol. *38*, 1482-1487.

Stinear, T.P., Jenkin, G.A., Johnson, P.D., and Davies, J.K. (2000). Comparative genetic analysis of *Mycobacterium ulcerans* and *Mycobacterium marinum* reveals evidence of recent divergence. J. Bacteriol. *182*, 6322-6330.

Stinear, T.P., Mve-Obiang, A., Small, P.L., Frigui, W., Pryor, M.J., Brosch, R., Jenkin, G.A., Johnson, P.D., Davies, J.K., Lee, R.E., Adusumilli, S., Garnier, T., Haydock, S.F., Leadlay, P.F., and Cole, S.T. (2004). Giant plasmid-encoded polyketide synthases produce the macrolide toxin of *Mycobacterium ulcerans*. Proc. Natl. Acad. Sci. U S A. *101*, 1345-1349.

Stinear, T., Ross, B.C., Davies, J.K., Marino, L., Robins-Browne, R.M., Oppedisano, F., Sievers, A., and Johnson, P.D. (1999). Identification and characterization of IS2404 and IS2606: two distinct repeated sequences for detection of *Mycobacterium ulcerans* by PCR. J. Clin. Microbiol. *37*, 1018-1023

Talaat, A.M., Reimschuessel, R., Wasserman, S.S., and Trucksis, M. (1998). Goldfish, *Carassius auratus*, a novel animal model for the study of *Mycobacterium marinum* pathogenesis. Infect. Immun. *66*, 2938-2942.

Taillard, C., Greub, G., Weber, R., Pfyffer, G.E., Bodmer, T., Zimmerli, S., Frei, R., Bassetti, S., Rohner, P., Piffaretti, J.C., Bernasconi, E, Bille, J., Telenti, A., and Prod'hom, G. (2003). Clinical implications of *Mycobacterium kansasii* species heterogeneity: Swiss National Survey. J. Clin. Microbiol. 41, 1240-4.

Tchornobay, A.M., Claudy, A.L., Perrot, J.L., Levigne, V., and Denis, M. (1992). Fatal disseminated *Mycobacterium marinum* infection. Int. J. Dermatol. *31*, 286-287.

Teelken, M.A., Stienstra, Y., Ellen, D.E., Quarshie, E., Klutse, E., van der Graaf, W.T., and van der Werf, T.S. (2003). Buruli ulcer: differences in treatment outcome between two centres in Ghana. Acta. Trop. *88*, 51-56.

Tonjum, T., Welty, D.B., Jantzen, E., and Small, P.L. (1998). Differentiation of *Mycobacterium ulceran*, , *M. marinum*, and *M. haemophilum*: mapping of their relationships to *M. tuberculosis* by fatty acid profile analysis, DNA-DNA hybridization, and 16S rRNA gene sequence analysis. J. Clin. Microbiol. *36*, 918-925.

Torrelles, J.B., Chatterjee, D., Lonca, J. G., Manterola, J.M. Ausina, V.R., and Brennan, P. (2000). Serovars of *Mycobacterium avium* complex isolated from AIDS and non-AIDS patients in Spain. J. Appl. Microbiol. 88, 266-279.

Tortoli, E. (2003). *Mycobacterium kansasii*, species or complex? Biomolecular and epidemiological insights. Kekkaku. *78*, 705-709.

Tortoli, E., Simonetti, M.T., Lacchini, C., Penati, V., and Urbano, P. (1994). Tentative evidence of AIDS-associated biotype of *Mycobacterium kansasii*. J. Clin. Microbiol. *32*, 1779-1982.

Tortoli, E. (2004). Impact of genotypic studies on mycobacterial taxonomy: the new mycobacteria of the 1990s. Clin. Micro. Rev. 16, 319-354.

Trott, K.A., Stacy, B.A., Lifland, B.D., Diggs, H.E., Harland, R.M., Khokha, M.K., Grammer, T.C., and Parker, J.M. (2004). Characterization of a *Mycobacterium ulcerans*-like infection in a colony of African tropical clawed frogs (*Xenopus tropicalis*). Comp Med. 54, 309-317.

Tzen, C.Y., Chen, T.L., Wu, T.Y., Yong, D.I., and Lee, H.C. (2001). Disseminated cutaneous infection with *Mycobacterium kansasii*: genotyping versus phenotyping. J. Am. Acad. Dermatol. *45*, 620-624.

Ucko, M., Colorni, A., Kvitt, H., Diamant, A., Zlotkin, A., and Knibb, W.R. (2002). Strain variation in *Mycobacterium marinum* fish isolates. Appl. Environ. Microbiol. *68*, 5281-5287.

Ullrich, H.J., Beatty, W.L., and Russell, D.G. (2000). Interaction of *Mycobacterium avium*-containing phagosomes with the antigen presentation pathway. J. Immunol. *165*, 6073-6080.

van der Sar, A.M., Abdallah, A.M., Sparrius, M., Reinders, E., Vandenbroucke-Grauls, C.M., and Bitter, W. (2004). *Mycobacterium marinum* strains can be divided into two distinct types based on genetic diversity and virulence. Infect. Immun. *72*, 6306-6312.

van der Werf, T.S., Stinear, T., Stienstra, Y., van der Graaf, W.T., and Small, P.L. (2003). Mycolactones and *Mycobacterium ulcerans* disease. Lancet *362*, 1062-1064.

van der Werf, T, S., van der Graaf, W.T., Tappero, J.W., and Asiedu, K. (1999). *Mycobacterium ulcerans* infection. Lancet. *354*, 1013-1018.

Vieira, O.V., Harrison, R.E., Scott, C.C., Stenmark, H., Alexander, D., Liu, J., Gruenberg, J., Schreiber, A.D., and Grinstein, S. (2004). Acquisition of Hrs, an essential component of phagosomal maturation, is impaired by mycobacteria. Mol. Cell Biol. *24*, 4593-4604.

Vignal, C., Guerardel, Y., Kremer, L., Masson, M., Legrand, D., Mazurier, J., and Elass, E. (2003). Lipomannans, but not lipoarabinomannans, purified from *Mycobacterium chelonae* and *Mycobacterium kansasii* induce TNF-alpha and IL-8 secretion by a CD14-toll-like receptor 2-dependent mechanism. J. Immunol. *171*, 2014-2023.

Wallace, R.J. Jr., Zhang, Y., Brown, B.A., Dawson, D., Murphy, D.T., Wilson, R., and Griffith, D.E. (1998). Polyclonal *Mycobacterium avium* complex infections in patients with nodular bronchiectasis. Am. J. Respir. Crit. Care Med. *158*, 1235-1244.

Wayne, L.G. (1962). Two varieties of *Mycobacterium kansasii* with different clinical significance. Am. Rev. Respir. Dis. *86*, 651-959.

Weir E. (2002). Buruli ulcer: the third most common mycobacterial infection. CMAJ. *166*, 1691.

Zenone, T., Boibieux, A., Tigaud, S., Fredenucci, J.F., Vincent, V., Chidiac, C., and Peyramond, D. (1999). Non-tuberculous mycobacterial tenosynovitis: a review. Scand. J. Infect. Dis. *31*, 221-228.

Zhang, Y., Mann, L.B., Wilson, R.W., Brown-Elliott, B.A., Vincent, V., Iinuma, Y., and Wallace, R.J. Jr. (2004). Molecular analysis of *Mycobacterium kansasii* isolates from the United States. J. Clin. Microbiol. *42*, 119-125.

Chapter 7

The Stress Response

G. R Stewart, I. Papatheodorou and D.B. Young

ABSTRACT

Heat shock provides a convenient and well-characterised model to study the response of cells exposed to environmental stress. The major stress proteins induced by heat shock are molecular chaperones which play an essential role in the protein-folding activities of the cell under all conditions, but are particularly important when the protein complement of the cell begins to denature and aggregate during exposure to stress. The dramatic increase in the expression of chaperones during infection identifies the stress response as an important element of mycobacterial pathogenesis. Recent genome-wide analyses allow us to view the regulation and function of the major stress proteins in the context of the whole cell response to infection. The prominence of conserved stress proteins in the microbial response to infection presents a convenient target for recognition by the host immune system, triggering innate immune signalling as well as providing antigens for adaptive immunity. In this review, we describe advances in understanding the biology of mycobacterial stress proteins from the perspective both of the microbe and of the infected host.

INTRODUCTION

The need to be able to survive exposure to environmental stress provides one of the driving forces of evolution that have shaped the structure and mechanisms of cellular life. A major component of the stress response is the elevated expression of molecular chaperones that help to maintain the integrity of proteins and protein complexes within the cell. This response, and the chaperones themselves, have been highly conserved throughout evolution.

In their natural habitats most bacteria have to survive in conditions that are suboptimal for growth. For example, saprophytic mycobacteria that inhabit the soil must cope with a deficient supply of organic matter from which they must extract an energy source. In addition to carbon and organic nitrogen limitation, they may also have to cope with low levels of inorganic nutrients such as iron and phosphorus. On top of nutrient limitations, bacteria may have to cope with variations in oxygen availability, osmotic stress, desiccation, low or high temperatures, oxidative stress, UV irradiation, and acid or alkaline conditions. It is clear that bacteria must possess flexible adaptive systems to survive these conditions.

For pathogenic mycobacteria, the stress response is a fundamental element of survival in the challenging and hostile environment encountered during infection (Fig. 1). *M. tuberculosis* can face dessication, low temperatures and UV irradiation during transmission from host to host. Having reached the alveolar surface in the lung, it is subjected to the antimicrobial action of surfactants which may damage the bacterial cell surface. Once phagocytosed by a macrophage, *M. tuberculosis* is exposed to a potentially lethal oxidative burst produced by NADPH oxidase, and if the macrophage has been activated, to reactive nitrogen species produced through the action of inducible nitric oxide synthase (iNOS) (Nathan and Shiloh, 2000). In an activated macrophage the phagosome is acidified by the acquisition of vacuolar ATPases (Schaible et al., 1998; Via et al., 1998), exposing the intracellular bacterium to low pH. Toxic proteins such as granulysin released by macrophages and natural killer (NK) cells (Dieli et al., 2001; Stenger et al., 1998), and toxic free fatty acids (Akaki et al., 1997) can also be damaging or lethal to the bacterium. The phagosomal environment may also exert nutritional stress, requiring adaptation to particular carbon sources (McKinney et al., 2000), and potential deficiencies in micronutrients. As granulomatous lesions develop, the bacteria take on a different appearance which is similar to that of *in vitro* starved bacilli suggesting a further decrease in nutrient availability (Nyka, 1974). In these lesions the bacteria are also subjected to hypoxic conditions which may represent one of the most important elements of *M. tuberculosis* persistence (Stewart et al., 2003).

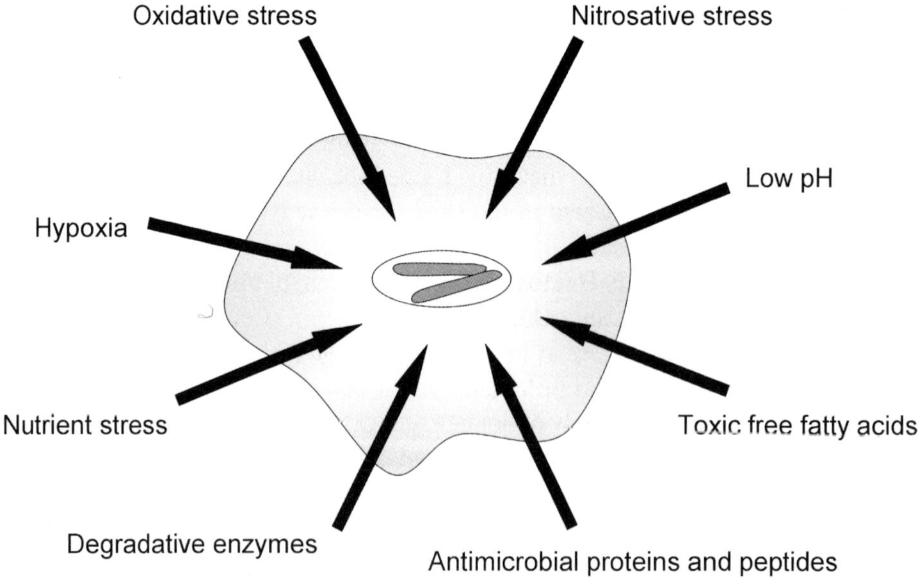

Fig. 1. The stressful life of pathogenic mycobacteria. Mycobacteria like *M. tuberculosis* residing within macrophages are exposed to many different stresses.

Each of these environmental threats is countered by a combination of general and specific defences. In this chapter we examine the biology of the major stress proteins that represent a common theme throughout many of these different stress responses.

THE HEAT SHOCK RESPONSE: A PARADIGM OF THE STRESSED CELL

Exposure to an abrupt increase in growth temperature induces the characteristic heat shock response. Although most organisms do not encounter such changes in temperature in their natural habitat, heat shock provides an experimentally robust means of studying what happens in a cell when proteins start to denature and aggregate. In this respect, the heat shock response represents a useful paradigm of the stressed cell.

The first heat shock studies performed on mycobacteria revealed the increased expression of a common subset of proteins in a range of different species (Lamb et al., 1990; Lathigra et al., 1991; Mehlert and Young, 1989; Patel et al., 1991; Shinnick et al., 1988). A combination of immunoblotting and sequence analysis identified several of these proteins as members of the highly conserved stress protein families DnaK/Hsp70, GroEL/Hsp60/Cpn60 and GroES/Hsp10/Cpn10. The process of identification was facilitated by the fact that these proteins were dominant antigens recognised by monoclonal antibodies generated by immunisation of mice with crude mycobacterial extracts (Young et al., 1988). The first attempts to examine the heat shock response at the mRNA level were made by Patel et al. (Patel et al., 1991) using Northern blots to demonstrate that *hsp60* and *hsp70* were both transcriptionally upregulated following heat shock in *M. bovis* BCG. More than 10 years later microarray technology made it possible to perform a whole genome inspection of transcription during heat shock in *M. tuberculosis* (Stewart et al., 2002). As expected, the chaperones were prominent members of the induced gene set, but it was clear that the heat shock response encompassed a genome-wide rearrangement of the transcriptome.

THE MAJOR STRESS PROTEINS: THE CHAPERONES

Chaperones are ubiquitous in cellular life and function primarily to promote the correct folding of proteins and to prevent the aggregation of misfolded proteins. They are also involved in the translocation of proteins across membranes and have adapted to a variety of additional specialised functions including transcriptional control (Xing et al., 2004), buffering of the consequences of mutations (Rutherford and Lindquist, 1998), and control of apoptosis (Takayama et al., 2003). They are among the most conserved of all proteins in nature and are expressed during normal cellular growth. However, their expression is increased markedly following exposure to many different environmental stresses including thermal stress, heavy metals, UV radiation, oxidative stress, ethanol and nutrient starvation (Lindquist,

1986), with accumulation of unfolded polypeptides representing the likely common stimulus (Narberhaus, 1999).

During heat shock of mycobacteria, increased expression was observed for members of the chaperone families GroEL, DnaK, DnaJ/Hsp40, ClpB/Hsp100, α-crystallin/Hsp20 and GroES (Lamb et al., 1990; Lathigra et al., 1991; Mehlert and Young, 1989; Patel et al., 1991; Shinnick et al., 1988; Stewart et al., 2002). These molecules form the vast majority of total *de novo* protein synthesis during heat shock. In *E. coli*, GroEL functions in a heptameric ring formation where two ring structures coordinate with heptameric associations of GroES to bind polypeptide substrates and promote their appropriate folding and complex formation (Langer et al., 1992; Viitanen et al., 1992). All bacteria possess at least one *groEL* gene; in common with several other bacterial genera, mycobacteria contain multiple copies. *M. tuberculosis, M. leprae* and *M. avium paratuberculosis* each have two *groEL* genes, and *M. smegmatis* has three (Fig. 2B). In *M. tuberculosis,* GroEL1 (Rv3417c) is encoded in a predicted operon together with *groES* (Rv3418c), and separated from the *groEL2* (Rv0440) gene. Only a single *groES* exists in all sequenced mycobacterial genomes. It is not clear whether all the GroEL paralogues partner this single GroES, nor whether they confer redundant functions or perhaps differ in their specificities for different subsets of polypeptide substrates. It is also possible that the mycobacterial GroEL proteins function independently of GroES, as a recent report suggests that the *M. tuberculosis* GroELs do not form heptameric ring structures at all but function as lower oligomers (Qamra et al., 2004).

DnaK (Rv0350) functions as a monomer in conjunction with its co-chaperones GrpE (Rv0351) and DnaJ (Rv0352). The three genes form an operon together with the HspR regulator (see below). Unlike GroEL, the primary function of DnaK is not directly to promote refolding of unfolded/misfolded polypeptides to their native state, but to bind to their hydrophobic domains and prevent their inappropriate aggregation (Langer et al., 1992; Schroder et al., 1993). With the help of the AAA+ chaperone ClpB, DnaK can disaggregate large aggregates of misfolded proteins (Goloubinoff et al., 1999; Mogk et al., 1999; Zolkiewski, 1999). The antiquity of the chaperones has allowed nature to engender many with functions additional to their primary role in protein folding. In *E. coli* for example, DnaK also plays a role in the regulation of DNA replication by interacting with chromosomal (Gamer et al., 1996; Liberek et al., 1995), phage (Alfano and McMacken, 1989; Wickner et al., 1991; Zylicz et al., 1989) and plasmid (Ezaki et al., 1989; Konieczny and Liberek, 2002) replication proteins. There is evidence that a similar role exists for the DnaK chaperone machine in controlling chromosomal DNA replication in *M. bovis* (Stewart et al., 2004). In contrast to *E. coli* which contains several *dnaK* paralogues, all mycobacteria possess just a single well-conserved *dnaK* (Fig. 2A).

DnaJ is a member of the Hsp40 family of co-chaperones and acts as a facilitator and a regulator of DnaK activity by assisting in the delivery of peptide substrates (Gamer et al., 1996; Schroder et al., 1993; Szabo et al., 1994) and by controlling

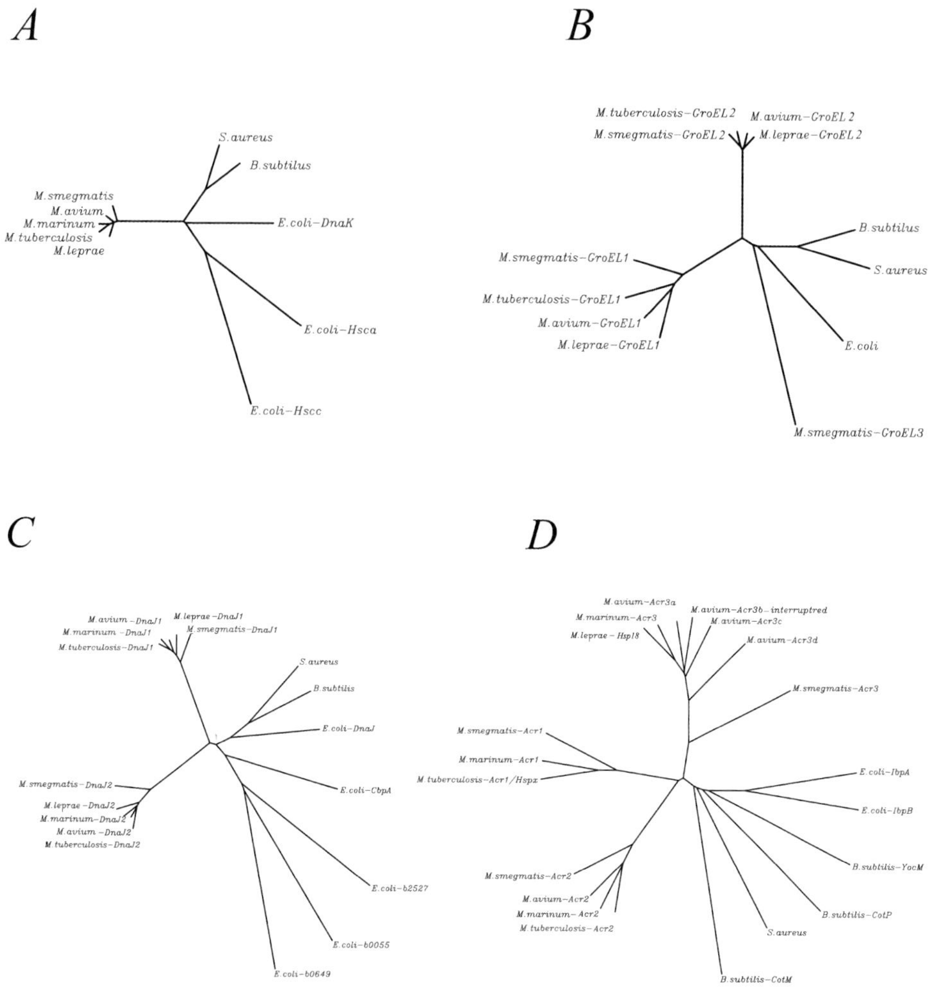

Fig. 2. Phylogenetic relationship of mycobacterial chaperones. Sequences were compared using ClustalW and displayed as an unrooted neighbour-joining dendrogram. All sequenced mycobacteria encode a single highly conserved DnaK (A). Other chaperones are present in multiple: there are 3 classes of GroEL family proteins (B), 2 classes of DnaJ proteins (C) and 3 classes of α-crystallins (D). *M. avium* is *M. avium paratuberculosis*.

the ATPase activity that mediates the folding reaction (Karzai and McMacken, 1996; Laufen et al., 1999; Wawrzynow et al., 1995). The diversification of function within chaperone families is again evidenced in *E. coli* which possesses five DnaJ paralogs, of which only three appear to function as chaperones, each associated with a specific DnaK partner. *M. tuberculosis* possesses two genes encoding DnaJ-like proteins (Fig. 2C). Rv0352 is part of the DnaK operon and is annotated as *dnaJ1*. The second gene, *dnaJ2* (Rv2373c), appears to lack a DnaK-like partner. A study examining the global transcriptional effect of over-expressing these *dnaJ* homologues in *M. bovis* BCG suggested that DnaJ1 is the co-chaperone partner of DnaK, involved in classical chaperone interactions with peptides as well as the role of DnaK in regulating DNA replication (Stewart et al., 2004). However, DnaJ2 appeared to have no functional overlap with DnaJ1. This is consistent with DnaJ homologues functioning in concert with particular DnaK partners. The function of DnaJ2 remains to be revealed.

The α-crystallin, or Hsp20, family of chaperones function to prevent the irreversible aggregation of proteins under stress conditions. In most bacteria the expression of α-crystallins during normal growth is negligible, however, during stress their expression can be dramatically elevated. To add further flexibility to the α-crystallin response, these proteins can exist as inactive multimers which instantaneously disaggregate upon stress into smaller functionally active configurations (Narberhaus, 2002). Most, though not all, bacteria possess an α-crystallin (Narberhaus, 2002). There are three classes of α-crystallins in mycobacteria (Fig. 2D). All mycobacteria examined so far, including the highly degenerate *M. leprae*, encode at least one member of this chaperone family, though there does not appear to be a necessity to encode a particular class of α-crystallin. Furthermore, there does not appear to be a dependency of pathogenicity on a particular α-crystallin class as *M. tuberculosis* contains an *acr1* (previously *hspX* or *acr*) and an *acr2*, *M. leprae* an *acr3*, and *M. avium paratuberculosis* one *acr2* and three functional *acr3* genes (Stewart et al., 2005). It is interesting to note that the mycobacteria which reside in environments with the greatest variability of temperature – *M. smegmatis*, *M. marinum* and *M. avium* – are those which possess the largest complement of α-crystallins. It has been postulated that diversity in α-crystallins provides an organism with the ability to respond to a wider range of environments (Narberhaus, 2002).

OTHER GENES INDUCED DURING HEAT SHOCK

Heat shock produces a massive insult to a bacterial cell and the result is a broad rearrangement of cellular activities, reflected in genome-wide changes in corresponding bacterial transcriptomes. In *M. tuberculosis,* like other bacteria (Helmann et al., 2001; Muffler et al., 2002), the chaperone response is dominant but several hundred other genes are up- or down-regulated following 45°C heat shock for 30 minutes. The distribution of heat-induced genes across functional categories is biased towards genes involved in adaptation and detoxification processes and away from genes involved in core growth activities such as cell wall synthesis (Stewart et al., 2002). Further emphasising the dramatic transcriptomic change, 15% of the upregulated genes encode regulatory genes, as compared to a genomic frequency of less than 5%. Details of these global transcriptional changes can be obtained from the Bacterial Microarray Group at St Georges database (http://bugs.sghms.ac.uk/) and in supplementary data to Stewart et al. (2002).

REGULATION OF THE HEAT SHOCK RESPONSE

Two forms of transcriptional regulation are used to control expression of the major chaperone systems in mycobacteria; release of repression, and induction by alternative sigma factors. Two repressor systems have been identified. The first involves the protein HspR, a repressor found also in *Streptomyces* (Bucca et al., 1997; Grandvalet et al., 1997) and *Helicobacter pylori* (Grandvalet et al., 1997). In mycobacteria this protein is encoded by the fourth and final gene of the *dnaK* operon. It interacts with DnaK itself to form the functional repressor, binding to inverted repeat motifs situated in the promoter region of the transcriptional unit (Stewart et al., 2001 and Stewart et al. unpublished data). The control system is coordinated with the stressed state of the cell by sequestration of DnaK by denatured proteins, releasing repression during heat shock. In *M. tuberculosis* there are two transcriptional start sites upstream of the *dnaK* operon and each is associated with an adjacent HspR Associated Inverted Repeat, or HAIR sequence. HAIR motifs are also present in the promoter regions of at least two other transcriptional units containing heat-inducible chaperone genes; *clpB* and the *acr2* operon (Rv0249c-251c) (Stewart et al., 2002).

The second heat shock repressor system involves the conserved HrcA protein and controls transcription of the GroE system. In *Bacillus* this system has been shown to work by a mechanism analogous to that described for HspR. The HrcA protein represses transcription by interacting with GroEL to bind inverted repeats in the promoter regions of target genes (Hecker et al., 1996). Transcriptional repression is released by sequestration of GroEL by denatured peptides. In mycobacteria, a similar system appears to be in operation as an *hrcA* deletion mutant constitutively over-transcribes *groEL1*, *groES* and *groEL2* (Stewart et al., 2002). The HrcA binding sites, Controlling Inverted Repeats of Chaperone Expression (CIRCE), are found upstream of the GroE transcriptional units. Two additional genes (Rv0990c and Rv0991c) are associated with CIRCE motifs and were upregulated in the HrcA deletion mutant and during heat shock; their function is unknown.

Three sigma factors, σ^B, σ^H and σ^E, have been shown to be heat-inducible (Fernandes et al., 1999; Manganelli et al., 1999; Stewart et al., 2001). σ^B (Rv2710) is a primary-like sigma factor that is encoded near to the principal sigma factor, σ^A (Rv2703). It is not essential for normal growth but is required for resistance against a variety of stresses including heat shock, oxidative stress and detergent-induced surface stress (Manganelli et al., 2004). σ^H and σ^E belong to the extracellular function (ECF) group of sigma factors. In addition to high temperature, σ^H is induced by redox stress generated by exposure to diamide, and a mutant strain is susceptible to both stresses (Manganelli et al., 2002; Raman et al., 2001). All of the HspR-regulated transcriptional units, the *dnaK* operon, *clpB* and the *acr2* operon, are also under the control of σ^H, as is the HrcA-regulated Rv0990c-Rv0991c unit (Manganelli et al., 2002; Raman et al., 2001). Like, σ^H, the ECF sigma factor σ^E is also induced by a number of environmental stresses including oxidative stress, detergent stress, and heat shock (Manganelli et al., 1999). Transcriptional profiles of a σ^E mutant demonstrate that σ^E confers a third level of transcriptional control on the *acr2* operon (Manganelli et al., 2001)(Fig. 3). In addition to control of the major stress proteins, many of the top 100 heat-inducible genes contain the σ^E and σ^H consensus promoter regions (Manganelli et al., 2001; Raman et al., 2001). The σ^B regulon and the σ^B recognition sequence have yet to be accurately identified, but it is likely that the σ^B regulon is well represented in the heat shock gene set.

The relationship between the σ^B, σ^H and σ^E regulons is a complex stimulus-dependent hierarchy which is reviewed in detail (Manganelli et al., 2004). Both σ^E and σ^H can promote transcription at the *sigB* promoter. In addition, σ^H promotes transcription of its own gene as well as that of *sigE*. As *sigE* and *sigH* are responsive to different environmental stimuli, this regulatory network allows for more flexibility in the gene expression response to different environmental conditions. A simple example of plasticity in gene expression is shown in Fig. 4, which is a directed graph comparison of the transcriptional response to exposure to heat shock, SDS, and diamide (Papatheodorou et al., 2004). Clearly there are subsets of genes that are only regulated by a single environmental condition, together with other genes which are induced by multiple stimuli. There are five transcriptional units that are

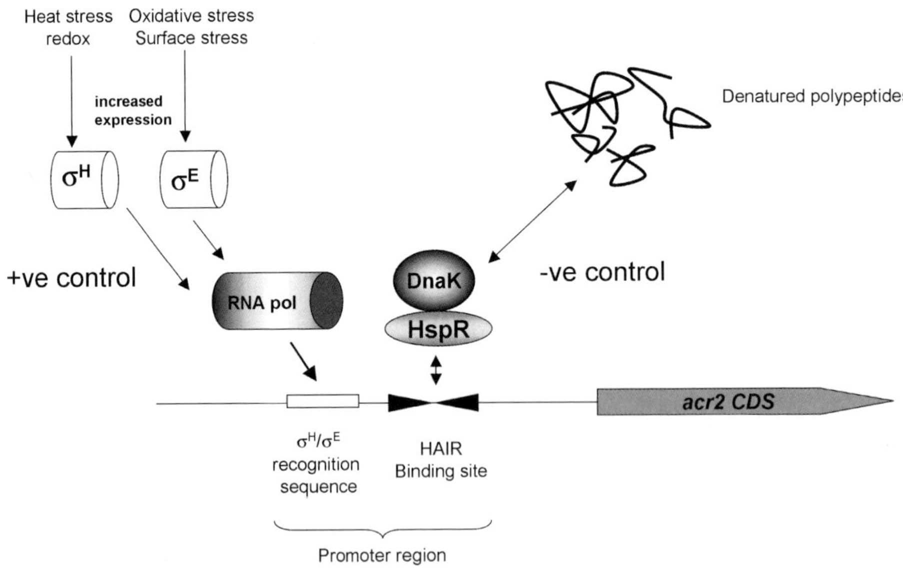

Fig. 3. Transcriptional control of the α-crystallin gene *acr2*. Positive control is achieved through the stress responsive sigma factors σ^H and σ^E which direct RNA polymerase holoenzyme to a dual σ^H and σ^E recognition sequence. Transcription is also negatively controlled by the HspR repressor protein. This protein acts with DnaK to form the functional repressor with sequestration of DnaK by denatured polypeptides promoting release of repression during stress.

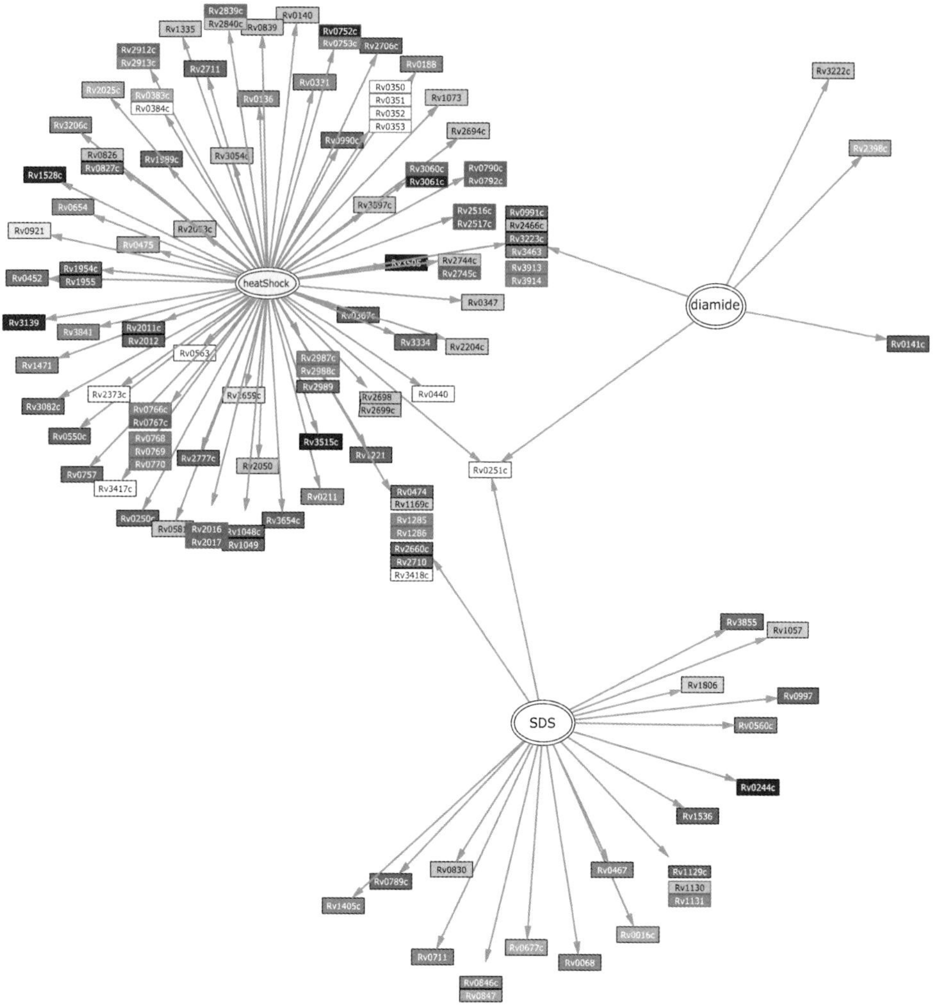

Fig. 4. Force directed graph showing overlap between the heat shock, diamide and SDS exposure stress stimulons (upregulated genes)(Papatheodorou et al., 2004).

upregulated both under heat shock and after exposure to diamide, and for several of these the mechanisms of this regulation are obvious. Rv0991c is regulated by the HrcA repressor protein (responsive to heat) and σ^H (Rv3223c, responsive to heat and diamide), and *trxB* (Rv3913) and *sigH* are regulated by σ^H itself. There are six transcriptional units upregulated by both heat shock and SDS; two of these, Rv1295-96 (*cysD*) and Rv1169c (*PE11*), are known to be regulated by the heat and surface stress-responsive σ^E (Manganelli et al., 2001). The *acr2* gene is induced by all three stimuli; this gene is controlled by the HspR repressor, σ^E and σ^H giving it perhaps the greatest range of conditions under which it can be induced.

THE IMPORTANCE OF MAJOR STRESS PROTEINS DURING INFECTION

The importance of chaperones for intracellular pathogens such as *M. tuberculosis*, *Listeria moncytogenes, Rickettsia prowazekii, Chlamydia pneumoniae* and *Salmonella* has been inferred from studies demonstrating the elevation of their expression following phagocytosis by host cells (Buchmeier and Heffron, 1990; Eriksson et al., 2003; Gahan et al., 2001; Gaywee et al., 2002; Haranaga et al., 2003; Lee and Horwitz, 1995; Monahan et al., 2001). In *M. tuberculosis*, the range of chaperone families induced in the intracellular environment includes α-crystallins (Hsp20), GroELs (Hsp60), DnaK (Hsp70) and ClpB (Hsp100) (Monahan et al., 2001; Schnappinger et al., 2003). Presumably their chaperone functions are essential to protect the bacteria against the stresses induced by the antimicrobial mechanisms of the macrophage.

To try to gain an insight to the conditions and stresses present in the macrophage phagosome, Schnappinger et al. *(Schnappinger* et al., *2003)* used the approach of comparing the transcriptional profile of *M. tuberculosis* induced following phagocytosis by resting and activated murine macrophages to different *in vitro* stimulons; heat shock (Stewart et al., 2002), SDS (Manganelli et al., 2001), low oxygen (Sherman et al., 2001), low pH (Fisher et al., 2002), hydrogen peroxide, nitric oxide (NO) (Voskuil et al., 2003), low iron (Rodriguez et al., 2002), palmitic acid and starvation (Betts et al., 2002). Subsets of the intraphagosomally regulated genes could be categorized as belonging to the different stress stimulons. A group of genes was also identified as the general stress subset, as these intraphagosomally upregulated genes were induced by more than one individual stress. Several chaperone genes are present in this list, and prominent among these is the *acr2* gene. Once again this appears to confirm the position of Acr2 as the utility stress protein. Two particular stimulons were induced specifically in the phagosomes of activated macrophages: the hypoxic/low dose-NO stimulon which is controlled by the DosR/DosS two component regulator (Sherman et al., 2001; Voskuil et al., 2003) and the oxidative/nitrosative (high NO) stimulon. The hypoxic dormancy regulon/stimulon does not include any genes that are part of other stress stimulons such as SDS, heat shock, nutrient deprivation and oxidative stress and thus forms a very specific adaptation which is not part of the general stress response. Perhaps to fill the functional void left by the absence of any general stress response chaperones, the hypoxic/NO response includes its own α-crystallin chaperone, *acr1*, whose expression is induced more than 200-fold during macrophage infection (Schnappinger et al., 2003; Yuan et al., 1998).

Extension of these findings to the analysis of bacterial stress response genes during whole animal infection has relied largely on the analysis of individual genes using RT-PCR. In mice, the expression of the NO-regulated *acr1* gene is induced by an impressive 800-fold compared to its level in broth culture (Schnappinger et al., 2003). Induction kinetics follow the profile of the host cellular immune response and expression of iNOS (Shi et al., 2003), further confirming its reliance on NO as a stimulus. Analogous studies in human tissues have not proved as conclusive in demonstrating increased expression *in vivo* (Timm et al., 2003). Although elevation of gene expression is indicative of importance, it is desirable to make a more direct assessment of functional importance by generation of mutant strains and observation of resultant phenotypes during infection. A strain of *M. tuberculosis* deleted of *acr1* was severely attenuated for survival in macrophage culture (Yuan et al., 1998). An *acr2* mutant strain was able to grow in macrophages and to establish a persistent infection in mice which was of equal bacterial load to that of the wild-type *M. tuberculosis*. However, there was a marked alteration in disease progression measured by reduced weight loss over a prolonged infection. This correlated with a reduced immune response to the *acr2* mutant and a reduction in immune related lung pathology (Stewart et al., 2005). These findings demonstrate that both α-crystallins contribute to persistent infection with *M. tuberculosis*, at least in the mouse model of tuberculosis, and suggest that manipulation of *acr* expression can influence the host response to infection.

Whole genome expression profiles of *M. tuberculosis* during murine infection again show a breadth of chaperone expression to be prominent (Talaat et al., 2004) but similar data from human studies are not presently available.

PROBLEMS FOR PATHOGENS EXPRESSING CHAPERONES

It is thus clear that the expression of chaperone stress proteins is of vital importance during infection. The host immune response has learned to take advantage of the ubiquity of chaperone molecules among microbes and the requirement for their elevated expression during infection. Several reports describe specific receptors on the surface of immune cells that recognise mycobacterial GroEL, DnaK and GroES (Huang et al., 2000; Lewthwaite et al., 2001; MacAry et al., 2004; Meghji et al., 1997; Peetermans et al., 1993; Peetermans et al., 1994). Ligation of these receptors initiates a rapid inflammatory response thus alerting the innate immune system to the presence of the pathogen. In addition to acting as a danger signal, the peptide-binding function of *M. tuberculosis* DnaK is utilized by the host as a convenient means to scavenge pathogen peptides for presentation to T-cells (Stewart and Young, 2004). DnaK is captured by receptors on antigen presenting cells, internalized, and any pathogen peptides to which it is bound are processed for MHC-class I presentation and the resultant generation of antigen specific CD8+ T cell responses (Srivastava, 2002). In addition, T-cell recognition of chaperones themselves is a strong feature of *M. tuberculosis* infection. Recognition of GroEL and DnaK has been exploited in experimental vaccination strategies (Lowrie et al., 1999; Tascon et al., 1996). Indeed the expression of Acr2 early after infection has been correlated with prominence of this stress protein as an early target of the immune response in cattle and humans making it an interesting candidate vaccine molecule (K. Wilkinson unpublished).

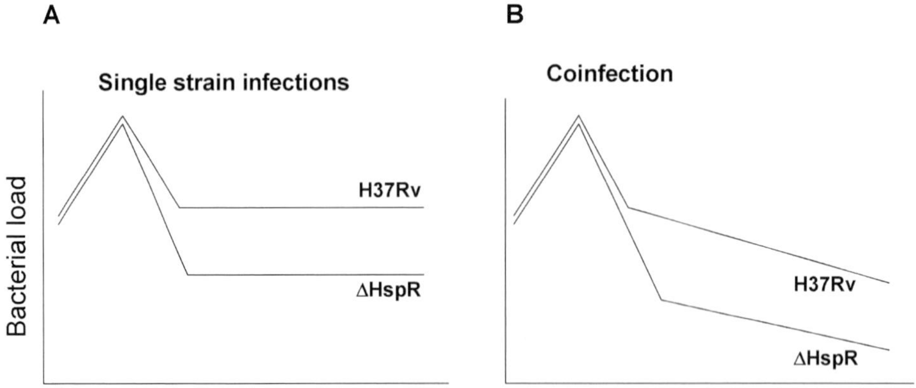

Fig. 5. Deregulation of the *M.tuberculosis* DnaK response by deletion of the HspR repressor leads to DnaK overexpression and attenuation of the bacteria during the chronic phase of murine infection. (A) Bacterial loads in spleens of mice infected with either wild-type or hspR-mutant *M. tuberculosis*. Attenuation is due to increased stimulation of the immune response by the *hspr*-mutant strain. (B) In mice coinfected with both wild-type and *hspr*-mutant strains, the HspR mutant causes a concomitant reduction of wild-type bacterial loads.

It is thus critical that the pathogen regulates chaperone expression, balancing the need to stabilize stressed proteins against the risk of activating the host immune response. In support of this are experiments which have demonstrated that a strain of *M. tuberculosis* deleted of the HspR heat shock repressor protein and constitutively overexpressing DnaK is impaired in its ability to persist during the chronic stage of murine infection (Stewart et al., 2001). One explanation for this is that the uncontrolled release of Hsp70 from mycobacterial cells promotes an increased MHC class I presentation of peptides. Indeed, it was shown that infection with a BCG strain deleted of HspR induces an increased number of CD8 γ-IFN secreting T-cells. Further evidence that this attenuation is immune mediated is provided by experiments in which HspR mutant *M. tuberculosis* concomitantly reduces the bacterial load of a coinfected wild-type strain (Fig. 5). Clearly manipulation of heat shock protein expression presents a potential route to improved infection control via novel vaccination or immunotherapy strategies.

CONCLUSION

Understanding the biology of model stress responses has made a significant contribution to our appreciation of how mycobacteria maintain homeostasis during stress conditions. It has also made a significant impact on our perception of how mycobacteria regulate the transcription of their genes to produce a plastic and adaptable response to environmental change. In addition, it has informed as to the nature of the intracellular environment and the host/pathogen interaction during infection with pathogenic mycobacteria. The role of stress proteins as both protectors of the bacterial cell and stimulators of innate and adaptive immune responses in the host presents an interesting dilemma for the bacterium. We anticipate that as we continue to learn more about the stress responses and stress proteins we will be further enlightened on the pathogenesis of mycobacteria.

WEB RESOURCES

The Bacterial Microarray Group at St Georges (Bugs) Home page and database is available at http://bugs.sghms.ac.uk/

The Graphgene software for drawing directed graph comparisons of microarray data is housed at http://zebrafish.doc.ic.ac.uk

REFERENCES

Akaki, T., Sato, K., Shimizu, T., Sano, C., Kajitani, H., Dekio, S., and Tomioka, H. (1997). Effector molecules in expression of the antimicrobial activity of macrophages against *Mycobacterium avium* complex: roles of reactive nitrogen intermediates, reactive oxygen intermediates, and free fatty acids. J. Leukoc. Biol. *62*, 795-804.

Alfano, C., and McMacken, R. (1989). Heat shock protein-mediated disassembly of nucleoprotein structures is required for the initiation of bacteriophage lambda DNA replication. J. Biol. Chem. *264*, 10709-10718.

Betts, J.C., Lukey, P.T., Robb, L.C., McAdam, R.A., and Duncan, K. (2002). Evaluation of a nutrient starvation model of *Mycobacterium tuberculosis* persistence by gene and protein expression profiling. Mol. Microbiol. *43*, 717-731.

Bucca, G., Hindle, Z., and Smith, C.P. (1997). Regulation of the *dnaK* operon of *Streptomyces coelicolor* A3(2) is governed by HspR, an autoregulatory repressor protein. J. Bacteriol. *179*, 5999-6004.

Buchmeier, N.A., and Heffron, F. (1990). Induction of Salmonella stress proteins upon infection of macrophages. Science *248*, 730-732.

Dieli, F., Troye-Blomberg, M., Ivanyi, J., Fournie, J.J., Krensky, A.M., Bonneville, M., Peyrat, M.A., Caccamo, N., Sireci, G., and Salerno, A. (2001). Granulysin-dependent killing of intracellular and extracellular *Mycobacterium tuberculosis* by Vgamma9/Vdelta2 T lymphocytes. J. Infect. Dis. *184*, 1082-1085.

Eriksson, S., Lucchini, S., Thompson, A., Rhen, M., and Hinton, J.C. (2003). Unravelling the biology of macrophage infection by gene expression profiling of intracellular *Salmonella enterica*. Mol. Microbiol. *47*, 103-118.

Ezaki, B., Ogura, T., Mori, H., Niki, H., and Hiraga, S. (1989). Involvement of DnaK protein in mini-F plasmid replication: temperature-sensitive seg mutations are located in the *dnaK* gene. Mol. Gen. Genet. *218*, 183-189.

Fernandes, N.D., Wu, Q.L., Kong, D., Puyang, X., Garg, S., and Husson, R.N. (1999). A mycobacterial extracytoplasmic sigma factor involved in survival following heat shock and oxidative stress. J. Bacteriol. *181*, 4266-4274.

Fisher, M.A., Plikaytis, B.B., and Shinnick, T.M. (2002). Microarray analysis of the *Mycobacterium tuberculosis* transcriptional response to the acidic conditions found in phagosomes. J. Bacteriol. *184*, 4025-4032.

Gahan, C.G., O'Mahony, J., and Hill, C. (2001). Characterization of the *groESL* operon in *Listeria monocytogenes*: utilization of two reporter systems (*gfp* and *hly*) for evaluating *in vivo* expression. Infect. Immun. *69*, 3924-3932.

Gamer, J., Multhaup, G., Tomoyasu, T., McCarty, J.S., Rudiger, S., Schonfeld, H.J., Schirra, C., Bujard, H., and Bukau, B. (1996). A cycle of binding and release of the DnaK, DnaJ and GrpE chaperones regulates activity of the *Escherichia coli* heat shock transcription factor sigma32. Embo J. *15*, 607-617.

Gaywee, J., Radulovic, S., Higgins, J.A., and Azad, A.F. (2002). Transcriptional analysis of *Rickettsia prowazekii* invasion gene homolog (*invA*) during host cell infection. Infect. Immun. *70*, 6346-6354.

Goloubinoff, P., Mogk, A., Zvi, A.P., Tomoyasu, T., and Bukau, B. (1999). Sequential mechanism of solubilization and refolding of stable protein aggregates by a bichaperone network. Proc. Natl. Acad. Sci. U S A *96*, 13732-13737.

Grandvalet, C., Servant, P., and Mazodier, P. (1997). Disruption of *hspR*, the repressor gene of the *dnaK* operon in *Streptomyces albus* G. Mol Microbiol *23*, 77-84.

Haranaga, S., Yamaguchi, H., Ikejima, H., Friedman, H., and Yamamoto, Y. (2003). *Chlamydia pneumoniae* infection of alveolar macrophages: a model. J. Infect. Dis. *187*, 1107-1115.

Hecker, M., Schumann, W., and Volker, U. (1996). Heat-shock and general stress response in *Bacillus subtilis*. Mol. Microbiol. *19*, 417-428.

Helmann, J.D., Wu, M.F., Kobel, P.A., Gamo, F.J., Wilson, M., Morshedi, M.M., Navre, M., and Paddon, C. (2001). Global transcriptional response of *Bacillus subtilis* to heat shock. J. Bacteriol. *183*, 7318-7328.

Huang, Q., Richmond, J.F., Suzue, K., Eisen, H.N., and Young, R.A. (2000). In vivo cytotoxic T lymphocyte elicitation by mycobacterial heat shock protein 70 fusion proteins maps to a discrete domain and is CD4(+) T cell independent. J. Exp. Med. *191*, 403-408.

Karzai, A.W., and McMacken, R. (1996). A bipartite signaling mechanism involved in DnaJ-mediated activation of the *Escherichia coli* DnaK protein. J. Biol. Chem. *271*, 11236-11246.

Konieczny, I., and Liberek, K. (2002). Cooperative action of *Escherichia coli* ClpB protein and DnaK chaperone in the activation of a replication initiation protein. J. Biol. Chem. *277*, 18483-18488.

Lamb, F.I., Singh, N.B., and Colston, M.J. (1990). The specific 18-kilodalton antigen of *Mycobacterium leprae* is present in *Mycobacterium habana* and functions as a heat-shock protein. J. Immunol. *144*, 1922-1925.

Langer, T., Lu, C., Echols, H., Flanagan, J., Hayer, M.K., and Hartl, F.U. (1992). Successive action of DnaK, DnaJ and GroEL along the pathway of chaperone-mediated protein folding. Nature *356*, 683-689.

Lathigra, R.B., Butcher, P.D., Garbe, T.R., and Young, D.B. (1991). Heat shock proteins as virulence factors of pathogens. Curr. Top. Microbiol. Immunol. *167*, 125-143.

Laufen, T., Mayer, M.P., Beisel, C., Klostermeier, D., Mogk, A., Reinstein, J., and Bukau, B. (1999). Mechanism of regulation of Hsp70 chaperones by DnaJ cochaperones. Proc. Natl. Acad. Sci. U S A *96*, 5452-5457.

Lee, B.Y., and Horwitz, M.A. (1995). Identification of macrophage and stress-induced proteins of *Mycobacterium tuberculosis*. J. Clin. Invest. *96*, 245-249.

Lewthwaite, J.C., Coates, A.R., Tormay, P., Singh, M., Mascagni, P., Poole, S., Roberts, M., Sharp, L., and Henderson, B. (2001). *Mycobacterium tuberculosis* chaperonin 60.1 is a more potent cytokine stimulator than chaperonin 60.2 (Hsp 65) and contains a CD14-binding domain. Infect. Immun. *69*, 7349-7355.

Liberek, K., Wall, D., and Georgopoulos, C. (1995). The DnaJ chaperone catalytically activates the DnaK chaperone to preferentially bind the sigma 32 heat shock transcriptional regulator. Proc. Natl. Acad. Sci. U S A *92*, 6224-6228.

Lindquist, S. (1986). The heat-shock response. Annu. Rev. Biochem. *55*, 1151-1191.

Lowrie, D.B., Tascon, R.E., Bonato, V.L., Lima, V.M., Faccioli, L.H., Stavropoulos, E., Colston, M.J., Hewinson, R.G., Moelling, K., and Silva, C.L. (1999). Therapy of tuberculosis in mice by DNA vaccination. Nature *400*, 269-271.

MacAry, P.A., Javid, B., Floto, R.A., Smith, K. G., Oehlmann, W., Singh, M., and Lehner, P.J. (2004). HSP70 peptide binding mutants separate antigen delivery from dendritic cell stimulation. Immunity *20*, 95-106.

Manganelli, R., Dubnau, E., Tyagi, S., Kramer, F.R., and Smith, I. (1999). Differential expression of 10 sigma factor genes in *Mycobacterium tuberculosis*. Mol. Microbiol. *31*, 715-724.

Manganelli, R., Provvedi, R., Rodrigue, S., Beaucher, J., Gaudreau, L., and Smith, I. (2004). Sigma factors and global gene regulation in *Mycobacterium tuberculosis*. J. Bacteriol. *186*, 895-902.

Manganelli, R., Voskuil, M.I., Schoolnik, G.K., Dubnau, E., Gomez, M., and Smith, I. (2002). Role of the extracytoplasmic-function sigma factor sigma(H) in *Mycobacterium tuberculosis* global gene expression. Mol. Microbiol. *45*, 365-374.

Manganelli, R., Voskuil, M.I., Schoolnik, G.K., and Smith, I. (2001). The *Mycobacterium tuberculosis* ECF sigma factor sigmaE: role in global gene expression and survival in macrophages. Mol. Microbiol. *41*, 423-437.

McKinney, J.D., Honer zu Bentrup, K., Munoz-Elias, E.J., Miczak, A., Chen, B., Chan, W.T., Swenson, D., Sacchettini, J.C., Jacobs, W.R., Jr., and Russell, D.G. (2000). Persistence of *Mycobacterium tuberculosis* in macrophages and mice requires the glyoxylate shunt enzyme isocitrate lyase. Nature *406*, 735-738.

Meghji, S., White, P.A., Nair, S.P., Reddi, K., Heron, K., Henderson, B., Zaliani, A., Fossati, G., Mascagni, P., Hunt, J.F., et al. (1997). *Mycobacterium tuberculosis* chaperonin 10 stimulates bone resorption: a potential contributory factor in Pott's disease. J. Exp. Med. *186*, 1241-1246.

Mehlert, A., and Young, D.B. (1989). Biochemical and antigenic characterization of the *Mycobacterium tuberculosis* 71kD antigen, a member of the 70kD heat-shock protein family. Mol. Microbiol. *3*, 125-130.

Mogk, A., Tomoyasu, T., Goloubinoff, P., Rudiger, S., Roder, D., Langen, H., and Bukau, B. (1999). Identification of thermolabile *Escherichia coli* proteins: prevention and reversion of aggregation by DnaK and ClpB. Embo J. *18*, 6934-6949.

Monahan, I.M., Betts, J., Banerjee, D.K., and Butcher, P.D. (2001). Differential expression of mycobacterial proteins following phagocytosis by macrophages. Microbiology *147*, 459-471.

Muffler, A., Bettermann, S., Haushalter, M., Horlein, A., Neveling, U., Schramm, M., and Sorgenfrei, O. (2002). Genome-wide transcription profiling of *Corynebacterium glutamicum* after heat shock and during growth on acetate and glucose. J. Biotechnol. *98*, 255-268.

Narberhaus, F. (1999). Negative regulation of bacterial heat shock genes. Mol. Microbiol. *31*, 1-8.

Narberhaus, F. (2002). Alpha-crystallin-type heat shock proteins: socializing minichaperones in the context of a multichaperone network. Microbiol. Mol. Biol. Rev. *66*, 64-93.

Nathan, C., and Shiloh, M.U. (2000). Reactive oxygen and nitrogen intermediates in the relationship between mammalian hosts and microbial pathogens. Proc. Natl. Acad. Sci. U S A *97*, 8841-8848.

Nyka, W. (1974). Studies on the effect of starvation on mycobacteria. Infect. Immun. *9*, 843-850.

Papatheodorou, I., Sergot, M., Randall, M., Stewart, G.R., and Robertson, B.D. (2004). Visualization of microarray results to assist interpretation. Tuberculosis (Edinb) *84*, 275-281.

Patel, B.K., Banerjee, D.K., and Butcher, P.D. (1991). Characterization of the heat shock response in *Mycobacterium bovis* BCG. J. Bacteriol. *173*, 7982-7987.

Peetermans, W.E., Langermans, J.A., van der Hulst, M.E., van Embden, J.D., and van Furth, R. (1993). Murine peritoneal macrophages activated by the mycobacterial 65-kilodalton heat shock protein express enhanced microbicidal activity *in vitro*. Infect. Immun. *61*, 868-875.

Peetermans, W.E., Raats, C.J., Langermans, J.A., and van Furth, R. (1994). Mycobacterial heat-shock protein 65 induces proinflammatory cytokines but does not activate human mononuclear phagocytes. Scand. J. Immunol. *39*, 613-617.

Qamra, R., Srinivas, V., and Mande, S.C. (2004). *Mycobacterium tuberculosis* GroEL homologues unusually exist as lower oligomers and retain the ability to suppress aggregation of substrate proteins. J. Mol. Biol. *342*, 605-617.

Raman, S., Song, T., Puyang, X., Bardarov, S., Jacobs, W.R., Jr., and Husson, R.N. (2001). The alternative sigma factor SigH regulates major components of oxidative and heat stress responses in *Mycobacterium tuberculosis*. J. Bacteriol. *183*, 6119-6125.

Rodriguez, G.M., Voskuil, M.I., Gold, B., Schoolnik, G.K., and Smith, I. (2002). ideR, An essential gene in *Mycobacterium tuberculosis*: role of IdeR in iron-dependent gene expression, iron metabolism, and oxidative stress response. Infect. Immun. *70*, 3371-3381.

Rutherford, S.L., and Lindquist, S. (1998). Hsp90 as a capacitor for morphological evolution. Nature *396*, 336-342.

Schaible, U.E., Sturgill-Koszycki, S., Schlesinger, P.H., and Russell, D.G. (1998). Cytokine activation leads to acidification and increases maturation of *Mycobacterium avium*-containing phagosomes in murine macrophages. J. Immunol. *160*, 1290-1296.

Schnappinger, D., Ehrt, S., Voskuil, M.I., Liu, Y., Mangan, J.A., Monahan, I.M., Dolganov, G., Efron, B., Butcher, P.D., Nathan, C., and Schoolnik, G.K. (2003). Transcriptional adaptation of *Mycobacterium tuberculosis* within macrophages: insights into the phagosomal environment. J. Exp. Med. *198*, 693-704.

Schroder, H., Langer, T., Hartl, F.U., and Bukau, B. (1993). DnaK, DnaJ and GrpE form a cellular chaperone machinery capable of repairing heat-induced protein damage. Embo J. *12*, 4137-4144.

Sherman, D.R., Voskuil, M., Schnappinger, D., Liao, R., Harrell, M.I., and Schoolnik, G.K. (2001). Regulation of the *Mycobacterium tuberculosis* hypoxic response gene encoding alpha -crystallin. Proc. Natl. Acad. Sci. U S A *98*, 7534-7539.

Shi, L., Jung, Y.J., Tyagi, S., Gennaro, M.L., and North, R.J. (2003). Expression of Th1-mediated immunity in mouse lungs induces a *Mycobacterium tuberculosis* transcription pattern characteristic of nonreplicating persistence. Proc. Natl. Acad. Sci. U S A *100*, 241-246.

Shinnick, T.M., Vodkin, M.H., and Williams, J.C. (1988). The *Mycobacterium tuberculosis* 65-kilodalton antigen is a heat shock protein which corresponds to common antigen and to the *Escherichia coli* GroEL protein. Infect. Immun. *56*, 446-451.

Srivastava, P. (2002). Roles of heat-shock proteins in innate and adaptive immunity. Nat. Rev. Immunol. *2*, 185-194.

Stenger, S., Hanson, D.A., Teitelbaum, R., Dewan, P., Niazi, K.R., Froelich, C.J., Ganz, T., Thoma-Uszynski, S., Melian, A., Bogdan, C., et al. (1998). An antimicrobial activity of cytolytic T cells mediated by granulysin. Science *282*, 121-125.

Stewart, G.R., Newton, S.M., Wilkinson, K.A., Humphreys, I., Murphy, H.M., Robertson, B.D., Wilkinson, R.J., and Young, D. (2005). The stress responsive chaperone alpha-crystallin 2 is required for pathogenesis of *Mycobacterium tuberculosis*. Mol. Microbiol. *55*, 1127-1137.

Stewart, G.R., Robertson, B.D., and Young, D.B. (2003). Tuberculosis: a problem with persistence. Nat. Rev. Microbiol. *1*, 97-105.

Stewart, G.R., Robertson, B.D., and Young, D.B. (2004). Analysis of the function of mycobacterial DnaJ proteins by overexpression and microarray profiling. Tuberculosis (Edinb) *84*, 180-187.

Stewart, G.R., Snewin, V.A., Walzl, G., Hussell, T., Tormay, P., O'Gaora, P., Goyal, M., Betts, J., Brown, I.N., and Young, D.B. (2001). Overexpression of heat-shock proteins reduces survival of *Mycobacterium tuberculosis* in the chronic phase of infection. Nat. Med. *7*, 732-737.

Stewart, G.R., Wernisch, L., Stabler, R., Mangan, J.A., Hinds, J., Laing, K.G., Young, D.B., and Butcher, P.D. (2002). Dissection of the heat-shock response in *Mycobacterium tuberculosis* using mutants and microarrays. Microbiology *148*, 3129-3138.

Stewart, G.R., and Young, D.B. (2004). Heat shock proteins and the host pathogen interaction during bacterial infection. Current Opinion in Immunology *16*, 506-510.

Szabo, A., Langer, T., Schroder, H., Flanagan, J., Bukau, B., and Hartl, F. . (1994). The ATP hydrolysis-dependent reaction cycle of the *Escherichia coli* Hsp70 system DnaK, DnaJ, and GrpE. Proc. Natl. Acad. Sci. U S A *91*, 10345-10349.

Takayama, S., Reed, J.C., and Homma, S. (2003). Heat-shock proteins as regulators of apoptosis. Oncogene *22*, 9041-9047.

Talaat, A.M., Lyons, R., Howard, S.T., and Johnston, S.A. (2004). The temporal expression profile of *Mycobacterium tuberculosis* infection in mice. Proc. Natl. Acad. Sci. U S A *101*, 4602-4607.

Tascon, R.E., Colston, M.J., Ragno, S., Stavropoulos, E., Gregory, D., and Lowrie, D.B. (1996). Vaccination against tuberculosis by DNA injection. Nat. Med. *2*, 888-892.

Timm, J., Post, F.A., Bekker, L.G., Walther, G.B., Wainwright, H.C., Manganelli, R., Chan, W.T., Tsenova, L., Gold, B., Smith, I., et al. (2003). Differential expression of iron-, carbon-, and oxygen-responsive mycobacterial genes in the lungs of chronically infected mice and tuberculosis patients. Proc. Natl. Acad. Sci. U S A *100*, 14321-14326.

Via, L.E., Fratti, R.A., McFalone, M., Pagan-Ramos, E., Deretic, D., and Deretic, V. (1998). Effects of cytokines on mycobacterial phagosome maturation. J. Cell. Sci. *111*, 897-905.

Viitanen, P.V., Gatenby, A.A., and Lorimer, G.H. (1992). Purified chaperonin 60 (groEL) interacts with the nonnative states of a multitude of *Escherichia coli* proteins. Protein. Sci. *1*, 363-369.

Voskuil, M.I., Schnappinger, D., Visconti, K.C., Harrell, M.I., Dolganov, G.M., Sherman, D.R., and Schoolnik, G.K. (2003). Inhibition of respiration by nitric oxide induces a *Mycobacterium tuberculosis* dormancy program. J. Exp. Med. *198*, 705-713.

Wawrzynow, A., Banecki, B., Wall, D., Liberek, K., Georgopoulos, C., and Zylicz, M. (1995). ATP hydrolysis is required for the DnaJ-dependent activation of DnaK chaperone for binding to both native and denatured protein substrates. J. Biol. Chem. *270*, 19307-19311.

Wickner, S., Hoskins, J., and McKenney, K. (1991). Monomerization of RepA dimers by heat shock proteins activates binding to DNA replication origin. Proc. Natl. Acad. Sci. U S A *88*, 7903-7907.

Xing, H., Mayhew, C.N., Cullen, K.E., Park-Sarge, O.K., and Sarge, K.D. (2004). HSF1 modulation of Hsp70 mRNA polyadenylation via interaction with symplekin. J. Biol. Chem. *279*, 10551-10555.

Young, D., Lathigra, R., Hendrix, R., Sweetser, D., and Young, R.A. (1988). Stress proteins are immune targets in leprosy and tuberculosis. Proc. Natl. Acad. Sci. U S A *85*, 4267-4270.

Yuan, Y., Crane, D.D., Simpson, R.M., Zhu, Y.Q., Hickey, M.J., Sherman, D.R., and Barry, C.E., 3rd (1998). The 16-kDa alpha-crystallin (Acr) protein of *Mycobacterium tuberculosis* is required for growth in macrophages. Proc. Natl. Acad. Sci. U S A *95*, 9578-9583.

Zolkiewski, M. (1999). ClpB cooperates with DnaK, DnaJ, and GrpE in suppressing protein aggregation. A novel multi-chaperone system from *Escherichia coli*. J. Biol. Chem. *274*, 28083-28086.

Zylicz, M., Ang, D., Liberek, K., and Georgopoulos, C. (1989). Initiation of lambda DNA replication with purified host- and bacteriophage-encoded proteins: the role of the *dnaK, dnaJ* and *grpE* heat shock proteins. Embo J. *8*, 1601-1608.

Chapter 8

Mycobacterial Dormancy and Its Relation to Persistence

Michael Young, Galina V. Mukamolova and Arseny S. Kaprelyants

ABSTRACT

Latent tuberculosis is a long-term, asymptomatic infection caused by the persistence of *Mycobacterium tuberculosis* within the human body. "The question of how bacteria survive for decades in immunologically educated hosts without causing disease has puzzled microbiologists for a century" (Stewart et al., 2003). Several models have been established to mimic presumed or deduced features of the persistent state. Many, though not all of them, are associated with reduced bacterial culturability. Since our understanding of the precise nature of the organisms that persist during TB latency is so rudimentary, we do not know which models are most relevant. In this chapter we consider the proposition that persistence of *M. tuberculosis* during TB latency is associated with the adoption of a dormant state, in which the bacteria lose culturability, sometimes for protracted periods. We suggest that the persisting organisms may be a dynamic and mixed population, in which individual bacteria are able to adopt or cycle between different physiological states, some of which may be characterised by impaired culturability.

INTRODUCTION

In common with several other bacterial pathogens, *Mycobacterium tuberculosis*, the causative agent of human tuberculosis, has evolved mechanisms to ensure its survival inside macrophages and establish a long-term, asymptomatic infection of its host. The features that allow *M. tuberculosis* to remain within the body for extended periods without causing disease are not known. Apart from host immunity, which keeps the organism in check, the potential ability of *M. tuberculosis* to produce specialised forms that remain intact in spite of a vigorous immune response may contribute to the maintenance of a chronic, asymptomatic infection. The subject has been extensively reviewed, especially in recent years (Boshoff and Barry, 2005; Flynn and Chan, 2001b; Gomez and McKinney, 2004; Honer zu Bentrup and Russell, 2001; Manabe and Bishai, 2000; Parrish et al., 1998; Stewart et al., 2003; Wayne, 1994; Wayne and Sohaskey, 2001; Zahrt, 2003). This chapter, like a recent review by Zhang (2004), will focus on "microbiological" aspects of the interaction between *M. tuberculosis* and its human host during the establishment and maintenance of long-term asymptomatic infections.

The terms latency, persistence, dormancy and non-culturability, which are used throughout, require some initial explanation. Latency is a clinical term used to describe the asymptomatic, chronic human infection caused by *M. tuberculosis*. Following primary infection, the immune system usually controls but typically fails to eradicate the pathogen, leaving residual organisms somewhere within the body where they can remain for many years. Reactivation, or post-primary tuberculosis can occur when the host immune system senesces or decays during old age, or as a result of infection with HIV, for example (Flynn, 2004; Flynn and Chan, 2001a; Flynn and Chan, 2001b). There have been several recent well-documented cases of the activation of latent tuberculosis in patients receiving immunosuppressive treatment against other diseases, such as rheumatoid arthritis (reviewed by Zhang, 2004). Significant improvements in the diagnosis and preventative therapy of latent tuberculosis will be required to deal with this unwelcome side effect of drug treatment. The bacteria present in latently infected individuals show persistence (Wayne and Sohaskey, 2001). Very little is known about these persistent organisms, except that they are difficult to locate and to eradicate. *M. tuberculosis* also shows persistence during chemotherapy, which has to be prolonged in order to be effective. Complete clearance of the organism may take six months or more, suggesting that there is heterogeneity in the bacterial population, which contains some organisms in a phenotypic state that renders them insensitive to the action of anti-mycobacterial drugs (Mitchison, 1980). This provides an important clue as to the possible nature of these particular persisting organisms, since those that are actively growing are readily removed, but those with low metabolic activity are not. Treatment with antibiotics may even drive a fraction of the bacterial population into a persistent state. Moreover, it is possible (perhaps even likely) that the persistent organisms associated with naturally occurring latent tuberculosis are physiologically quite distinct from those associated with anti-tubercular chemotherapy. Persisting organisms may be in a state of dormancy, which has been defined as "a reversible state of low metabolic activity, in which cells can persist for extended periods without division" (Kaprelyants et al., 1993). Dormant organisms are only able to resume growth following a period of resuscitation. Dormancy is often associated with non-culturability. This is an operational term describing an ill-defined range of physiological states, in which organisms temporarily lose the ability to form colonies on agar-solidified medium, though they may retain the ability to multiply in a liquid medium, which promotes their resuscitation (Mukamolova et al., 2003).

The huge numbers of latently infected, asymptomatic individuals that harbour a persistent *M. tuberculosis* infection present a formidable challenge to current efforts aimed at substantially reducing the global burden of tuberculosis (http://www.nap.edu/books/0309070287/html/). These individuals form a vast reservoir for future infection associated with senescence of the immune system. Persistence is now the subject of intensive research. Although documented many decades ago and worked on quite extensively since, it is a sobering thought that we still understand very little about the organisms that persist in the human body in latent infections. In fact, our ignorance could hardly be more profound. Apart from the simple fact that persisting organisms do exist (the compelling evidence for this is summarised below) little more can be said with any degree of certainty. How many organisms are likely to persist in a latent infection? Where are they located? Are they metabolically active? Do they undergo multiplication? There has been much speculation about these and other related matters but despite decades of research, summarised in the reviews cited above, satisfactory answers are still unavailable. A variety of persistence models have been established, both *in vivo* using experimental animals, and *in vitro* to circumvent the obvious problems associated with locating, isolating and studying what may be small numbers of persisting organisms in the human host. Much of this chapter is devoted to a consideration of these models. Since we lack fundamental understanding of precisely what it is we are trying to model, great caution must be exercised in extrapolating the results obtained to persistence in the human host. It is even possible (though not likely in our opinion), that this considerable body of work may have little direct relevance to the phenomenon of persistence in the human body. Nevertheless, we have learned a great deal about the biology and the capabilities of *M. tuberculosis* from its behaviour in these models of persistence. Much of this knowledge is certainly relevant to the infection process, though not necessarily to persistence *per se*. New methods and approaches are required to study persistence of *M. tuberculosis* within the human body.

EVIDENCE FOR *M. TUBERCULOSIS* PERSISTENCE IN HUMANS

HISTOLOGY

Histological investigations of autopsy material have provided clear evidence for the presence of persistent organisms. Whilst *M. tuberculosis* is readily cultured from young tuberculous lesions, these become fibrous and calcified with time and it is increasingly difficult to recover viable bacteria from them (Vandivière et al., 1956). This may reflect death of the bacteria originally present, but the possibility that some organisms can persist in old tubercles, having adopted a non-culturable form, should not be excluded. There is direct evidence that organisms can persist outside tuberculous lesions, since they have been found in autopsy samples taken from asymptomatic individuals who have died of causes unrelated to TB (reviewed by Stewart et al., 2003). In some cases detection was by direct culture, in others samples were employed to infect experimental animals. For example, Opie and Aronson (1927) were able to culture bacteria from healthy lung tissue and lymph nodes and they suggested that organisms situated outside primary lesions were the most likely cause of post-primary disease. This would account for the location of post-primary lesions in the apical regions of the lung where primary lesions are not usually found (Balasubramanian et al., 1994).

IMMUNOLOGY

In spite of its well-known limitations, as discussed, for example by Bishai and colleagues (Parrish et al., 1998), skin testing using tuberculin, also known as purified protein derivative (PPD), is widely used to identify persons infected with *M. tuberculosis* and to assess cell-mediated immune responses to the organism. Most patients with active tuberculosis are PPD-positive, although there are some exceptions (Delgado et al., 2002). Approximately two billion people (*i.e.* one third of the entire world population) are PPD-positive (Dye et al., 1999). Some of these individuals show a positive skin test as a result of BCG vaccination, whereas others may have had previous contact with harmless environmental mycobacteria (Enarson, 2004; Fine et al., 1994), but the great majority are believed to have encountered *M. tuberculosis* itself. These individuals are likely to harbour a persistent infection, which can be maintained until death (Tufariello et al., 2003). More accurate indicators of previous exposure to the tubercle bacillus have recently been developed, based on the production of interferon γ by peripheral blood lymphocytes in response to exposure to specific *M. tuberculosis* antigens (Ewer et al., 2003; Mazurek et al., 2001), but the essential picture remains the same; the human population harbours a huge reservoir of persistent *M. tuberculosis*.

PCR

M. tuberculosis DNA has been found in nucleic acid extracted from apparently healthy lung tissue derived from individuals who died of causes other than tuberculosis. DNA was detected by conventional PCR, using as target the multiple copies of IS*6110* that are found in the bacterial genome. Confirmation was obtained by *in situ* PCR, which indicated that *M. tuberculosis* DNA, and presumably the organism itself, was associated with a variety of cell types in addition to macrophages (Hernández-Pando et al., 2000).

TB REACTIVATION

It would be very difficult to account for the substantial numbers of individuals that develop tuberculosis as a result of senescence of the immune system or HIV infection, especially in areas where the risk of exposure to active tuberculosis is extremely low, were it not the result of the recrudescence of a cryptic reservoir of infection within the body. In one particularly striking case, molecular typing methods showed that a man developed TB as a result of reactivation of the same organism that had infected his father some 33 years previously (Lillebaek et al., 2002). It has been estimated that reactivation occurs in between 5 % and 10 % of infected individuals, leading to active tuberculosis. Reactivation provides clear evidence for the presence of persisting organisms in a substantial proportion of the human population, but it is not yet clear where they are. One likely location is the lung, either within pre-existing granulomatous lesions, or at secondary sites, elsewhere (see above). However, some persistently infected patients go on to develop extra-pulmonary tuberculosis in the absence of any new pulmonary lesions, suggesting that bacilli may potentially be widely disseminated around different sites within the body (Parrish et al., 1998).

ATYPICAL FORMS

The possibility that the persisting *M. tuberculosis* cells associated with latent tuberculosis are non-culturable or even dormant is rather controversial. If they do exist, they would be of obvious significance, since their resuscitation could be responsible for reactivation disease. Their existence was originally proposed to account for the presence of acid-fast, but non-culturable bacteria in closed pulmonary lesions (*i.e.* those lacking any clear bronchial connection) (Medlar et al., 1952). These organisms might be dead, or they might be injured and potentially culturable following a period of resuscitation. The introduction of careful washing procedures and extended incubation times in liquid medium led to greater success in culturing bacteria from such lesions (Hobby et al., 1954; Vandivière et al., 1956), raising the interesting possibility that they are in a state of transient non-culturability or possibly even dormant and requiring appropriate conditions for their resuscitation.

Khomenko and colleagues have undertaken a detailed analysis of non-culturable forms in the human host and in experimentally infected animals. They established the existence of filterable or mini-forms of *M. tuberculosis* that remain within the tissues of guinea pigs after curing them with anti-tubercular drugs. After several months of therapy, viable *M. tuberculosis* could no longer be detected by standard plating procedures, although microscopic examination of organ homogenates filtered through 0.2 - 0.7-micron filters revealed the presence of electron-dense forms with a rounded shape and an average diameter of 0.25 microns. When administered to guinea pigs, these filtrates induced the development of tuberculosis and after several passages *M. tuberculosis* was isolated by standard culture methods. The authors suggested that their filterable forms could represent persistent organisms, which are able to convert to actively growing cells under appropriate conditions (Khomenko and Golyshevskaya, 1984). However, the nature of the infectious agent was not unequivocally established in these studies because pure cultures of the "mini forms" of *M. tuberculosis* were not isolated. In later work, this group reported the presence of filterable forms of *M. tuberculosis* in broncho-alveolar lavage of patients with tuberculosis. The morphology of these cells was similar to that of the cells found in guinea pigs (coccoid cells with a diameter of 0.2 - 1.5 microns with an electron dense cell wall) (Khomenko, 1987). These forms induced tuberculosis in 42 % of the guinea pigs tested and in 23 % of them pure cultures of *M. tuberculosis* were subsequently isolated (Khomenko et al., 1994).

It is also noteworthy that some instances of sarcoidosis may be associated with latent *M. tuberculosis* infection (Almenoff et al., 1996; Khomenko et al., 1994); although the organism cannot be cultured from sarcoid tissue it is sometimes detectable by PCR (Eishi et al., 2002; Popper et al., 1994). The aetiology of sarcoidosis remains unclear but this disease is characterised by the formation of granulomas similar to those associated with tuberculosis. In some cases sarcoidosis accompanies typical tuberculosis. Nevertheless, patients with sarcoidosis do not show a positive tuberculin skin test and anti-tubercular drugs are ineffective (Khomenko et al., 1994). In one study, 51 % of patients with sarcoidosis contained mini-forms of mycobacteria whereas only 3.5 % harboured normal bacilli. Again, administration of such forms to guinea pigs resulted in 20 % of the animals developing changes in their organs that are typical of *M. tuberculosis* infection and cultivation of organ homogenates revealed viable, colony-forming *M. tuberculosis*. The chronic type of sarcoidosis was accompanied predominantly by mini-forms of mycobacteria, whereas sampling of broncho-alveolar lavage from patients with acute exacerbated sarcoidosis revealed the presence of normal tubercle bacilli. The authors suggested that sarcoidosis could be a form of persistent tuberculosis (Khomenko et al., 1994).

Finally, cells with a substantially altered envelope (L-forms and organisms that are no longer acid-fast), representing an adaptation to the environment *in vivo* and especially to antibiotic treatment, have also been reported (Chandrasekhar and Ratnam, 1992; Khomenko et al., 1980; Seiler et al., 2003). Mycobacteria with a substantially altered cell envelope may require exacting conditions in order to bring them into laboratory culture (Mattman, 1970).

MODELLING PERSISTENCE OF *M. TUBERCULOSIS*

"The question of how bacteria survive for decades in immunologically educated hosts without causing disease has puzzled microbiologists for a century" (Stewart et al., 2003). Since it has been so difficult to pin down the precise location of persistent organisms within the human body, and they would in any case be extremely difficult to study *in situ*, a variety of model systems have been established, based on suppositions and predictions about the environment in which they may be located and their possible physiological state. None of the models (Table 1) has been rigorously validated as reproducing conditions akin to those experienced by the persistent organisms associated with latency. We shall consider the *in vivo* and *in vitro* models in turn.

IN VIVO MODELS

The main advantage of *in vivo* models is that the immune system of the host, known to play a very important role in regulating mycobacterial activity in latent infections, is an integral component. However, different organisms have different immune systems and that of the mouse, which is the most widely used experimental host, is quite different from our own. Not surprisingly, therefore, the pattern of disease progression in the mouse is very different from that in man (Gomez and McKinney, 2004; Stewart et al., 2003). Although there is no widely accepted animal model of persistence of *M. tuberculosis* in the human host, the mouse remains the most

Table 1. Persistence models.

Model	Antibiotic treatment required	Culturability (CFU)	Number of non-culturable cells	Proteomic or transcriptomic data (reference)
In vivo models				
Cornell (McCune et al., 1957; Scanga et al., 1999)	Yes	Zero	Not known	Not currently applicable
Chronic, low dose (Orme and Collins, 1994; Phyu et al., 1998)	No	0 - 10^6 per organ	10^4-10^6 per organ	Not currently applicable
Artificial granuloma	No	ca. 10^3 per granuloma	Not known	(Karakousis et al., 2004)
Ex vivo models				
Cultured macrophages (Paul et al., 1996) (Biketov et al., 2000; Zhang et al., 1998)	No	10^3 - 10^6 per ml lysate	Up to 10^5 per ml lysate	Not currently applicable
Splenocytes from BCG-vaccinated mice (Turner et al., 2002)	No	10^3 per ml lysate	Not known	Not currently applicable
In vitro models				
Wayne (Wayne and Hayes, 1996) Wayne and Sohaskey, 2001)	No	High	Close to 0	(Boon et al., 2001; Rosenkrands, 2002; Voskuil et al., 2004; Muttucumaru et al., 2004; Starck et al., 2004; Boshoff et al., 2004)
Prolonged stationary phase (Hu et al., 2000)	Yes	Low, close to zero	10^3 per ml	(Hu et al., 2001)
Prolonged stationary phase (Shleeva et al., 2004; Shleeva et al., 2002)	No	Low, close to zero	10^6 - 10^7 per ml	(work in progress) (Hampshire et al., 2004)*
Prolonged stationary phase (Sun and Zhang, 1999; Zhang et al., 2001)	No	10^5 per ml	Present but not enumerated	(Yuan et al., 1996)
Starvation in phosphate buffer (Betts et al., 2002)	No	10^7 per ml	Probably close to zero	(Betts et al., 2002)
Amino acid starvation of a proline auxotroph (Parish, 2003)$	No	ca. 10^8 per ml	Probably close to zero	Not currently available

*These authors used a prolonged stationary phase model similar to that of Shleeva et al. (2002), except that the oxygen tension was maintained at 50 % saturation. The bacteria lost culturability (CFU), but the number of non-culturable cells (difference between MPN and CFU counts) was not reported.

$In this investigation, two other amino acid auxotrophs (requiring tryptophan and histidine) failed to survive following amino acid starvation, although all three auxotrophs maintained viability as well as the wild type when transferred to water.

convenient and most widely used organism for modelling purposes. Alternatives that are likely to assume greater importance in the future are guinea pigs, rabbits and monkeys, in which the progression of the disease more closely resembles that in the human host. Animal models have been used extensively to evaluate the effects of defined genetic lesions on the ability of *M. tuberculosis* to colonise and cause disease (see below). A drawback of the two *in vivo* models that have been most widely used, is that sufficient numbers of persisting cells of *M. tuberculosis* cannot be isolated to explore their physiology and biochemistry by standard methods, although quantitative, real time RT-PCR has been employed to investigate the expression of selected genes (see below). However, a new model introduced very recently, has effectively overcome this limitation (Karakousis et al., 2004).

CORNELL MODEL OF TUBERCULOSIS

This animal model of latency, established at Cornell University, USA, in the1950s, is based on the treatment of *M. tuberculosis*-infected mice with a mixture of antibiotics. After treatment, the mice attain a superficially sterile condition during which no viable bacteria are detectable by organ (lung and spleen) culture. However, during a further 3-month period without antibiotics, about one third of the mice will proceed to develop tuberculosis (McCune et al., 1957) and the number of animals showing relapse is further increased by treatment with immunosuppressive drugs (*e.g.* corticosteroids) (McCune et al., 1966a; McCune et al., 1966b). There are numerous variants of this model, depending on the precise experimental procedure, *e.g.* mouse genotype and lineage, size of inoculum and how it is prepared and administered, and length of the drug-free rest period, to name but a few. In general, the model is difficult to reproduce and there is no standard established protocol (Scanga et al., 1999). Moreover, its relevance to latent tuberculosis in humans is not clear (see discussion below). Nevertheless, it has been widely exploited and has proved particularly useful for characterising those branches of the immune system that are involved in controlling *M. tuberculosis* reactivation – see, for example, (van Pinxteren et al., 2000).

There have been many failed attempts to isolate viable organisms (using both solid and liquid media and a range of culture conditions) during the apparently sterile phase of the Cornell model. The absence of culturable organisms has promoted speculation concerning the origin of the viable organisms that can be isolated from animal organs during the subsequent relapse phase. Either the mice contain very small numbers of viable *M. tuberculosis* that have so far escaped detection, or the organisms present are somehow altered (*e.g.* injured) and unable to form colonies (*i.e.* they are non-culturable). This dilemma was partially resolved by the detection of DNA equivalent to ca. 10^5 bacteria per organ (lung or spleen) using PCR (de Wit et al., 1995). Of course, this might simply represent residual DNA resulting from the lysis of dead cells, but the presence of non-culturable organisms has been confirmed by the detection of transcripts corresponding to several different genes as well as 16S RNA (Hu et al., 2000; Pai et al., 2000). This is significant evidence indicating that the superficially sterile mouse tissues contain non-culturable, but metabolically active cells of *M. tuberculosis*. Since metronidazole has no activity, either in the initial sterilising phase or in the subsequent sterile state in the Cornell model (Brooks et al., 1999; Dhillon et al., 1998), it may be inferred that (at least some of) the bacteria present do not have an anaerobic type of metabolism as would be expected were they in a hypoxic micro-environment. *M. tuberculosis* reactivation was prevented by treatment with a combined course of isoniazid, pyrazinamide and rifampin (Dhillon et al., 1996), but not by rifapentine monotherapy (Miyazaki et al., 1999). It is possible that the residual non-culturable bacilli in the Cornell model are injured cells or L-forms, resulting from the action of the antibiotics used to establish the apparently sterile state, as has been suggested for human patients undergoing anti-tubercular chemotherapy (Khomenko et al., 1980). Alternatively, prolonged antibiotic therapy may select a sub-population of metabolically inactive persisters that have lost culturability. Irrespective of their precise nature, about which we can only speculate at present, these persisters may require exacting conditions in order to recover or resuscitate. However, we still do not know how many non-culturable cells actually retain the capacity for *in vivo* resuscitation. One estimate suggests that reactivation disease in the mouse may be caused by the resuscitation of less than 10 organisms (LeCoeur et al., 1989), which may account for the negative results obtained when attempting to cultivate them.

CHRONIC MOUSE MODEL OF TUBERCULOSIS

This *in vivo* model seems less complicated than the Cornell model but its relevance to tuberculosis latency in humans is also questionable (see discussion below). To establish chronic tuberculosis, animals are infected with a low dose of *M. tuberculosis* via various different routes - see Phyu et al. (1998) and references therein. For example, following intravenous administration of 10^1-10^3 CFU of *M. tuberculosis*, viable counts (CFU) of bacteria in organs show an initial lag followed by a transient increase of the bacterial load (acute phase) after which, a steady-state level of CFU (that is significantly lower than the maximum) is established (chronic phase) (Brown et al., 1995). The mice survive the chronic infection for a long time with bacteria in their organs in the absence of significant pathology. However, if the level of adrenocorticotropic hormone is increased experimentally to suppress immunity, there is a significant increase in the bacterial load and the development of active disease (Brown et al., 1995). This is reminiscent of the reactivation of tuberculosis in the Cornell model (see above). The results following intravenous infection are highly dependent on the infection dose (which varies significantly in different studies), the way the inoculum is prepared, and the mouse line. More stable results are obtained when a small number of organisms are administered in the form of aerosol particles (Orme and Collins, 1994; Rhoades et al., 1997). This permits long-term survival of the mice with relatively high and essentially stable numbers of bacteria in the lungs and spleen (Dahl et al., 2003). Although less frequently used, intra-peritoneal administration of *M. tuberculosis* results in a moderate and stable CFU load in the organs over a period of 40- 50 weeks post-infection (Phyu et al., 1998, Dhillon et al., 2004).

The chronic infection model has been extensively used to study the significance of host immunity in the establishment of a chronic infection and in subsequent reactivation of the disease. In particular, the significance of interferon-gamma production and the roles of CD4+ and CD8+ T-cells in controlling bacterial infection were established (Gomez and McKinney, 2004).

ARTIFICIAL GRANULOMAS

Very recently, Bishai and colleagues have developed a novel *in vivo* model of persistence based on the formation of artificial granulomas in mice (Karakousis et al., 2004). Mycobacteria encapsulated into hollow polyvinylidene fluoride fibres that permit the diffusion of molecules up to 500 kD have been implanted in mice. The subsequent formation of granulomatous lesions around these "capsules" containing mycobacteria permitted concomitant study of bacterial adaptation and the host response. In spite of the fact that granulomas form outside the lung, this technique represents a huge advance compared with other *in vivo* models, since mycobacteria persisting in the granulomas can be investigated readily. However, like other mouse models, there are significant differences in disease progression compared with the human host. The authors found that encapsulated mycobacteria *in vivo* (in artificial granulomas) but not *in vitro* entered an altered physiological state. These bacteria did not lose culturability, but they had very low metabolic activity (as judged by measurement of luciferase activity from a *lux* fusion). They also had a pattern of antibiotic-sensitivity that is characteristic of bacteria in persistent infections. Interestingly, the cells persisting in these artificial granulomas needed more time to form colonies than similarly encapsulated cells grown *in vitro*. The authors proposed that their low metabolic activity might be a reason for this difference, but it is also possible that the cells maintained *in vivo* were injured in some way. The experiments were terminated after 28 days, so it is not clear whether the surviving cells would have become non-culturable after more prolonged persistence in the granuloma environment.

EX VIVO MODELS OF PERSISTENCE IN MACROPHAGES

M. tuberculosis infects and initially lives within mononuclear phagocytes where its behaviour has been extensively documented (Deretic and Fratti, 1999; McDonough et al., 1993; Moulder, 1985; Russell, 2001). Murine macrophages have been widely used for this purpose, as have human monocyte-derived cell lines such as THP-1. Various methods, including microarrays, have been employed to examine the gene expression profiles of mycobacteria isolated from infected macrophages (Dubnau et al., 2002; Graham and Clark-Curtiss, 1999; Schnappinger et al., 2003). These have revealed that the organism induces genes characteristic of hypoxia and nitrosative and oxidative stress, as well as genes for β-oxidation of fatty acids, and that there is some remodelling of the cell envelope. These studies have defined the primary interactions that occur between the tubercle bacillus and the cells that confront them when they invade the human or animal host. Virulent strains of *M. tuberculosis* resist killing by reactive oxygen and nitrogen intermediates (Chan et al., 1992; Flesch and Kaufmann, 1991; Moulder, 1985) by preventing phagosome maturation (Russell, 2001) and although they may suffer transient injury, they multiply within the phagocyte (Falcone et al., 1994; North and Izzo, 1993). Avirulent strains generally proliferate less well in phagocytes than virulent strains, although they can persist (or even multiply slowly) in human monocytes (Silver et al., 1998), human macrophages (Zhang et al., 1998) murine splenocytes (Turner et al., 2002) and murine macrophages (McDonough et al., 1993; Paul et al., 1996). Of course, any macrophage-based model of persistence must be regarded with extreme care; the persistence time within the macrophage is only a matter of days (limited by the *in vitro* life time of the cell culture) as compared with months for the *in vivo* models discussed above, and years for latent tuberculosis of humans.

There have been very few studies to date concerning the physiological properties of mycobacteria obtained from macrophages. In one such report (Paul et al., 1996), measurements were made of both total count (microscopic examination) and viable count (CFU) of *M. tuberculosis* within human monocyte-derived macrophages six days post-infection. A significant proportion (between 70 % and 90 %) of the bacteria obtained from some macrophage cultures were unable to form colonies. These non-culturable bacteria might be injured, or moribund, or even dead following persistence within the macrophage environment. Alternatively, they could be in a non-culturable or dormant state. Evidence in favour of the latter was obtained in another investigation of the culturability of *M. tuberculosis* following persistence in murine peritoneal macrophages (Biketov et al., 2000). Each phagocyte contained between two and four acid-fast bacteria and although the total count of acid-fast bacteria isolated from the macrophages between 4 and 14 d post-infection was quite uniform (ca. 10^6 organisms ml^{-1}), the number of CFU in these suspensions was extremely variable from one experiment to another (from zero to 10^4 CFU ml^{-1}). The total count and the viable count differed by between two and six orders of magnitude in different experiments. The great majority of these organisms were unable to form colonies and these populations showed no detectable respiratory activity. Moreover, these bacteria had atypical morphology (short rods or ovoid cells) and they had lost the ability to adsorb specific bacteriophages indicative of alterations to the cell envelope. The persistence of these organisms within macrophages was associated with the adoption of a non-culturable (possibly dormant) state, since significant resuscitation occurred when the viable count was estimated in a liquid medium (MPN assay), especially when the initial number of CFU was very low. Addition of a protein called Rpf (see below) to the culture medium further enhanced viability, sometimes by several orders of magnitude. The population of persisting organisms was heterogeneous, since only a few percent of the total was recoverable/resuscitable.

IN VITRO MODELS

The main advantage of *in vitro* models is that the bacteria are available in quantity and are readily accessible to experimental manipulation. As a result, much is known about the physiology and biochemistry of these organisms, especially those that persist in the Wayne model and, to a lesser extent, the stationary phase models (see below). However, they do have major disadvantages, some of which have been alluded to previously. We do not know how accurately they reflect aspects of persistence *in vivo*. To illustrate the need for caution in this regard, it has been shown that a class of genes specifically up-regulated in amphibian granulomas caused by *Mycobacterium marinum* were not activated *in vitro* in response to various conditions thought to be operative in the granuloma environment (Chan et al., 2002). Moreover, *in vitro* models cannot mimic the immune system of the host, which plays a cardinal role in regulating mycobacterial activity in latent infections. A further potential drawback, which applies to all models but is more readily noticeable *in vitro*, is the problem of heterogeneity within bacterial populations that frequently contain mixtures of culturable/non-culturable and viable/dead cells, the proportions of the different types varying according to the culture conditions (Davey and Kell, 1996).

WAYNE'S MICROAEROPHILIC MODEL OF PERSISTENCE

Wayne's model of "non-replicating persistence" (see Wayne and Sohaskey (2001) for a review) is based on the premise that organisms surviving *in vivo*, possibly within closed cavities in the lung, are likely to experience an oxygen deficit. There is certainly ample evidence that oxygen starvation restricts mycobacterial growth *in vivo*. Open lesions in the lung with direct connection to the airways are usually populated with culturable, acid-fast organisms, whereas closed lesions with no such direct connection are not. TB was commonly treated during the 1930s –1950s by collapse therapy effectively reducing the oxygen supply to the infected tissue and restricting further active multiplication of the bacillus. Direct analysis of gas samples in the two types of cavities showed a markedly reduced oxygen tension in closed cavities as compared with open cavities (Haapanen et al., 1959). The genome sequence (http://genolist.pasteur.fr/TubercuList/) indicates that *M. tuberculosis* is capable of using alternative terminal electron acceptors such as nitrate and fumarate instead of oxygen (Cole et al., 1998). Although normally regarded as an aerobe, when given time to adjust to gradual oxygen depletion, *M. tuberculosis* can persist, but not apparently multiply, in a state of anaerobiosis *in vitro* for protracted periods (Corper and Cohn, 1933). Wayne (1976) initially used stationary cultures in liquid medium and showed that re-suspension and aeration of the settled organisms led to the immediate resumption of growth, indicating that the mycobacteria were in a state of oxygen deficit-induced growth arrest. Wayne and Lin (1982) demonstrated that abrupt oxygen depletion in axenic liquid cultures resulted in rapid cell death. In contrast, under conditions of slow oxygen depletion, which permitted the gradual adjustment of cellular metabolism, the bacteria adapted by entering a stable non-replicating, persistent (NRP) state. Bacteria growing exponentially in sealed flasks with a defined culture to headspace ratio and with slow stirring, consume the available oxygen and generate a temporal oxygen gradient. After a brief microaerophilic state (NRP-1) they enter an anoxic stationary phase (NRP-2) during which DNA synthesis is very low indeed but ATP levels are sufficient to ensure the maintenance of viability (CFU) for many days (Wayne and Hayes, 1996). When oxygen was introduced into the anaerobic culture, the bacilli immediately started to multiply, exhibiting synchronous division during first 30 h of growth. Although the term "dormancy" is widely used in the literature in connection with the Wayne model, we consider it inappropriate, since the mycobacteria are in a state of oxygen starvation-induced growth arrest from which they do not require resuscitation in order to resume multiplication.

Later it was shown that under these conditions, the bacteria shut down protein synthesis, and that it re-starts after oxygen is introduced (Hu et al., 1998). The authors considered the switching off of protein synthesis as an important component of the overall mechanism for the adoption of a non-replicating state. Very similar results have been obtained with *Mycobacterium bovis* (Lim et al., 1999). Remarkably, the non-pathogenic, rapidly growing organism, *Mycobacterium smegmatis*, also undergoes a similar transition to a non-replicating state, including arrest of DNA and RNA synthesis, under the conditions described above. Synchronous cell division occurs after oxygen availability is restored, suggesting that growth arrest occurs at a defined stage of the bacterial cell cycle (Dick et al., 1998). A simple method permitting the establishment of a non-replicating state for *M. smegmatis* using solid medium (agar plates) in anoxic jars has also been described (Lim and Dick, 2001). Maintenance of the bacteria anaerobically under these conditions did not permit colony formation after 10 days. However, 90 % of the CFU initially plated were recovered within 3 days when the jars were re-opened to re-introduce oxygen. Very similar results have been obtained using *Micrococcus luteus*, another non-pathogenic member of the actinobacteria (A.S. Kaprelyants, unpublished).

Wayne's microaerophilic model of TB persistence has important medical implications, as organisms in the NRP-2 state were comparatively insensitive to antibiotics such as isoniazid and rifampin (Wayne and Sramek, 1994), which are routinely employed to treat tuberculosis. This is not surprising, since these antibiotics are inactive against non-growing organisms in general (which again stresses the difficulties associated with eradication of any non-replicating cells of *M. tuberculosis* doing anti-tubercular therapy). Metronidazole, a drug that has been used successfully for the treatment of anaerobic infections (*e.g. Helicobacter pylori*), was active against *M. tuberculosis* in the oxygen starvation-induced non-replicating state, though it has no effect on bacteria growing exponentially in the presence of oxygen (Wayne and Sramek, 1994). Similarly, non-replicating cells of *M. bovis* and *M. smegmatis* induced by oxygen-starvation were also sensitive to metronidazole (Dick et al., 1998; Lim and Dick, 2001). These findings suggested that metronidazole might be employed therapeutically against persistent organisms *in vivo*. However, attempts to use it to kill persistent *M. tuberculosis* in mice (Cornell model) were unsuccessful (Dhillon et al., 1998), indicating that these organisms were probably not in a sufficiently hypoxic state. Interestingly, other instances where mycobacteria were resistant to rifampin, isoniazid and metronidazole have been reported *e.g.* under conditions of nutrient starvation in phosphate buffer (Betts et al., 2002).

PERSISTENCE IN PROLONGED STATIONARY PHASE

The multiplication pattern of *M. tuberculosis* in mouse organs can be divided into several phases, which are typical of those associated with the growth of the organism in batch cultures *in vitro* (lag, exponential and stationary phases). Some of the characteristics of cells in stationary phase *in vitro* seem to resemble those of persistent cells *in vivo* and this evident similarity has led to the development of stationary phase models of persistence. One salient feature that they share is the formation of non-culturable (and in some cases dormant) cells.

The first documented demonstration of persistence *in vitro* was probably that of Corper and Cohn (1933) who showed that organisms could be revived from cultures established 12 years previously and maintained without agitation in sealed vessels thereafter. Following a brief initial period of growth, these organisms presumably survived under anaerobic conditions in stationary phase. In detailed studies, conducted much later, prolonged incubation (up to two years) of *M. smegmatis* cells in aerobic stationary phase did not result in a transition to dormancy at least for the majority of cells, which remained metabolically active and able to divide. Because the CFU number of this culture decreased slightly over the period of measurement, the opposing processes of cell multiplication and cell death were presumably taking place simultaneously (Smeulders et al., 1999). The behaviour of *M. tuberculosis* cells in prolonged stationary phase under aerated conditions was similar to that of *M. smegmatis*: the number of culturable bacteria did not change significantly over a period of 8 months (Shleeva et al., 2002) indicating that the majority of cells in stationary phase cultures are active and not dormant.

The imposition of microaerophilic conditions using several different experimental approaches led to the formation of apparently dormant cells of *M. tuberculosis* after prolonged stationary phase. Hu et al. (2000) established microaerophilic conditions by keeping *M. tuberculosis* cells in stationary phase without shaking to achieve a gradient of oxygen. After 100 days in stationary phase, viable cells were killed by the addition of a high dose of rifampin to the culture. After several days of incubation with the antibiotic the CFU count was zero. However, estimates of the number of viable cells in liquid medium using MPN methods revealed the presence of 2×10^3 cells per ml, indicating the presence of a significant sub-population of bacteria that are non-culturable on plates but able to resuscitate in liquid medium. Because the rifampin-resistance of these cells was phenotypic, the authors concluded that they represent a persistent sub-population. The persisting organisms incorporated uridine (at a reduced rate) and they contained mRNA and rRNA. The authors concluded that the persisting cells of *M. tuberculosis* "may exist in some physiological forms in which limited metabolism occurs" and that this accounted for their antibiotic tolerance (Hu et al., 2000). While this approach clearly shows the formation of a defined sub-population of cells with a persistent phenotype, the possibility that this population arose after antibiotic treatment and did not pre-exist in the stationary phase population was not excluded.

Shleeva et al. (2002) used filtration to separate non-culturable cells from viable cells in static *M. tuberculosis* cultures kept in tubes for several months without oxygen input. These cultures were passed through a series of filters of decreasing pore size (20 - 0.45 μm). It was expected that cells with a non-culturable phenotype would have a reduced size, as had been observed previously with *M. luteus* and *R. rhodochrous* (Mukamolova et al., 1995a; Shleeva et al., 2002). Indeed, the culturability of the cells (estimated as the ratio CFU/total count) decreased with decreasing pore size of the filter. The 1.5 μm filtrate was unable to form colonies on agar despite the fact that microscopy revealed the presence of large numbers of cells (total count 4×10^7 cells ml^{-1}). Non-culturable cells in the 1.5 μm filtrate were a mixture of ovoid cells with a length of ca. 1.1 μm and small individual cocci with a diameter of 0.5-0.7 μm. Both types of cells had an intact membrane barrier and lacked detectable respiratory activity. The culturability of the cells in the 1.5 μm filtrate was time-dependent: it was minimal ca. 4 months post-inoculation, after which the CFU number gradually increased and approached the total count after 8 months of cultivation. The transient character of non-culturable cell formation in these experiments could reflect the re-growth of a small number of initially viable cells or the resuscitation of non-culturable cells or a mixture of both processes in the highly heterogeneous bacterial population (Shleeva et al., 2002). It is not known whether these filterable forms of *M. tuberculosis* obtained *in vitro* resemble the previously mentioned "mini forms" observed *in vivo* following antitubercular drug therapy (Khomenko and Golyshevskaya, 1984; Khomenko, 1987; Khomenko et al., 1994).

To avoid problems associated with heterogeneity, *M. tuberculosis* cultures were grown in sealed flasks and shaken at 200 rpm for up to 8 months. Under these conditions, the cell population contained many of the typical ovoid and coccoid cells after 4 months of cultivation. These cultures contained zero CFU, even after incubation of the plates for a 2-month period. In contrast to cells grown in static cultures this state of non-culturability was maintained for the remainder of the experimental period (4 - 8 months post-inoculation) (Shleeva et al., 2002). However, since only approximately 10^5 cells per ml were able to resuscitate, it was not possible to assess their metabolic status.

Thus, it seems that cultivation of *M. tuberculosis* in a prolonged "anoxic" stationary phase without antibiotic treatment leads to the formation of a persistent population of cells characterised by non-culturability on solid medium. This is a reversible state that disappears following a period of resuscitation (see below), indicating that these cells may be considered dormant. The proportion of such cells in the bacterial population is not constant and depends on the cultivation conditions. Generally, conditions that decrease heterogeneity and cryptic growth and synchronize the culture result in uniform populations of non-culturable cells.

The procedures so far devised for producing persistent populations of non-culturable cells of *M. tuberculosis* involve protracted exposure to hypoxic conditions, but we suspect that it will be possible to find conditions under which aerobically grown cells of *M. tuberculosis* can also become dormant, as has been demonstrated for other actinobacteria. In *M. smegmatis*, for example, variously modified media were used to support aerobic growth at a reduced rate and under these conditions, bacterial populations lost culturability very substantially (in some experiments completely) during stationary phase (Shleeva et al., 2004). Some residual metabolic activity was measurable in these non-culturable populations (Kuznetsov et al., 2004) suggesting that despite their non-culturability, these bacteria are not completely dormant. In contrast, conditions that supported aerobic growth at maximal rates did not lead to the formation of non-culturable cells in stationary phase and neither did the sudden imposition of starvation for either carbon or nitrogen or phosphorus. In starved cultures, cell lysis accompanied by cryptic growth occurred during stationary phase. The formation of dormant and non-culturable cells of *M. luteus* after aerobic cultivation in a chemostat at a very low dilution rate (Kaprelyants and Kell, 1992), and of *Rhodococcus rhodochrous* grown aerobically under poor nutrient conditions (Shleeva et al., 2002), suggests that this conclusion can be generalised to encompass many of the actinobacteria. It therefore seems likely that aerobically grown cells of *M. tuberculosis* can potentially form non-culturable cells in stationary phase under conditions that await discovery. Indeed, we have very recently shown that a substantial proportion of the bacteria in cultures of *M. tuberculosis* in a suitably modified Sauton's medium could adopt a transiently non-culturable state under aerobic conditions (M. O. Shleeva, personal communication). Two aerobic models of persistence in long-term stationary phase in Middlebrook 7H9 medium under oxygen-sufficient conditions have been reported (Hampshire et al., 2004; Sun and Zhang, 1999). In the latter, the authors showed that only 10^5 cells ml^{-1} of *M. tuberculosis* H37Ra could produce colonies on agar plates, but the addition of spent culture medium from early stationary phase significantly increased the viability as estimated by the MPN method (Sun and Zhang, 1999). Hampshire et al. (2004) showed that bacteria lost viability (CFU) substantially during the period 20 – 70 days post-inoculation, after which the number of CFU started to increase again. The authors did not make MPN measurements during the experiment and the number of non-culturable cells cannot therefore be assessed, but the general behaviour of these bacteria resembles that observed in the investigation, in which non-culturability was demonstrated.

M. TUBERCULOSIS CULTURABILITY IN THE *IN VITRO*, *EX VIVO* AND *IN VIVO* MODELS

If the models of mycobacterial persistence briefly described above reflect the transition of *M. tuberculosis* to the state of persistence in the human host, one may expect them to share similar properties. One such feature is that populations of mycobacteria in the different models tend to be heterogeneous, but this is probably true of bacterial populations in general, even those grown under carefully controlled conditions of continuous culture (Davey and Kell, 1996). Bacterial culturability in the various models shows very significant differences. The organisms that persist in the Cornell mouse model show very low culturability, whereas those that persist in the chronic mouse infection and artificial granuloma models retain substantial culturability. In the *in vitro* models of mycobacterial persistence there are similar disparities. Bacteria in the Wayne model show oxygen deficit-induced growth arrest at a low culture density and retain high viability by CFU, whereas those that persist in prolonged stationary phase can have very low culturability - sometimes CFU are undetectable, depending on the precise experimental conditions (Hu et al., 2000; Shleeva et al., 2002). In the *ex vivo* model of persistence in macrophages culturability is variable. Of course, the really important question is: are the organisms that persist in the human host culturable or not and if not, are they dormant? Unfortunately, we do not have a clear answer to this simple question and therefore, the relevance of culturability to the phenomenon of persistence remains controversial.

In the *in vivo* Cornell model, bacteria lose culturability following the period of antibiotic treatment. This correlates well with the poor viability of bacteria isolated from organs of humans with latent TB (Wayne, 1960), suggesting that non-culturability could be an intrinsic property of persisting cells of *M. tuberculosis* in the human host. The proportion of such cells in bacterial populations can vary substantially, and unless these cells represent a substantial majority of the total, it is very difficult to study them using modern analytical procedures, like proteomics and transcriptomics.

In the context of dormancy, we may reiterate the widely accepted definition of the phenomenon as a "*reversible state of low metabolic activity, in which cells can persist for extended periods without division*" (Kaprelyants et al., 1993). If we consider Wayne's model in the light of this definition it is noteworthy that the persisting organisms develop sensitivity to metronidazole as they become anaerobic (Wayne and Sramek, 1994) indicating that they remain metabolically active. Moreover, they do not require a period of resuscitation in order to resume active growth. Therefore, they should not be considered as dormant. In contrast, the non-culturable cells that persist in prolonged stationary phase either have substantially reduced metabolic activity (Hu et al., 2000; Kuznetsov et al., 2004) or no detectable metabolic activity (Shleeva et al., 2004; Shleeva et al., 2002) and their culturability is restored, at least partially, as a result of resuscitation, suggesting that they are in a state approaching dormancy.

To estimate the proportion of dormant cells in a bacterial population, they must first be resuscitated to restore their ability to grow, which is required for conventional counting procedures (CFU). A method commonly used for resuscitating bacteria is to incubate them in a liquid medium (see below for more details). In the case of dormant cells formed in prolonged stationary phase, this results in the detection (by MPN) of significantly more viable cells than are detected by direct plate counts. Sometimes, resuscitation can lead to differences of several orders of magnitude between MPN and CFU measurements. Unfortunately, the number of *ex vivo* or *in vivo* experiments in which culturability has been estimated using liquid medium (MPN) is very small. Some of them have revealed a significant increase in viable counts over CFU measurements, as was found for *M. tuberculosis* isolated from murine peritoneal macrophages (Biketov et al., 2000).

What is urgently required is a careful analysis of the culturability of cells of *M. tuberculosis* isolated from the organs of mice in the two *in vivo* models of persistence (Cornell model and the chronic infection model), with measurements of viability using both CFU and MPN procedures. Recently, Dhillon et al. (2004) examined the viability of cells of *M. tuberculosis* isolated from the organs of mice that had been chronically infected for 10 months. They compared the viability of cells by CFU and MPN measurements and found that the viable count by plating was only 1 to 3 % of the MPN, suggesting the presence of significant numbers of resuscitable bacteria, possibly in a dormant state. In some mice the CFU in organs was below the detection limit, whereas the MPN assay revealed the presence of 10^3 - 10^4 viable cells per ml in the same population (Dhillon et al., 2004). As far as we are aware, there has only been one report suggesting zero viability by both CFU and MPN measurement in the Cornell model (Hu et al., 2000). However, estimation of the number of viable cells using a liquid medium is very dependent on the precise conditions employed; those that permit resuscitation need to be established more or less empirically for each particular case (Mukamolova et al., 1998b, 2003; Shleeva et al., 2002). As might be expected, *in vitro* experiments indicate that the longer organisms stay in a dormant state, the more difficult it is to resuscitate them (M. Shleeva, unpublished observations). This may account for the fact that Hu et al. (2000) failed to resuscitate bacteria from the apparently sterile state in the Cornell model.

DORMANCY, NON-CULTURABILITY, AND THE SLOWLY REPLICATING STATE

M. tuberculosis contains 13 genes encoding sigma factors, some of which resemble the sporulation-specific sigma factors found in bacilli and clostridia (DeMaio et al., 1996). Genome sequencing has revealed that it lacks most of the other genes specifically associated with endospore formation in bacilli and clostridia, so it therefore seems very unlikely that *M. tuberculosis* can produce structures similar to profoundly dormant bacterial endospores. Nevertheless, there is a substantial body of evidence indicating that some non-sporulating bacteria can adopt a state similar to dormancy (reviewed by Barer and Harwood (1999) and Mukamolova et al. (2003). Essentially, the transition of non-sporulating cells to a dormant state is induced by deterioration of the environment such that conditions are no longer able to sustain active growth. This could potentially result in cell death, or starvation state survival or the formation of dormant forms. Under some starvation conditions cell death and lysis may cause the release of nutrients into the surrounding medium and this may provoke the renewed multiplication of surviving cells (cryptic growth). Although previously considered as a spontaneous and stochastic process, it is now clear that bacterial cell death can be genetically determined (Lewis, 2000; Nystrom, 2003) and that it is involved in the dissemination of biofilm organisms (Webb et al., 2003) and in facilitating cell multiplication in logarithmic phase (Voloshin et al., 2004). Bacterial cell death and associated cryptic growth may occur in some of the *M. tuberculosis* persistence models motioned above. For example, it could account for the maintenance of high viability and a constant number of CFU throughout protracted aerobic stationary phase *in vitro* or in the chronic mouse model *in vivo*.

The starvation survival response differs from dormancy because the bacteria retain culturability throughout, which enables them to resume growth as soon as favourable conditions are restored. It differs from cell-death associated cryptic growth, because there is a shut down of metabolic processes concerned with cell multiplication and growth. The transition from active metabolism (characteristic of dividing cells) to maintenance metabolism is an important feature of starvation survival. Valuable resources are conserved but the organism retains sufficient metabolic activity to support core metabolism and repair processes to maintain cell integrity and culturability (Morita, 1990). Starvation survival has been observed in *M. tuberculosis* following transfer to phosphate buffer or water and also in response to amino acid deprivation (Loebel, 1933; Betts et al., 2002; Parish, 2003). Wayne's anaerobic (or microaerophilic) model of the non-replicating persistent state fits neatly into this conceptual framework and may be considered as an example of an oxygen starvation-survival response.

Apart from low metabolic activity and growth arrest, a characteristic property of the dormant forms of non-sporulating bacteria is their non-culturability (Kaprelyants et al., 1993; Mukamolova et al., 2003). Opinions differ concerning the origin, mechanism and especially the significance of the non-culturable state (reviewed by Mukamolova et al., 2003). It is well documented that many bacteria can produce so called "viable but non-culturable" (VBNC) forms (Roszak and Colwell, 1987). Since the terminology used to describe these forms is both confused and confusing, as has been pointed out previously (Barer and Harwood, 1999; Kell et al., 1998), the term non-culturable has been used in a purely operational sense here (meaning not culturable on solid media). The available data suggest that non-culturability can be a consequence of a special survival strategy (Mukamolova et al., 2003). Non-culturable cells show some resemblance to the more specialised survival structures produced by sporulating bacteria, since they have altered morphology and are adapted for survival under stressful conditions. However only cells that satisfy several different criteria, including integrity of cellular components (*i.e.* not damaged), low metabolic activity and, most importantly, capacity to resuscitate, should be considered as having adopted the non-culturable state. Since the necessary measurements have not yet been made, we still do not know whether VBNC cells, obtained during starvation at low temperatures, persisting cells in latent infections *in vivo* and non-culturable cells arising in stationary-phase or chemostat cultures, are different manifestations of the same underlying process. Non-culturable but metabolically active (*i.e.* damaged or injured) cells should be excluded from this category of specialised surviving forms. Bearing all these criteria in mind, we may accept that among the various persistence models described above, only those mycobacterial cells with a non-culturable phenotype, characterised by a low metabolic activity and capable of resuscitation are truly dormant. According to these criteria, bacteria showing zero CFU during prolonged stationary phase and cells present in the apparently sterile phase of the Cornell model may be considered as dormant.

In order to become non-culturable, bacteria must first adapt to a low metabolic cost existence and as a consequence of this they then lose culturability. Shleeva et al. (2002) have previously suggested that Wayne's model may represent a step towards the emergence of non-culturable cells of *M. tuberculosis* that need to be resuscitated before they can resume active growth. Currently, anaerobic conditions are required to produce such cells during prolonged stationary phase, which might reflect a mechanistic link between Wayne's model and the stationary phase model. However, Shleeva et al. (2004) were able to induce a non-culturable state for the fast growing organism, *M. smegmatis*, under oxygen-sufficient conditions and similar results have also been obtained with other representatives of the actinobacteria like *M. luteus* and *R. rhodochrous* (Kaprelyants and Kell, 1993; Shleeva et al., 2002). Oxygen might be one of many possible inducers of non-culturability in *M. tuberculosis* but the induction of this state may also be possible aerobically, under as yet undefined conditions. It is interesting that attempts to use metronidazole as a drug to cure persistent *M. tuberculosis* infections in mice (Brooks et al., 1999), including the Cornell model (Dhillon et al., 1998) were unsuccessful, indicating that at least some persisting cells *in vivo* are not metabolising anaerobically. It is also known that cells persisting *in vivo* (unlike bacteria in anaerobic model) retain sensitivity to rifampin. Moreover, non-replicating persistence under hypoxia is not only observed in pathogenic *M. tuberculosis*; it is also a feature of the life style of fast growing, non-pathogenic organisms like *M. smegmatis* (Dick et al., 1998) and *M. luteus* (G. V. Mukamolova, unpublished).

Unlike profoundly dormant bacterial endospores, non-culturable forms of actinobacteria cannot survive for years. In this respect they resemble *Streptomyces* spores (generally accepted as specialised survival forms), which also have limited longevity. Moreover, they are much more sensitive to hostile environments than are the highly resistant bacterial endospores. We have suggested that actively growing cells, non-culturable forms, *Streptomyces* spores and endospores can be considered to represent different points on a continuum, where survival is favoured as metabolic activity is gradually reduced (Mukamolova et al., 2003). Persistent forms of *M. tuberculosis* localised in animal tissues also have a limited lifetime and their very persistence (for years) could actually arise from the spontaneous resuscitation of some cells followed by limited growth and then a transition to dormancy. Persistence may be a dynamic process involving cell multiplication, death and non-culturability. During the multiplication phase these organisms would become susceptible to anti-tubercular drugs, whereas during the non-culturable or dormant phase they would be refractory. This could account for the well-known fact that persistent organisms are only cleared from the body by protracted treatment with antibiotics.

Finally, stationary phase gene expression in *Escherichia coli* is essentially orchestrated by two proteins, the RpoS sigma factor and the leucine-responsive regulator, Lrp (Hengge-Aronis, 1993; Hengge-Aronis, 2002; Tani et al., 2002). Collectively, they activate a program of gene expression concerned with starvation-survival, which may eventually result in loss of culturability on solid medium (Ericsson et al., 2000). Presumably, adaptation to stationary phase in *M. tuberculosis* involves similar regulatory mechanisms, but whether or not there is a specific regulator of the persistent state *in vivo* remains an open question (see below).

ARE THE CURRENT MODELS ADEQUATE?

From the above discussion it is clear that the different approaches developed (either *in vitro* or *in vivo*) to model the latent state of human tuberculosis, possibly reflect different parts or different phases of establishment of persistence and none of them adequately covers all aspects of the phenomenon. Indeed, as we already discussed, viability/culturability of *M. tuberculosis* cells are very different in the various models. In some cases, moreover, persistence is established after antibiotic treatment, which may be more relevant to the phenomenon of persistence following incomplete anti-tubercular drug therapy than persistence following natural control of a primary infection by the host immune system. An evident disadvantage of all the *in vitro* models is the lack of any potential pressure that may be exerted by the host immune system leading to adoption of a non-culturable/dormant state. The arguably more relevant Cornell model produces such small numbers of non-culturable/dormant organisms that they cannot readily be analysed using modern biochemical or genetic methods (*e.g.* proteomics or transcriptomics). Table 1 summarises all the models indicating their positive and negative features.

CHANGES ASSOCIATED WITH THE TRANSITION TO THE PERSISTENT STATE

MORPHOLOGICAL CHANGES

In many previous studies in which bacterial cultures were subjected to nutrient starvation, the organisms changed their morphology to become smaller and more rounded in shape (Kaprelyants et al., 1993). For example, *M. luteus* has a coccoid morphology during active growth and the dormant cells formed during prolonged stationary phase have a reduced size and an unusually thick cell wall (Mukamolova et al., 1995b). Non-culturable cells of *M. tuberculosis* after several months of persistence in stationary phase *in vitro* become ovoid or coccoid in shape (Fig. 1), similar to the morphology of non-culturable cells of *R. rhodochrous* (Shleeva et al., 2002). In Wayne's model of persistence, *M. tuberculosis* cells are similarly characterised by the occurrence of cell wall thickening during NRP-1 (Cunningham and Spreadbury, 1998). Later, during NRP-2, a novel pigment appears in the cell envelope that is not observed under oxygen-sufficient conditions (Cunningham et al., 2004).

Less is known about morphological changes associated with persistence in animal models and in patients with latent tuberculosis. Persisting cells in patients gradually lose their acid-fast staining with the Ziehl-Neelsen reagent and this may be connected with the occurrence of alterations in the structure of the cell envelope (Seiler et al., 2003). The filterable forms extensively characterised by Khomenko and colleagues (Khomenko and Golyshevskaya, 1984; Khomenko, 1987; Khomenko

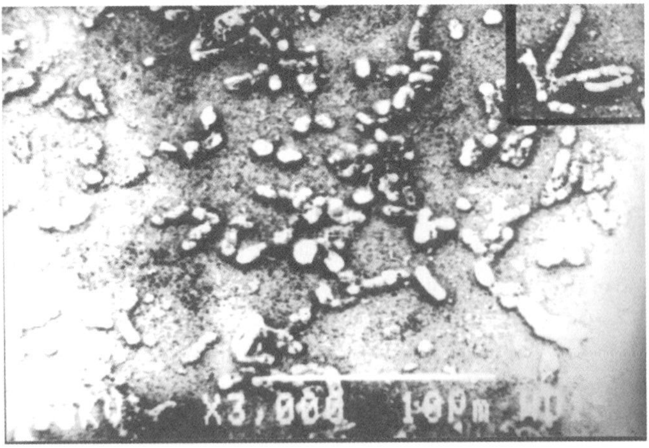

Fig. 1. Non-culturable cells of *M. tuberculosis*. *M. tuberculosis* cells were grown for 4 months in Sauton's medium and filtered through a 2 μm filter. These organisms fail to produce colonies when plated on agar-solidified medium, but they can be resuscitated in liquid medium under MPN conditions. Reproduced from Shleeva et al. (2002) with permission.

et al., 1994) have been discussed previously. The persisting cells present in the apparently sterile state of the Cornell model have not yet been identified so their morphological characteristics are unknown.

BIOCHEMICAL CHANGES

M. tuberculosis cells in a non-replicative state in Wayne's model of persistence have been quite extensively characterised. One of the most conspicuous changes is the appearance of the α-crystallin chaperone (Wayne and Hayes, 1998; Yuan et al., 1998). Alpha crystallin (HspX, Acr) is also induced in response to other stresses such as exposure to reactive nitrogen species (Garbe et al., 1999), or certain antibiotics (Boshoff, 2004; O. A. Turapov, unpublished observations) or, more generally, during stationary phase (Manabe et al., 1999; Yuan et al., 1996) and during persistence in macrophages (see below). Another characteristic change is the induction of enzymes concerned with nitrate reduction. In particular, NarX, a putative fused nitrate reductase, was strongly up regulated in the anaerobic NRP-2 stage of Wayne's model (Hutter and Dick, 1999). Wayne and Hayes have reported significant nitrite production by *M. tuberculosis* under anaerobic conditions (Wayne and Hayes, 1998) suggesting that in the absence of oxygen, the organism uses nitrate, when available, as an alternative terminal electron acceptor. This may have significance *in vivo*; although *M. bovis* BCG lacking nitrate reductase persisted normally in the spleen of immune-competent mice, it was unable to persist in the lungs, liver and kidneys (Fritz et al., 2002).

In addition to these characteristic changes in energy metabolism, cells of *M. tuberculosis* change their carbon metabolism significantly in order to survive under microaerophilic/anaerobic conditions. In the Wayne model, the glyoxylate bypass operates as an alternative to the complete tricarboxylic acid cycle and the activity of isocitrate lyase, the first enzyme of this shunt, increases five-fold (Wayne and Lin, 1982). Accordingly, the cells start to use fatty acids as a carbon source under anaerobic conditions (Wayne and Sohaskey, 2001). There is also a marked increase in the activity of glycine dehydrogenase, which could regenerate NAD from NADH by reductively aminating glyoxylate to form glycine (Wayne and Lin, 1982; Wayne and Sohaskey, 2001). An endogenous source of fatty acids that may be used during persistence in the Wayne model has recently been proposed. Daniel et al. (2004) have shown that triacylglycerol accumulates in response to hypoxia or treatment with nitric oxide. The authors also identified a cohort of 15 potential triacylglycerol synthases, which were up regulated under these conditions. They postulated that triacylglycerol plays a vital role in carbon and energy storage during long-term persistence (dormancy) of *M. tuberculosis*, as has been documented in other, completely unrelated, dormant organisms such as hibernating animals and oily seeds (Daniel et al., 2004). Interestingly, some of the bacilli in sputum samples from patients with active tuberculosis contain lipophilic inclusion bodies (Garton, 2002) that may represent deposits of triacylglycerol. Mycobacteria may use fatty acids released from the degrading host tissues to synthesise endogenous triacylglycerol (Daniel et al., 2004). Alternatively, free fatty acids could originate from the pathogen itself, since microarray studies (see below) show characteristic up regulation of genes encoding enzymes that participate in the β-oxidation of fatty acids under conditions of non-replicative persistence in the Wayne model (Muttucumaru et al. 2004; Voskuil et al., 2004). Genes involved in the β-oxidation of fatty acids as well as *icl* are also up regulated in mycobacteria exposed to acidic conditions (Fisher et al., 2002). Moreover, starvation in phosphate buffer resulted in a rapid shutdown of aerobic metabolism and significant up-regulation of a putative glycine dehydrogenase and fumarate reductase (Betts et al., 2002). It is possible that the transition from active aerobic metabolism to an alternative "anaerobic" pathway is part of a universal mechanism that provides a basis for "maintenance" metabolism permitting survival under conditions of starvation and other stresses.

Although the inhibition of some functions is almost certainly as important as the activation of others for achieving persistence, much less is known about the enzymes and pathways that are down regulated under conditions presumed to mimic persistence in the human host. It is well established that the transition of *M. tuberculosis*, *M. bovis* BCG and *M. smegmatis* to a non-replicative state in Wayne's model resulted in an inhibition of DNA and protein synthesis (Wayne and Sohaskey, 2001). Similarly, the transition of *M. luteus*, *R. rhodochrous* and *M. smegmatis* to a non-culturable state during prolonged stationary phase, led to almost complete inhibition of respiration under oxygen-sufficient conditions (Mukamolova et al., 1995b; Shleeva et al., 2004; Shleeva et al., 2002). In the case of dormant *M. luteus* this was connected with the inhibition of some of the respiratory chain enzymes (Mukamolova et al., 1995b). Shleeva et al. (ms. in preparation) have recently found that free fatty acids accumulate in the membranes of *M. smegmatis* during the transition to non-culturability and this could be a reason for the observed inhibition of respiration under these oxygen-sufficient conditions. Voskuil et al. (2003) showed that inhibition of *M. tuberculosis* respiration by nitric oxide *in vitro* resulted in the induction of the DosR regulon comprising some 48 genes (see below). Inhibition of respiration mediated by nitric oxide produced *in vivo* by macrophages, or more generally within the granuloma environment, could be an important event that provokes growth arrest and further changes may then ensue, resulting eventually in the assumption of a dormant state.

M. smegmatis also adopts a non-replicative persistent state similar to that of *M. tuberculosis* in the Wayne model (Dick et al., 1998). In this organism, the transition to non-replicative persistence was associated with increased activity of alanine dehydrogenase, which might participate in NADH regeneration by reductive amination of pyruvate under conditions of oxygen limitation (Hutter and Dick, 1998). The level of a 27-kD histone-like protein, whose role in the model remains to be established, also increased (Lee et al., 1998).

Summarizing this section, we should stress that the majority of the changes in cellular metabolism described above were observed using *in vitro* models (especially Wayne's model of persistence in response to hypoxia), which have serious drawbacks, since it is not known to what extent they mimic aspects of persistence of *M. tuberculosis* in the human host. Moreover, the observed changes are not specific for any particular model of persistence; rather they are part of general strategy of survival. Mutants unable to produce the 27-kD histone-like protein (Lee et al., 1998), α-crystallin (Yuan et al., 1998) and isocitrate lyase (McKinney et al., 2000) grew as well as the wild type under hypoxic conditions. Nevertheless, a strain unable to produce α-crystallin showed a reduced ability to multiply within macrophages (Yuan et al., 1998), whereas a strain unable to produce isocitrate lyase showed reduced persistence in activated macrophages (though it behaved as the wild type in resting macrophages) (McKinney et al., 2000). The observed differences between the behaviour of these strains *in vitro* and in macrophages serve to highlight the limitations of the Wayne hypoxic model of *M. tuberculosis* persistence *in vitro*, which, as we have suggested above, could mimic early events along one or more converging pathways of events culminating in the establishment of a dormant state.

GENES IMPLICATED IN PERSISTENCE

The response of the *M. tuberculosis* transcriptome to a variety of stresses *in vitro* has now been quite extensively characterised – see, for example, (Boshoff et al., 2004; Schnappinger et al., 2003; Wilson et al., 2001; Wilson et al., 1999) and gene expression profiles in various persistence models have been reported (Bacon et al., 2004; Hampshire et al., 2004; Kendall et al., 2004; Muttucumaru et al. 2004; Voskuil et al., 2004; Betts et al., 2002). The response of the *M. tuberculosis* transcriptome to macrophage infection has also been documented (Schnappinger et al., 2003), but no data are currently available for either the Cornell model or the chronic persistence model in the mouse.

As indicated above, the DosSR two-component system regulates the hypoxic response in mycobacteria (Boon and Dick, 2002; Boon et al., 2001; Sherman et al., 2001) and is also induced by nitric oxide (Voskuil et al., 2003). Transcriptional profiling of a *M. tuberculosis* strain in which expression of *dosSR* was abolished, established that most of the genes strongly up regulated in response to hypoxia are under DosR control (Park et al., 2003; Sherman et al., 2001). This regulon is also significantly up regulated in the artificial granuloma model in mice (Karakousis et al., 2004) and is associated with the expression of Th1-mediated immunity in mouse lungs (Shi et al., 2003). One very prominent member of the regulon is *hspX*, encoding the α-crystallin chaperone (Narberhaus, 2002), which is up regulated in response to a variety of environmental stresses in addition to hypoxia.

Global gene expression profiles during the Wayne model of persistence have been determined by Muttucumaru et al. (2004) and Voskuil et al. (2004). In one investigation, single time points representative of the NRP-1 (day 7) and NRP-2 (day 14) states were used, whereas in the other a kinetic analysis was made, with many more time points extending to 80 days post-inoculation. Both studies revealed that gene expression in the microaerophilic NRP-1 state differs from that in the anaerobic NRP-2 state. Some of the genes up regulated were known markers of the hypoxic response, including *dosSR* and members of the previously characterised DosR regulon including *hspX*, *narX* and *narK2*. The *narX* gene encodes a fused nitrate reductase that is co-transcribed with *narK2*, encoding its associated nitrite extrusion protein. Both are up regulated under similar conditions in *M. bovis* BCG (Hutter and Dick, 1999; Hutter and Dick 2000) and the former is also expressed *in vivo* (Fenhalls et al., 2002a). Both groups also reported up regulation of *fdxA* (ferredoxin), *metC* (homocysteine synthase), *lpqS* (lipoprotein of unknown function) as well as PPE17, PE11, *fprB* (probable NADPH reductase), *ahpD* (alkyl peroxidase) and *alkB* (fatty acid metabolism). The down regulation of many genes concerned with growth, biosynthesis and aerobic, oxidative metabolism was also documented.

Voskuil (2004), in a very useful compilation of the results of these array experiments and several others dealing with persistence *in vitro* under other somewhat related conditions, has remarked that there are many genes whose expression altered significantly in one study but not in the other. The discrepancies are probably attributable to differences in the procedures employed for undertaking the experiments and/or the methods used to analyse the data and the reader should consult the original publications for further details. Genes up regulated in one study, but not the other, include *gcvB* (potential glycine dehydrogenase), *ald* (secreted alanine dehydrogenase) and (*aceA*/*icl*) encoding isocitrate lyase. The activities of all three enzymes are known to increase in response to hypoxia (Andersen et al., 1992; Hutter and Dick, 1998; Raynaud et al., 1998; Wayne and Lin, 1982) and isocitrate lyase is expressed in, and required for persistence of *M. tuberculosis* in macrophages and mice (Graham and Clark-Curtiss, 1999; McKinney et al., 2000). Only eight genes were up regulated after persistence for 80 days under the hypoxic conditions of the Wayne model. These were *acr2* encoding a heat shock protein, *mez* encoding a malate oxidoreductase and six additional genes of unknown function. The functions of many of the most strongly up-regulated genes in the Wayne model are currently unknown and these are obvious targets for further analysis. The up regulation of *rpfA* during NRP-1 but not NRP-2 (Muttucumaru et al. 2004) is of interest in the context of this review, since this gene encodes a protein that helps to restore culturability to non-culturable organisms (see below).

Microarray analysis of gene expression in an aerobic model of persistence in long-term stationary phase in Middlebrook 7H9 medium under oxygen-sufficient conditions (Hampshire et al., 2004) showed the up regulation of many genes previously reported to be associated with growth *in vivo*. Several of these are located within the RD1 region, which is deleted from the attenuated strain, BCG

Pasteur (Behr et al., 1999; Gordon et al., 1999). Genes involved in the metabolism of fatty acids were also up regulated. Some of these genes were also up regulated in an independent study in which sudden starvation was imposed for a period of 96 hours, after which changes in gene expression were monitored using proteomic and transcriptomic analysis (Betts et al., 2002). These experiments corroborate the results of biochemical analyses indicating that there is modification of the lipid envelope and adaptation to fatty acids as an alternative energy source during stationary phase and they suggest that the *in vitro* models do have some relevance to what the organism experiences *in vivo*.

Schnappinger et al. (2003) isolated primary bone marrow-derived macrophages from wild type and NOS2-deficient mice and studied gene expression in the resting and interferon-γ-activated state following infection with *M. tuberculosis*. These experiments have revealed that adaptation to the intra-phagosomal environment occurs within 24 h of infection, with little change in the gene expression profile over the next 24 h; about 11 % of *M. tuberculosis* genes were up regulated and about 4 % were down regulated. The observed changes were validated by quantitative real time-PCR (qRT-PCR) analysis of selected genes and, moreover, the results obtained correlated well with similar analyses of RNA isolated from the lungs of mice 21 and 56 days after infection with *M. tuberculosis*. Many of the genes showing differential regulation in macrophages responded similarly *in vitro* when bacteria were exposed one or more of a variety of stresses, including nitric oxide, hypoxia, iron-limitation, nutrient starvation, or heat shock. The authors identified 68 genes that responded specifically to stimulation of the macrophages with interferon-γ, none of which responded in the NOS2 macrophages, indicating that the oxidative stress response of *M. tuberculosis* within the phagosome is induced by nitric oxide. Nitric oxide-independent up-regulation of several genes involved in repair of DNA damage was also documented.

Microarray data are also available for *M. tuberculosis* under a range of other environmental conditions *in vitro* including a steady state model under reduced oxygen tension (Bacon et al., 2004), a static culture with reduced oxygen tension (Kendall et al., 2004) and a long-term stationary phase model at high oxygen tension (Hampshire et al., 2004). Comparative analysis of gene expression profiles obtained from the Wayne model (Muttucumaru et al. 2004; Voskuil et al., 2004), cells in prolonged stationary phase (Voskuil et al., 2004) and nutrient depleted cells (Hampshire et al., 2004), *i.e.* conditions in which *M. tuberculosis* proliferation is strictly limited, thereby mimicking one important aspect of persistence, revealed some overlap of up regulated genes (see Table 1 in Voskuil, 2004). As more data sets are obtained and carefully analysed this type of approach may help to reveal potential candidate genes such as *acr2* (heat shock protein) whose expression may be important for persistence.

The behaviour of several defined mutant strains has been investigated *in vivo* during infection of the mouse. Many investigations have focused on whether or not

Table 2. Genes associated with persistence.

Gene	Function	Association with persistence	References
INFORMATION PATHWAYS			
sigC rv2069	RNA polymerase σ factor	Time-to-death phenotype in mice	(Sun et al., 2004)
sigE rv1221	RNA polymerase σ factor	Time-to-death phenotype in mice	(Ando et al., 2003)
sigD rv3414c	RNA polymerase σ factor	Time-to-death phenotype in mice	(Raman et al., 2004)
sigF rv3286c	RNA polymerase σ factor	Time-to-death phenotype in mice, important in late stage disease	(Chen et al., 2000) (Geiman et al., 2004)
sigH rv3223c	RNA polymerase σ factor	Reduced pathology, time-to-death phenotype	(Kaushal et al., 2002)
dnaE2 rv3370c	DNA-polymerase	Reduced persistence *in vivo*; time-to-death phenotype in mice	(Boshoff et al., 2003)
REGULATORY PROTEINS			
whiB3 rv3416	Transcriptional regulator	Reduced pathology, time-to-death phenotype	(Steyn et al., 2002)
hspR rv0353	Transcriptional repressor	Over-expression of heat-shock proteins, reduced persistence in mice	(Stewart et al., 2001)
mprA rv0981	Two-component regulator	Reduced persistence in mice	(Zahrt and Deretic, 2001)
devR/dosR rv3132c	Two-component regulator	Increased lethality in immunodeficient mice	(Parish et al., 2003a)
kdpDE rv1027/8c	Two-component regulator	Increased lethality in immunodeficient mice	(Parish et al., 2003a)
trcS rv1032	Two-component regulator	Increased lethality in immunodeficient mice	(Parish et al., 2003a)
tcrXY rv3464/5c	Two-component regulator	Increased lethality in immunodeficient mice	(Parish et al., 2003a)
INTERMEDIARY METABOLISM AND RESPIRATION			
icl rv0467	Isocitrate lyase	Metabolic defect, reduced persistence in mice	(McKinney et al., 2000)
relA rv2583c	GTP pyrophosphate kinase	Reduced survival under hypoxia and nutrient stress. Impaired ability to sustain chronic infection.	(Primm et al., 2000) (Dahl et al., 2003)
narGHJI rv1161-4	Nitrate reductase	Reduced persistence of BCG in mouse model	(Fritz et al., 2002)

DETOXIFICATION AND ADAPTATION			
katG rv1908	Catalase-peroxidase	Reduced persistence in mice. Reduced survival in mice between 2 and 4 weeks after infection.	(Li et al., 1998; Ng et al., 2004; Pym et al., 2001)
treS rv0126	Trehalose synthesis	Time-to-death phenotype in mice	(Stewart et al., 2003)

LIPID METABOLISM			
pks1/15 rv2946/7c	Phenolic glycolipid synthesis	Time-to-death phenotype in mice	(Stewart et al., 2003)
pks2 rv3825c	Sulpholipid synthesis	Time-to-death phenotype in mice or no difference with the wild type	(Stewart et al., 2003) (Rousseau et al., 2003)
mmpL8 rv3823c	Lipid transport	Time-to-death phenotype in mice	(Domenech et al., 2004)
pcaA rv0470c	Cyclopropane synthase	Altered cell wall, reduced persistence in mice	(Glickman et al., 2000)
PE PROTEINS			
mag24-1 rv1651c/3812	PE-PGRS	Reduced persistence of *M. marinum* in frog granuloma	(Ramakrishnan et al., 2000)

the mutants show altered virulence but some have specifically addressed the issue of persistence or at least noted phenotypic changes that appear to be connected with altered persistence. The results of these investigations are summarised in Table 2. Although seemingly obvious, it is worthwhile considering the various possible phenotypes one might expect to observe in the chronic mouse persistence models that have predominantly been employed to analyse these mutants. We may distinguish the phenotype associated with genes specifically required for the establishment of persistence from that associated with genes specifically required for its maintenance, whilst noting that some genes may be required for both establishment and maintenance. In the chronic mouse infection model, a strain lacking a gene specifically concerned with maintenance of the persistent state (but not with the active growth phase required for its establishment) would show normal initial growth, but the subsequent plateau phase would be replaced by a phase of declining bacillary load (Stewart et al., 2003). This is precisely the pattern observed with a *relA* mutant of *M. tuberculosis* (Dahl et al., 2003). This mutant, defective in the stringent response (Cashel et al., 1996), grows normally in macrophages and also on citrate or phospholipids (though not on some other carbon sources) *in vitro*, but it cannot maintain viability (*i.e.* CFU; MPN measurements were not reported) during either stationary phase or anaerobic incubation (Primm et al., 2000). Microarray analysis revealed that the *relA* mutant shows altered expression of many genes, two of which, *rpfA* (up regulated) and *rpfC* (down regulated), encode proteins that resemble the resuscitation-promoting factor of *M. luteus*. The available data indicate that *relA* is required for persistence during hypoxia, starvation, and prolonged stationary phase, and also in the chronic mouse infection and artificial granuloma models *in vivo*, making it a prime candidate for involvement in persistence in the human host. Incidentally, the presence of cognate phenotypes *in vitro* and *in vivo* lends some credence to the notion that the various models do have some commonality in the challenges that they pose to the persisting organism.

Altered (particularly increased) persistence may be an indirect effect of reduced virulence. This seems to be the case for several mutants affected in some of the sigma factors found in *M. tuberculosis* that have been studied extensively by Bishai and colleagues, as well as others, and they are denoted as having a "time-to-death" phenotype in Table 2. *M. tuberculosis* contains genes encoding thirteen different sigma factors, one or more of which could potentially control the expression of genes specifically required for persistence. Two of them, SigF and SigJ, are up regulated in stationary phase cultures of *M. tuberculosis*, which led to speculation that they may play a role in orchestrating gene expression in persistent organisms *in vivo* (DeMaio et al., 1996; Hu and Coates, 2001). However, in neither instance has this proved to be true, at least in the mouse. SigF controls the expression of the *acr* gene and is responsible for the high level of expression of its product, the alpha crystallin chaperone during stationary phase (Manabe et al., 1999). A *sigF* mutant of *M. tuberculosis* showed reduced expression of many genes during stationary phase *in vitro* including several involved in cell envelope *e.g.* sulpholipid biosynthesis. Initially the mutant proliferated in mice as well as the wild type, but the number of CFU in the organs stabilised sooner and at a lower level, giving rise to reduced histopathology and increased survival of the host (Chen et al., 2000; Geiman et al., 2004). Enhanced expression of *sigJ* is maintained *in vitro* even after stationary phase bacteria have been treated with rifampin (Hu and Coates, 2001). Surprisingly, a *sigJ* mutant showed no difference (compared with the wild type) in survival both in a microaerophilic stationary phase model and during long-term persistence in mice (Hu et al., 2004).

A *sigC* mutant of *M. tuberculosis* grew as well as the wild type strain in macrophages and established a similar bacterial load in the organs of laboratory mice (Sun et al., 2004). However, the mutant caused substantially reduced immunopathology; infection with the wild type strain caused 100 % mortality within 235 days whereas mice infected with the *sigC* mutant showed no mortality after 300 days (*i.e. increased* persistence). Microarray analysis of the *sigC* mutant *in vitro* showed that the SigC regulon includes several genes associated with virulence, including *hspX* (α-crystallin), *senX3* and *mtrA*, (both of which encode two-component regulators), *mpt83* and *fbpC* (a lipoprotein and a fibronectin-binding protein associated with host immune recognition), all of which were down regulated in the *sigC* mutant. A mutation completely inactivating the *senX3-regX3* gene pair impaired growth of *M. tuberculosis* in macrophages and in mice (Parish et al., 2003b), whereas *mtrA* is induced following entry of *M. bovis* BCG into macrophages (expression is constitutive in *M. tuberculosis*) and the gene is essential for growth *in vitro* (Zahrt and Deretic, 2000). Sun et al. (2004) propose that the *sigC* strain is attenuated as a result of a faulty adaptive response to the *in vivo* environment.

A *sigD* mutant of *M. tuberculosis* also shows a "time to death" phenotype in mice, associated with a reduced bacterial load in the lung and spleen. This sigma factor is present throughout growth and stationary phase and it is up regulated in response to starvation in a *relA*-dependent manner (Betts et al., 2002; Dahl et al., 2003). One of the three genes whose expression level dropped most dramatically in the *sigD* mutant was *rpfC* (as observed in a *relA* mutant – see above), which appears to be directly controlled by SigD. This was compensated in part by increased expression of *rpfA* and *rpfE* (Raman et al., 2004).

A similar *increased* ability to persist, indicated by a time-to-death phenotype in Table 2, is associated with a mutation in *sigE*, which encodes a sigma factor involved in the adaptive response of *M. tuberculosis* to heat shock, oxidative stress and low pH (Manganelli et al., 2001). The mutant was less able to persist within macrophages (Manganelli et al., 2001). Somewhat paradoxically, in a mouse infection model bacterial numbers in organs were unaffected, and survival was prolonged (Ando et al., 2003). A further example is provided by *sigH*, which encodes a sigma factor involved in the responses to oxidative and redox stress (Kaushal et al., 2002; Manganelli et al., 2002). A *sigH* mutant showed normal responses following infection of macrophages (Kaushal et al., 2002) but it had substantially reduced virulence in a mouse infection model (Manganelli et al., 2002) and consequently, it persisted for much longer than the wild type strain. A *sigF* mutant, which showed reduced expression of many genes during stationary phase *in vitro* including several involved in cell envelope biosynthesis, initially proliferated in mice as well as the wild type, but the number of CFU in the organs stabilised sooner and at a lower level, giving rise to reduced histopathology and increased survival of the host (Chen et al., 2000; Geiman et al., 2004). Similar reduced immunopathology is also associated with mutation of *whiB3* (Steyn et al., 2002) and with deletion of the RD1 region of the bacterial chromosome (Lewis et al., 2003; Pym et al., 2002).

A *hspR* strain of *M. tuberculosis* that constitutively hyper-expresses the Hsp70 family of heat shock proteins was significantly affected in its ability to persist during the chronic stage of infection in mice, although it showed similar behaviour to the wild type in the initial acute stage of the infection (Stewart et al., 2001). The authors suggest that the mutant may be inappropriately alerting the immune system to its presence by hyper-production of these stress-associated proteins.

Parish et al. (2003a) investigated the phenotypes associated with inactivation of six of the 11, paired two-component systems in *M. tuberculosis*. Interestingly, in four cases, *devR*(*dosR*), *tcrXY*, *trcS* and *kdpDE*, the mutants had enhanced virulence compared with the wild type in the immune-deficient SCID mouse. This finding with respect to *dosR* is most unexpected, in view of the induction of the *dosR* regulon in the persistence models (see above). The authors suggest that these genes somehow limit bacterial multiplication *in vivo* and increase the susceptibility of the organism to killing by the host. These genes could be considered as part of the repertoire required to establish a persistent infection, since their activity is required in order to terminate the initial acute phase. Another two-component regulator, *mprA*, is repressed in *M. tuberculosis* but not in *M. bovis* BCG during growth in macrophages and a *mprA* mutant of *M. tuberculosis* showed enhanced growth and survival in resting (though not activated) macrophages (Zahrt and Deretic, 2001). The authors used growth competition experiments *in vivo*, with a mixture of *mprA*$^+$ and *mprA* strains, and observed interesting tissue-specific differences in persistence. The *mprA* mutant was defective in both the establishment and the maintenance of persistence in the spleen, and in the maintenance of persistence in the lung. However, there was no significant difference in survival in the liver between the wild type and the mutant.

The importance of *dnaE2* polymerase for long-term persistence of mycobacteria *in vivo* reflects the complexity of the host/mycobacterial interaction under hostile conditions (Boshoff et al., 2003). Highly efficient DNA repair together with an increased mutation rate allows the pathogen to adapt to changing conditions *in vivo*, including those that pertain under antibiotic chemotherapy. A counter example is provided by a mutant of *M. bovis* BCG lacking *recA* (another gene important for DNA repair, the absence of which causes increased sensitivity to metronidazole), which had the same pattern of survival as the wild type in both the Wayne model and a mouse infection model (Sander et al., 2001).

Mutants affected in their ability to adapt to hypoxia also show reduced persistence in mice. An *icl* mutant, defective in the production of isocitrate lyase, showed reduced virulence and persistence in immune-competent mice as well as reduced survival in activated but not in resting macrophages (McKinney et al., 2000). This phenotype was not observed in interferon-γ knockout mice, establishing a link between the immune status of the host and the requirement for isocitrate lyase. Similarly, *M. bovis* BCG lacking the anaerobic nitrate reductase encoded by the *narGHJI* gene cluster was unable to persist in the lungs, liver and kidneys of immune-competent mice whereas in immune-deficient SCID mice the mutant was able to establish a chronic infection with a reduced bacterial loading in these organs (Fritz et al., 2002). This phenotype was tissue-specific; the *narGHJI* mutant showed no impairment of its ability to colonise and grow in the spleen in either mouse strain.

A mutant defective in *katG* showed reduced virulence and ability to persist in mice and guinea pigs (Li et al., 1998). This gene encodes a catalase/peroxidase/peroxinitritase, whose major role during the infection process is to remove peroxides generated by the phagocyte NADPH oxidase; in knockout mice, lacking one subunit of the NADPH oxidase, the *katG* mutant proliferated as well as the wild type strain (Ng et al., 2004). This gene could be required for both establishment and maintenance of persistence.

Prolonged survival (time-to-death phenotype) has been reported for mutants variously impaired in the production of cell envelope lipids (Table 2), showing that they are important virulence determinants. Sulpholipids found in the mycobacterial cell envelope have been implicated in mycobacterial virulence (Converse et al., 2003). However, using a *pks2* mutant (Sirakova et al., 2001), defective in the production of sulpholipids, Rousseau et al. (2003) showed that growth, virulence and persistence of *M. tuberculosis* were unaffected in macrophages, mice and guinea pigs. A *M. tuberculosis* strain lacking *mmpL8* failed to incorporate an anionic sulpholipid lipid (SL-1) into its envelope (Converse et al., 2003). Although the initial stages of infection following aerosol administration of the organism were the same as in the wild type (replication rates and containment within organs) there was attenuation as evidenced by a time-to-death phenotype (Domenech et al., 2004). Another mutant, defective in the production of phenolic glycolipids that have an inhibitory effect on the innate immune response (Reed et al., 2004), also showed a time-to-death phenotype (Stewart et al., 2003). As with the sigma factor mutants considered above, the prolonged persistence of these strains is probably an indirect effect of a reduction in virulence. However, this is probably not the case for mutants lacking *pcaA*, a gene required for mycolic acid cyclopropane ring synthesis in the cell wall of *M. tuberculosis*. A *pcaA* mutant was defective in cord formation and unable to persist within and kill infected mice, despite showing normal replication during the initial phase of the infection. This mutant seems to show the characteristics of a gene required for the maintenance of a persistent infection (Glickman et al., 2000) and it would be of interest to explore its behaviour in some of the *in vitro* models of persistence.

Two genes that encode members of the repetitive glycine-rich PE-PGRS protein family have been inactivated in *Mycobacterium marinum* and the resulting strains were unable to replicate in macrophages and showed decreased persistence in frog granulomas (Ramakrishnan et al., 2000). It would be of interest to examine the persistence phenotypes of strains lacking the corresponding *M. tuberculosis* genes, *Rv1651c* and *Rv3812*.

Finally, a family of five genes is found in *M. tuberculosis* whose products resemble the Rpf (resuscitation-promoting factor) of *M. luteus*. These proteins have activity in various *in vitro* assays based on the resuscitation of bacteria from a non-culturable state (Mukamolova et al. 2002b; Zhu et al. 2003). Single mutants lacking any one of the five *M. tuberculosis* genes show no apparent persistence phenotype either *in vitro* or *in vivo* (Downing et al., 2004; Tufariello et al., 2004). However, Downing et al. (2005) have shown that two different triple mutants, each lacking three of the five *rpf*-like genes, are unable to resuscitate from the non-culturable state *in vitro* and they are also attenuated for growth *in vivo*, indicating that the products of these genes are at least partially redundant. The behaviour of these strains in a chronic mouse infection model is currently being explored.

REACTIVATION (RESUSCITATION) OF NON-CULTURABLE/DORMANT FORMS *M. TUBERCULOSIS IN VITRO*

The evidence summarised in the previous sections leads to the conclusion that in the various models developed to mimic aspects of persistence in the human body, *M. tuberculosis* persists as specialised forms whose metabolism is characterised by a slow or negligible turnover rate. An important feature of some cells is their non-culturable phenotype, which is generally connected with adaptation to a dormant state. The main criterion for the presence of non-culturable cells is that exacting conditions are required to coax them back into a state where growth and division are resumed. The ability to resuscitate serves unequivocally to distinguish cells that are non-culturable from those that are dead. Resuscitation is probably a very complex process, requiring restoration of many cellular functions, leading finally to restoration of culturability (Mukamolova et al., 2003). The main problem with any model is to find conditions suitable for resuscitation. These cells may be "metabolically injured" (Postgate, 1976) and components of a medium that is harmless for normal cells can be toxic for recovering cells, as observed in substrate accelerated death (Calcott and Postgate, 1972). For example, a very rich medium did not permit the resuscitation of non-culturable cells of *M. luteus* (Mukamolova et al., 1998b). Alternatively, non-culturable cells may be temporarily unable to synthesise compounds required for the initiation of division, which must then be supplied in the resuscitation medium. In the case of *M. luteus* and *R. rhodochrous*, small amounts of yeast extract improved resuscitation dramatically (Mukamolova et al., 1998b; Shleeva et al., 2002). Some non-culturable cells may require highly specialised environments for resuscitation. For non-culturable forms of pathogenic bacteria, resuscitation may only occur *in vivo*, or in artificial conditions that mimic those encountered *in vivo* (Mukamolova et al., 2003). For example, hormones or lymphokines from higher organisms stimulate the growth of some bacteria (reviewed by Kaprelyants and Kell, 1996).

Of particular interest here is whether special resuscitation-inducing factors exist for non-culturable/dormant forms of *M. tuberculosis*. In the case of *M. luteus* (a related actinobacterium), the addition of sterile culture supernatant was necessary for restoration of culturability to non-culturable cells, suggesting that metabolically active cells secrete a compound that promotes resuscitation. This active compound, Rpf (resuscitation-promoting factor), is a secreted protein, which is active at picomolar concentrations (Mukamolova et al., 1998a). In addition to resuscitative activity, Rpf also stimulated multiplication of normal, viable bacteria. Disruption of the *rpf* gene was not possible in *M. luteus* in the absence of a second functional copy, strongly suggesting essentiality (Mukamolova et al., 2002a). Moreover, the protein must be secreted (or provided exogenously) to be biologically active (Mukamolova et al., 2002a). In contrast, deletion of both of the *rpf*-like genes found in *Corynebacterium glutamicum* did not significantly affect the growth of active bacteria (Hartmann et al., 2004). However, the double mutant showed slower growth and an extended lag-phase when it was inoculated into fresh growth medium after prolonged storage. This phenotype suggests that the Rpf-like proteins of *C. glutamicum* play a role in cell recovery or possibly resuscitation. Recent evidence suggests that Rpf may be involved in remodelling of the cell envelope (Cohen-Gonsaud et al., 2004, 2005; Finan, 2003; Kazarian et al., 2003; Ravagnani et al., 2005).

Rpf is a member of a protein family that is found throughout the actinobacteria. There are several representatives in most organisms (Kell and Young, 2000) and the protein family now contains more than 40 members. The Rpf domain is annotated as a transglycosylase-like domain in the PFAM database. *M. tuberculosis* contains five *rpf*-like genes, whose products, RpfA-E, expressed as recombinant proteins in *E. coli*, have similar biological activity to that of *M. luteus* Rpf (Mukamolova et al., 2002b; Zhu et al., 2003). Unlike *M. luteus* Rpf, none of the Rpf-like proteins of *M. tuberculosis* are essential for viability (Downing et al., 2004; Tufariello et al., 2004). Analysis of the gene expression profiles of mutant strains, each lacking one of the five *rpf*-like genes, revealed that their products have overlapping functions (Downing et al., 2004). Some of them are probably secreted, whereas others may be anchored in the cytoplasmic membrane. All Rpf-like proteins share one highly conserved 70-residue domain that is primarily responsible for biological activity (Mukamolova et al., 2002a).

The presence of *rpf*-like genes in *M. tuberculosis* raises the possibility that their products may control bacterial growth and resuscitation. Indeed, addition of recombinant Rpf to the liquid medium for MPN assay allowed the resuscitation of non-culturable cells of *M. smegmatis* and maximum resuscitation was obtained by co-cultivation of non-culturable bacteria with viable cells of *M. luteus* (Shleeva et al., 2004) (Fig. 2). A similar procedure has been applied to resuscitate non-culturable cells of *M. tuberculosis* obtained in prolonged stationary phase (Shleeva et al., 2002). In contrast, *M. tuberculosis* resuscitated spontaneously in liquid medium. A culture completely unable to form colonies on plates (zero CFU) had a viable count of 10^5 in the MPN assay and the addition of Rpf further increased viability in the MPN assay (Shleeva et al., 2004) (Fig. 3). In accordance with these results, Dhillon et al. (2004) have recently reported the presence of substantial numbers of non-culturable cells of *M. tuberculosis* in the organs of chronically infected mice; the MPN values for bacterial suspensions were substantially greater than the number of CFU. Perhaps this type of spontaneous resuscitation is somehow connected with the characteristic slow growth of pathogenic mycobacteria; dormant cells of non-pathogenic organisms like *M. luteus* (Mukamolova et al., 1998a) and *R. rhodochrous* (Shleeva et al., 2002) that grow more rapidly were unable to resuscitate unless Rpf was provided exogenously. Nevertheless, the importance of the Rpf-like gene products of *M. tuberculosis* is shown in experiments where anti-Rpf

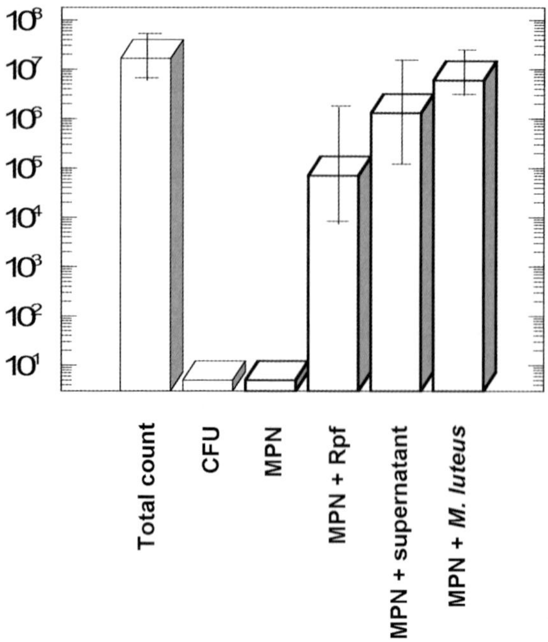

Fig. 2. Resuscitation of non-culturable cells of *M. smegmatis*. Bacteria were harvested from the point of minimum culturability during incubation in a modified form of Hartman's-de Bont medium. MPN assays (columns with a bold outline) were performed with and without recombinant Rpf (125 pM), in the presence of supernatant taken from an actively growing culture of *M. tuberculosis* and during co-culture with *M. luteus* actively producing Rpf. Reproduced from Shleeva et al. (2004) with permission.

antibodies significantly suppressed the growth of the organism *in vitro* following prolonged incubation in stationary phase (Mukamolova et al., 2002b). Moreover, when added exogenously to similar cultures the Rpf-like proteins of *M. tuberculosis* caused significant growth stimulation. Secreted Rpf may also be responsible for the reported stimulation of the growth rate of *M. tuberculosis* isolated from clinical specimens - supernatant taken from exponentially grown *M. luteus* enhanced the detection of *M. tuberculosis* in a rapid liquid culture system (Freeman et al., 2002). The significance of the Rpf-like proteins of *M. tuberculosis* was further underlined in a study of the role of the stringent response program controlled by Rel_{mtb} for persistence of the organism in a chronic mouse infection model. The Rel regulon is essential for maintenance of a chronic infection and RpfC (Rv1884c) is strongly up regulated in a Rel-dependent manner (Dahl et al., 2003).

Zhang and colleagues also found compounds in the supernatant of logarithmic phase cultures of *M. tuberculosis* that resuscitated aged non-culturable cells of the organism. They showed that supernatant from growing cultures contained phospholipids, which increased the number of CFU in aged cultures when added to plates and permitted growth from low inocula in liquid medium. Another active

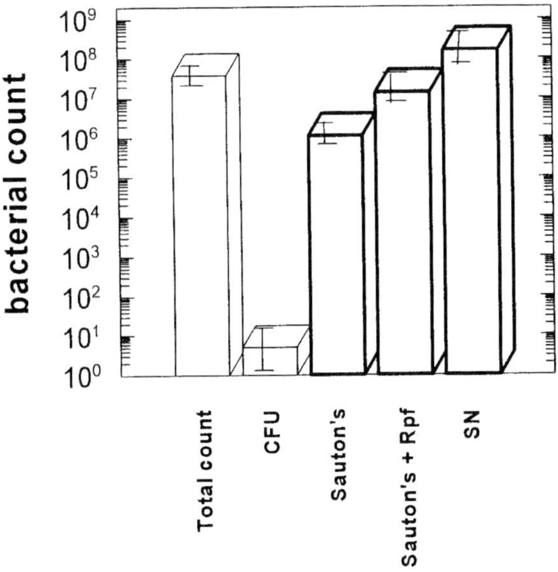

Fig. 3. Resuscitation of "non-culturable" cells of *M. tuberculosis*. The bacterial population at 4 months was passed though a 2 μm filter. This removes almost all of the viable CFU from the culture. MPN assays (columns with a bold outline) were performed with and without recombinant Rpf (125 pM) and also in the presence of supernatant taken from an actively growing culture of *M. tuberculosis*. Reproduced from Shleeva et al. (2002) with permission.

compound isolated from culture supernatants was an 8-kD protein (Rv1174c). Chemically synthesised peptides corresponding to three different segments of Rv1174c resuscitated *M. tuberculosis* cells when added at μM concentrations (Sun and Zhang, 1999; Zhang et al., 2001).

CONCLUSION

The problem of latent TB and *M. tuberculosis* persistence has been known for many decades, but a rapid and effective treatment for this global health problem remains elusive. It has proved extremely difficult to pin down the precise nature of the persisting organisms. Many attempts have been made to model the persistence of mycobacteria *in vivo* using mainly mouse models and *in vitro*. Detailed characterisation of these models in recent years has contributed significantly to our understanding of the biology of *M. tuberculosis*. However, we do not know whether any of the current models mimic persistence in the human host, since none of them can be properly validated. Clearly, what is urgently needed is reliable information to replace current suppositions and predictions concerning the defining characteristics of the persisting organisms associated with latent disease. This is not a trivial problem to address; it will require considerable ingenuity to locate and characterise the organisms within the human host and if, as seems likely, a rather small population of bacteria is implicated, new technological developments will be required in order to study them.

Modern molecular technologies including transcriptional profiling using microarrays, proteomic analyses and real time PCR have been employed to characterise *M. tuberculosis* in the various persistence models currently in use, and have revealed both similarities and differences between them. These analyses have confirmed that Wayne's model of persistence at low cell density essentially represents gradual adaptation of *M. tuberculosis* to hypoxia rather than to dormancy. Adaptation to hypoxia may be an important factor that must be satisfied to produce dormancy (or at least non-culturability) both *in vivo* and *in vitro*. Other important factors identified from *in vitro* experiments are the onset of stationary phase in shaken cultures and the presence of a high cell density. Proteomic and transcriptomic analyses to identify genes whose up- or down-regulation is required for persistence should help to reveal which environmental conditions are important for the establishment of this state.

A similar question mark applies to chronic *in vivo* models of persistence, where high numbers of viable cells of *M. tuberculosis* may persist in the animals, in contrast to the apparent disappearance of viable organisms in the human host. This latter behaviour is mimicked in the Cornell mouse model, but comparatively small numbers of organisms persist, and they are difficult to characterise biochemically using current proteomic and transcriptomic methods. Future developments, making these methods applicable to small numbers of organisms, could lead to important new advances in our understanding of the nature of persistence. Nevertheless, in view of important differences between the immunopathology of the disease in the mouse model and the human host (and the fact that persistence in the Cornell model is antibiotic-induced), its relevance to persistence in the human host is questionable.

From a microbiological point of view the central question is: are specialised dormant forms of the *M. tuberculosis* responsible for persistence? The answer to this question is probably no, since future disease in tuberculin-positive individuals, *i.e.* probable reactivation of latent tuberculosis, can be reduced by the prophylactic use of anti-microbial drugs (Comstock et al., 1979). The alternative suggestion that a very small population of fully viable bacteria persists somewhere within the body is also untenable, since these bacteria should be fully susceptible to antibiotics (the possibility that they reside in a microenvironment inaccessible to antibiotics seems unlikely). Several *in vitro* models, *i.e.* prolonged stationary phase in either shaken sealed flasks (Shleeva et al., 2002) or standing tubes (Sun and Zhang, 1999), or a prolonged microaerophilic stationary phase followed by rifampin treatment (Hu et al., 2000), show that viable cells of *M. tuberculosis* have the capacity to adopt a non-culturable state and that culturability can be restored by resuscitation. This resembles the situation *in vivo*, in the Cornell model and also in the chronic mouse model, although a significant proportion of the cell population *in vivo* is probably

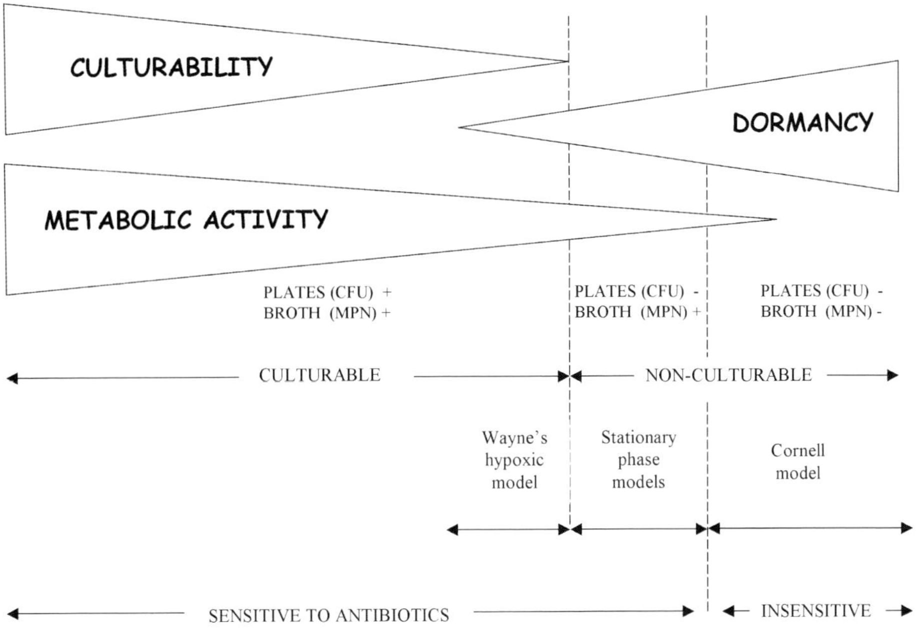

Fig. 4. Bacterial culturability, metabolic activity and dormancy: a continuum of physiological states. The phenotypic properties of organisms are indicated in relation to their position within the continuum.

not completely dormant. In view of the multitude of different microenvironments available within the human body, and the known heterogeneity of even ostensibly uniform bacterial cultures, we strongly suspect that persisting organisms represent a continuum of different types of cells with a range of different properties in respect of their culturability, metabolic activity and dormancy (Fig. 4). Some of them may be able to multiply slowly under the control of the host immune system whereas others may be in state of growth arrest with very low metabolic activity. We suggest, moreover, that individual bacteria are probably cycling between different states, which would account for the fact that protracted antibiotic treatment is necessary to effect complete clearance. A model of bacterial persistence based on the premise that persisting organisms are generally in an altered physiological state has recently been proposed (Balaban et al., 2004). Which particular cells, in which particular environments can be responsible for the reactivation of latent tuberculosis remains unclear. Accordingly, our strategy to combat latent tuberculosis must continue to be directed against the various different forms that the organism can adopt. Until more is known about the precise features of persistent organisms specifically associated with TB latency, detailed analysis of mycobacterial persistence in the various disparate models will continue.

FUTURE TRENDS

It will be evident from the forgoing discussion that the major lacuna holding up current progress is our lack of knowledge about the organisms that persist during human TB latency. Much of what we have inferred about these organisms comes from work undertaken several decades ago and there is now a really urgent need to revisit these experiments using modern analytical techniques. This will entail the development of methods for detecting and analysing the activity of (possibly individual) bacteria in tissue samples from asymptomatic patients *e.g.* those undergoing surgery as part of a cancer treatment. One particularly welcome development would be the ability to detect genes that are being actively transcribed in individual bacteria. If adequate sensitivity can be achieved in the future, an *in situ* hybridisation approach (Fenhalls et al., 2002a; Fenhalls et al., 2002b) following RT-PCR with *M. tuberculosis*-specific primers would prove extremely useful in this regard. If the persisting bacteria can be characterised better, a more rational approach to modelling could be adopted. In particular, the authors of this chapter would like to know whether long-term persistence necessitates loss of bacterial culturability.

A particularly exciting development is the artificial granuloma model recently published by Bishai and colleagues (Karakousis et al., 2004), which may provide vital clues concerning the nature of organisms persisting for long periods within the mouse. Anther approach that would help to clarify the roles of different genes in persistence in animal models would be an inducible antisense strategy. Tetracycline-inducible gene expression systems have recently been developed for *M. tuberculosis*, so that the effect of sudden loss or gain of function can now be assessed *in vivo* (Blokpoel et al., 2005; Ehrt et al., 2005). This would be particularly useful in identifying functions required for maintenance of persistence.

Publication of the *M. tuberculosis* genome sequence in 1998 (Cole et al., 1998) was a milestone in TB research and we can expect that further extensive use will be made of post-genomic approaches using the various *in vivo* and *in vitro* models and defined mutant strains of *M. tuberculosis* to gain insights into the nature of the persistent infection it causes.

In this chapter we have touched upon, but not in any way explored, the role of host genotype and host fitness in determining the course of an infection with *M. tuberculosis*. This is a very important area (Zhang, 2004) and it is possible that careful analysis of different mouse and human genotypes in relation to the ability of *M. tuberculosis* to persist may provide some useful clues about the nature of persistent bacteria.

We have raised a number of important questions in this review. For example, are persisting cells in the human host dormant or replicating slowly and are they culturable or non-culturable? Do they represent a homogeneous bacterial population or is there a continuum of different metabolic states? Do any of the *in vitro* and *in vivo* models adequately mimic events characteristic of persistence in the human host? How do cells change their metabolism during the transition to a persistent state? Can we distinguish genes that are required for the establishment of the persistent state from those that are required to maintain it? Unless we can better understand persistence, we are unlikely to find ways to prevent reactivation disease, which is a vitally important goal in current efforts to reduce the appalling global burden of tuberculosis.

REFERENCES

Almenoff, P.L., Johnson, A., Lesser, M., and Mattman, L.H. (1996). Growth of acid fast L forms from the blood of patients with sarcoidosis. Thorax *51*, 530-533.

Andersen, A.B., Andersen, P., and Ljungqvist, L. (1992). Structure and function of a 40,000-molecular-weight protein antigen of *Mycobacterium tuberculosis*. Infect. Immun. *60*, 2317-2323.

Ando, M., Yoshimatsu, T., Ko, C., Converse, P.J., and Bishai, W.R. (2003). Deletion of *Mycobacterium tuberculosis* sigma factor E results in delayed time to death with bacterial persistence in lungs of aerosol-infected mice. Infect. Immun. *71*, 7170-7172.

Bacon, J., James, B.W., Wernisch, L., Williams, A., Morley, K.A., Hatch, G.J., Mangan, J.A., Hinds, J., Stoker, N.G., Butcher, P.D., and Marsh, P.D. (2004). The influence of reduced oxygen availability on pathogenicity and gene expression in *Mycobacterium tuberculosis*. Tuberculosis (Edinb.) *84*, 205-217.

Balaban, N.Q., Merrin, J., Chait, R., Kowalik, L., and Leiber, S. (2004). Bacterial persistence as a phenotypic switch. Science *305*, 1622-1625.

Balasubramanian, V., Wiegeshaus, E.H., Taylor, B.T., and Smith, D.W. (1994). Pathogenesis of tuberculosis: pathway to apical localization. Tubercle Lung Dis. *75*, 168-178.

Barer, M.R., and Harwood, C.R. (1999). Bacterial viability and culturability. Adv. Microb. Physiol. *41*, 93-137.

Behr, M.A., Wilson, M.A., Gill, W.P., Salamon, H., Schoolnik, G.K., Rane, S., and Small, P.M. (1999). Comparative genomics of BCG vaccines by whole-genome DNA microarray. Science *284*, 1520-1523.

Betts, J.C., Lukey, P.T., Robb, L.C., McAdam, R.A., and Duncan, K. (2002). Evaluation of a nutrient starvation model of *Mycobacterium tuberculosis* persistence by gene and protein expression profiling. Mol. Microbiol. *43*, 717-731.

Biketov, S., Mukamolova, G.V., Potapov, V., Gilenkov, E., Vostroknutova, G., Kell, D.B., Young, M., and Kaprelyants, A.S. (2000). Culturability of *Mycobacterium tuberculosis* cells isolated from murine macrophages: a bacterial growth factor promotes recovery. FEMS Immunol. Med. Microbiol. *29*, 233-240.

Blokpoel, M.C.J., Murphy, H.N., O'Toole, R., Wiles, S., Runn, E.S.C., Stewart, G.R., Young, D.B., and Robertson, B.D. (2005). Tetracycline-inducible gene regulation in mycobacteria. Nucl. Acids Res. *33*, e22.

Boon, C., and Dick, T. (2002). *Mycobacterium bovis* BCG response regulator essential for hypoxic dormancy. J. Bacteriol. *184*, 6760-6767.

Boon, C., Li, R., Qi, R., and Dick, T. (2001). Proteins of *Mycobacterium bovis* BCG induced in the Wayne dormancy model. J. Bacteriol. *183*, 2672-2676.

Boshoff, H.I., and Barry, C.E., 3rd (2005). Tuberculosis: metabolism and respiration in the absence of growth. Nat. Rev. Microbiol. *3*, 70-80.

Boshoff, H.I., Myers, T.G., Copp, B.R., McNeil, M.R., Wilson, M.A., and Barry, C.E., 3rd (2004). The transcriptional responses of *Mycobacterium tuberculosis* to inhibitors of metabolism: novel insights into drug mechanisms of action. J. Biol. Chem. *279*, 40174-40184.

Boshoff, H.I., Reed, M.B., Barry, C.E., and Mizrahi, V. (2003). DnaE2 polymerase contributes to *in vivo* survival and the emergence of drug resistance in *Mycobacterium tuberculosis*. Cell *113*, 183-193.

Brooks, J.V., Furney, S.K., and Orme, I.M. (1999). Metronidazole therapy in mice infected with tuberculosis. Antimicrob. Agents Chemother. *43*, 1285-1288.

Brown, D.H., Miles, B.A., and Zwilling, B.S. (1995). Growth of *Mycobacterium tuberculosis* in BCG-resistant and BCG- susceptible mice: establishment of latency and reactivation. Infect. Immun. *63*, 2243-2247.

Calcott, P.H., and Postgate, J.R. (1972). On substrate-accelerated death in *Klebsiella aerogenes*. J. Gen. Microbiol. *70*, 115-122.

Cashel, M., Gentry, D.R., Hernandez, V.J., and Vinella, D. (1996). The stringent response. In *Escherichia coli* and *Salmonella thyphimurium*: cellular and molecular biology, F. C. Neidhardt, ed. (Washington, American Society for Microbiology), pp. 1458-1496.

Chan, J., Xing, Y., Magliozzo, R.S., and Bloom, B.R. (1992). Killing of virulent *Mycobacterium tuberculosis* by reactive nitrogen intermediates produced by activated murine macrophages. J. Exp. Med. *175*, 1111-1122.

Chan, K., Knaak, T., Satkamp, L., Humbert, O., Falkow, S., and Ramakrishnan, L. (2002). Complex pattern of *Mycobacterium marinum* gene expression during long-term granulomatous infection. Proc. Natl. Acad. Sci. U S A *99*, 3920-3925.

Chandrasekhar, S., and Ratnam, S. (1992). Studies on cell-wall deficient non-acid fast variants of *Mycobacterium tuberculosis*. Tubercle Lung Dis. *73*, 273-279.

Chen, P., Ruiz, R.E., Li, Q., Silver, R.F., and Bishai, W.R. (2000). Construction and characterization of a *Mycobacterium tuberculosis* mutant lacking the alternate sigma factor gene, *sigF*. Infect. Immun. *68*, 5575-5580.

Cohen-Gonsaud, M., Keep, N.H., Davies, A.P., Ward, J., Henderson, B., and Labesse, G. (2004). Resuscitation-promoting factors possess a lysozyme-like domain. Trends Biochem. Sci. *29*, 7-10.

Cohen-Gonsaud, M., Barthe, P., Bagneris, C., Henderson, B., Ward, J., Roumestand, C., and Keep, N.H. (2005). The structure of a resuscitation-promoting factor domain from *Mycobacterium tuberculosis* shows homology to lysozymes. Nat. Struct. Mol. Biol. *12*, 270-273.

Cole, S.T., Brosch, R., Parkhill, J., Garnier, T., Churcher, C., Harris, D., Gordon, S.V., Eiglmeier, K., Gas, S., Barry, C.E., et al. (1998). Deciphering the biology of *Mycobacterium tuberculosis* from the complete genome sequence. Nature *393*, 537-544.

Comstock, G.W., Baum, C., and Snider Jr., D.E. (1979). Isoniazid prophylaxis among Alaskan Eskimos: a final report of the bethel isoniazid studies. Am. Rev. Resp. Dis. *119*, 827-830.

Converse, S.E., Mougous, J.D., Leavell, M.D., Leary, J.A., Bertozzi, C.R., and Cox, J.S. (2003). MmpL8 is required for sulfolipid-1 biosynthesis and *Mycobacterium tuberculosis* virulence. Proc. Natl. Acad. Sci. U S A *100*, 6121-6126.

Corper, H.J., and Cohn, M.L. (1933). The viability and virulence of old cultures of tubercule bacille. Studies on twelve-year broth cultures maintained at incubator temperature. Am. Rev. Tuberc. *28*, 856-874.

Cunningham, A.F., Ashton, P.R., Spreadbury, C.L., Lammas, D.A., Craddock, R., Wharton, C.W., and Wheeler, P.R. (2004). Tubercle bacilli generate a novel cell wall-associated pigment after long-term anaerobic culture. FEMS Microbiol. Lett. *235*, 191-198.

Cunningham, A.F., and Spreadbury, C.L. (1998). Mycobacterial stationary phase induced by low oxygen tension: Cell wall thickening and localization of the 16-kilodalton α-crystallin homolog. J. Bacteriol. *180*, 801-808.

Dahl, J.L., Kraus, C.N., Boshoff, H.I., Doan, B., Foley, K., Avarbock, D., Kaplan, G., Mizrahi, V., Rubin, H., and Barry, C.E., 3rd (2003). The role of RelMtb-mediated adaptation to stationary phase in long-term persistence of *Mycobacterium tuberculosis* in mice. Proc. Natl. Acad. Sci. U S A *100*, 10026-10031.

Daniel, J., Deb, C., Dubey, V.S., Sirakova, T.D., Abomoelak, B., Moribondi, H.R., and Kolattukudy, P.E. (2004). Induction of a novel class of diacyl glycerol transferases and triacylglycerol accumulation in *Mycobacterium tuberculosis* as it goes into a dormancy-like state in culture. J. Bacteriol. *186*, 5017-5030.

Davey, H.M., and Kell, D.B. (1996). Flow cytometry and cell sorting of heterogeneous microbial populations: the importance of single-cell analysis. Microbiol. Rev. *60*, 641-696.

de Wit, D., Wootton, M., Dhillon, J., and Mitchison, D.A. (1995). The bacterial DNA content of mouse organs in the Cornell model of dormant tuberculosis. Tubercle Lung Dis. *76*, 555-562.

Delgado, J.C., Tsai, E.Y., Thim, S., Baena, A., Boussiotis, V.A., Reynes, J.M., Sath, S., Grosjean, P., Yunis, E.J., and Goldfeld, A.E. (2002). Antigen-specific and persistent tuberculin anergy in a cohort of pulmonary tuberculosis patients from rural Cambodia. Proc. Natl. Acad. Sci. U S A *99*, 7576-7581.

DeMaio, J., Zhang, Y., Ko, C., Young, D.B., and Bishai, W.R. (1996). A stationary-phase stress-response sigma factor from *Mycobacterium tuberculosis*. Proc. Natl. Acad. Sci. U S A *93*, 2790-2794.

Deretic, V., and Fratti, R.A. (1999). *Mycobacterium tuberculosis* phagosome. Mol. Microbiol. *31*, 1603-1609.

Dhillon, J., Allen, B.W., Hu, Y.M., Coates, A.R., and Mitchison, D.A. (1998). Metronidazole has no antibacterial effect in Cornell model murine tuberculosis. Int. J. Tuberc. Lung Dis. *2*, 736-742.

Dhillon, J., Dickinson, J.M., Sole, K., and Mitchison, D.A. (1996). Preventive chemotherapy of tuberculosis in Cornell model mice with combinations of rifampin, isoniazid, and pyrazinamide. Antimicrob. Agents Chemother. *40*, 552-555.

Dhillon, J., Lowrie, D.B., and Mitchison, D.A. (2004) *Mycobacterium tuberculosis* from chronic murine infections that grows in liquid but not on solid medium. BMC Infect. Dis. *4*, 51.

Dick, T., Lee, B.H., and Murugasu-Oei, B. (1998). Oxygen depletion induced dormancy in *Mycobacterium smegmatis*. FEMS Microbiol. Lett. *163*, 159-164.

Domenech, P., Reed, M.B., Dowd, C.S., Manca, C., Kaplan, G., and Barry, C.E., 3rd (2004). The role of MmpL8 in sulfatide biogenesis and virulence of *Mycobacterium tuberculosis*. J. Biol. Chem. *279*, 21257-21265.

Downing, K.J., Betts, J.C., Young, D.I., McAdam, R.A., Kelly, F., Young, M., and Mizrahi, V. (2004). Global expression profiling of strains harbouring null mutations reveals that the five *rpf*-like genes of *Mycobacterium tuberculosis* show functional redundancy. Tuberculosis *84*, 167-179.

Downing, K.J., Mischenko, V.V., Shleeva, M.O., Young, D.I., Young, M., Kaprelyants, A.S., Apt, A.S. and Mizrahi, V. (2005). Mutants of *Mycobacterium tuberculosis* lacking three of the five *rpf*-like genes are defective for growth *in vivo* and for resuscitation *in vitro*. Infect. Immun. *73*, in press.

Dubnau, E., Fontan, P., Manganelli, R., Soares-Appel, S., and Smith, I. (2002). *Mycobacterium tuberculosis* genes induced during infection of human macrophages. Infect. Immun. *70*, 2787-2795.

Dye, C., Scheele, S., Dolin, P., Pathania, V., and Raviglione, R.C. (1999). Consensus statement. Global burden of tuberculosis - estimated incidence, prevalence, and mortality by country. WHO Global Surveillance and Monitoring Project. JAMA *282*, 677-686.

Ehrt, S., Guo, X.V., Hickey, C.M., Ryou, M., Monteleone, M., Riley, L.W., and Schnappinger, D. (2005). Controlling gene expression in mycobacteria with anhydrotetracycline and Tet repressor. Nucl. Acids Res. *33*, e21.

Eishi, Y., Suga, M., Ishige, I., Kobayashi, D., Yamada, T., Takemura, T., Takizawa, T., Koike, M., Kudoh, S., Costabel, U., et al. (2002). Quantitative analysis of mycobacterial and propionibacterial DNA in lymph nodes of Japanese and European patients with sarcoidosis. J. Clin. Microbiol. *40*, 198-204.

Enarson, D.A. (2004). Use of the tuberculin skin test in children. Paediatr. Respir. Rev. *5 Suppl A*, S135-137.

Ericsson, M., Hanstorp, D., Hagberg, P., Enger, J., and Nystrom, T. (2000). Sorting out bacterial viability with optical tweezers. J. Bacteriol. *182*, 5551-5555.

Ewer, K., Deeks, J., Alvarez, L., Bryant, G., Waller, S., Andersen, P., Monk, P., and Lalvani, A. (2003). Comparison of T-cell-based assay with tuberculin skin test for diagnosis of *Mycobacterium tuberculosis* infection in a school tuberculosis outbreak. Lancet *361*, 1168-1173.

Falcone, V., Bassey, E.B., Toniolo, A., Conaldi, P.G., and Colllins, F.M. (1994). Differential release of tumor necrosis factor-α from murine peritoneal macrophages stimulated with virulent and avirulent species of mycobacteria. FEMS Immunol. Med. Microbiol. *8*, 225-232.

Fenhalls, G., Stevens, L., Moses, L., Bezuidenhout, J., Betts, J.C., van Helden, P., Lukey, P.T., and Duncan, K. (2002a). In situ detection of *Mycobacterium tuberculosis* transcripts in human lung granulomas reveals differential gene expression in necrotic lesions. Infect. Immun. *70*, 6330-6338.

Fenhalls, G., Stevens-Muller, L., Warren, R., Carroll, N., Bezuidenhout, J., van Helden, P., and Bardin, P. (2002b). Localisation of mycobacterial DNA and mRNA in human tuberculous granulomas. J. Microbiol. Methods *51*, 197-208.

Finan, C.L. (2003) Autocrine growth factors in streptomycetes. Ph.D. Thesis, University of Wales, Aberystwyth.

Fine, P.E., Sterne, J.A., Ponnighaus, J.M., and Rees, R.J. (1994). Delayed-type hypersensitivity, mycobacterial vaccines and protective immunity. Lancet *344*, 1245-1249.

Fisher, M.A., Plikaytis, B.B., and Shinnick, T.M. (2002). Microarray analysis of the *Mycobacterium tuberculosis* transcriptional response to the acidic conditions found in phagosomes. J. Bacteriol. *184*, 4025-4032.

Flesch, I.E., and Kaufmann, S.H.E. (1991). Mechanisms involved in mycobacterial growth inhibition by gamma interferon-activated bone marrow macrophages: role of reactive nitrogen intermediates. Infect. Immun. *59*, 3213-3218.

Fritz, C., Maass, S., Kreft, A., and Bange, F.C. (2002). Dependence of *Mycobacterium bovis* BCG on anaerobic nitrate reductase for persistence is tissue-specific. Infect. Immun. *70*, 286-291.

Flynn, J.L. (2004). Immunology of tuberculosis and implications in vaccine development. Tuberculosis (Edinb) *84*, 93-101.

Flynn, J.L., and Chan, J. (2001a). Immunology of tuberculosis. Annu. Rev. Immunol. *19*, 93-129.

Flynn, J.L., and Chan, J. (2001b). Tuberculosis: latency and reactivation. Infect. Immun. *69*, 4195-4201.

Freeman, R., Dunn, J., Magee, J., and Barrett, A. (2002). The enhancement of isolation of mycobacteria from a rapid liquid culture system by broth culture supernate of *Micrococcus luteus*. J. Med. Microbiol. *51*, 92-93.

Fritz, C., Maass, S., Kreft, A., and Bange, F.C. (2002). Dependence of *Mycobacterium bovis* BCG on anaerobic nitrate reductase for persistence is tissue-specific. Infect. Immun. *70*, 286-291.

Garbe, T.R., Hibler, N.S., and Deretic, V. (1999). Response to reactive nitrogen intermediates in *Mycobacterium tuberculosis*: induction of the 16-kilodalton alpha-crystallin homolog by exposure to nitric oxide donors. Infect. Immun. *67*, 460-465.

Garton, N.J., Christensen, H., Minnikin, D.E., Adegbola, R.A., and Barer, M.R. (2002) Intracellular lipophilic inclusion in mycobacteria *in vitro* and in sputum. Microbiology, *148*, 2951-2958.

Geiman, D.E., Kaushal, D., Ko, C., Tyagi, S., Manabe, Y.C., Schroeder, B.G., Fleischmann, R.D., Morrison, N.E., Converse, P.J., Chen, P., and Bishai, W.R. (2004). Attenuation of late-stage disease in mice infected by the *Mycobacterium tuberculosis* mutant lacking the SigF alternate sigma factor and identification of SigF-dependent genes by microarray analysis. Infect. Immun. *72*, 1733-1745.

Glickman, M.S., Cox, J.S., and Jacobs, W.R., Jr. (2000). A novel mycolic acid cyclopropane synthetase is required for cording, persistence, and virulence of *Mycobacterium tuberculosis*. Mol. Cell *5*, 717-727.

Gomez, J.E., and McKinney, J.D. (2004). *M. tuberculosis* persistence, latency, and drug tolerance. Tuberculosis (Edinb) *84*, 29-44.

Gordon, S.V., Brosch, R., Billault, A., Garnier, T., Eigelmeier, K., and Cole, S.T. (1999). Identification of variable regions in the genomes of tubercle bacilli using bacterial artificial chromosome arrays. Mol. Microbiol. *32*, 643-655.

Graham, J.E., and Clark-Curtiss, J.E. (1999). Identification of *Mycobacterium tuberculosis* RNAs synthesized in response to phagocytosis by human macrophages by selective capture of transcribed sequences (SCOTS). Proc. Natl. Acad. Sci. USA *96*, 11554-11559.

Haapanen, J.H., Kass, I., Gensini, G., and Middlebrook, E. (1959). Studies on the gaseous content of tuberculous cavities. Am. Rev. Resp. Dis. *80*, 1-5.

Hampshire, T., Soneji, S., Bacon, J., James, B.W., Hinds, J., Laing, K., Stabler, R.A., Marsh, P.D., and Butcher, P.D. (2004). Stationary phase gene expression of *Mycobacterium tuberculosis* following a progressive nutrient depletion: a model for persistent organisms? Tuberculosis (Edinb) *84*, 228-238.

Hartmann, M., Barsch, A., Niehaus, K., Puhler, A., Tauch, A., and Kalinowski, J. (2004). The glucosylated cell surface protein, Rpf2, containing a resuscitation-promoting factor motif, is involved in intercellular communication of *Corynebacterium glutamicum*. Arch. Microbiol. *182*, 299-312.

Hengge-Aronis, R. (1993). Survival of hunger and stress: the role of *rpoS* in early stationary phase gene regulation in *E coli*. Cell *72*, 165-168.

Hengge-Aronis, R. (2002). Signal transduction and regulatory mechanisms involved in control of the sigma(S) (RpoS) subunit of RNA polymerase. Microbiol. Mol. Biol. Rev. *66*, 373-395.

Hernández-Pando, R., Jeyanathan, M., Mengistu, G., Aguilar, D., Orozco, H., Harboe, M., Rook, G.A.W., and Bjune, G. (2000). Persistence of DNA from *Mycobacterium tuberculosis* in superficially normal lung tissue during latent infection. Lancet *356*, 2133-2138.

Hobby, G.L., Auerbach, O., Lenert, T.F., Small, M.J., and Comer, J.V. (1954). The late emergence of *M. tuberculosis* in liquid cultures of pulmonary lesions resected from humans. Am. Rev. Tuberc. *70*, 191-218.

Honer zu Bentrup, K., and Russell, D.G. (2001). Mycobacterial persistence: adaptation to a changing environment. Trends Microbiol. *9*, 597-605.

Hu, Y., and Coates, A.R. (2001). Increased levels of *sigJ* mRNA in late stationary phase cultures of *Mycobacterium tuberculosis* detected by DNA array hybridisation. FEMS Microbiol. Lett. *202*, 59-65.

Hu, Y.M., Butcher, P.D., Sole, K., Mitchison, D.A., and Coates, A.R.M. (1998). Protein synthesis is shutdown in dormant *Mycobacterium tuberculosis* and is reversed by oxygen or heat shock. FEMS Microbiol. Lett. *158*, 139-145.

Hu, Y., Kendall, S., Stoker, N.G., and Coates, A.R. (2004). The *Mycobacterium tuberculosis sigJ* gene controls sensitivity of the bacterium to hydrogen peroxide. FEMS Microbiol. Lett. *237*, 415-423.

Hu, Y.M., Mangan, J.A., Dhillon, J., Sole, K.M., Mitchison, D.A., Butcher, P.D., and Coates, A.R.M. (2000). Detection of mRNA transcripts and active transcription in persistent *Mycobacterium tuberculosis* induced by exposure to rifampin or pyrazinamide. J. Bacteriol. *182*, 6358-6365.

Hutter, B., and Dick, T. (1998). Increased alanine dehydrogenase activity during dormancy in *Mycobacterium smegmatis*. FEMS Microbiol. Lett. *167*, 7-11.

Hutter, B., and Dick, T. (1999). Up-regulation of *narX*, encoding a putative "fused nitrate reductase" in anaerobic dormant *Mycobacterium bovis* BCG. FEMS Microbiol. Lett. *178*, 63-69.

Hutter, B., and Dick , T. (2000). Analysis of the dormancy-inducible *narK2* promoter in *Mycobacterium bovis* BCG. FEMS Microbiol. Lett. *188*, 141-146.

Kaprelyants, A.S., Gottschal, J.C., and Kell, D.B. (1993). Dormancy in non-sporulating bacteria. FEMS Microbiol. Rev. *104*, 271-286.

Kaprelyants, A.S., and Kell, D.B. (1992). Rapid assessment of bacterial viability and vitality using rhodamine 123 and flow cytometry. J. Appl. Bacteriol. *72*, 410-422.

Kaprelyants, A.S., and Kell, D.B. (1993). Dormancy in stationary-phase cultures of *Micrococcus luteus*: flow cytometric analysis of starvation and resuscitation. Appl. Environ. Microbiol. *59*, 3187-3196.

Kaprelyants, A.S., and Kell, D.B. (1996). Do bacteria need to communicate with each other for growth? Trends Microbiol. *4*, 237-242.

Karakousis, P.C., Yoshimatsu, T., Lamichhane, G., Woolwine, S.C., Nuermberger, E.L., Grosset, J., and Bishai, W.R. (2004). Dormancy phenotype displayed by extracellular *Mycobacterium tuberculosis* within artificial granulomas in mice. J. Exp. Med. *200*, 647-657.

Kaushal, D., Schroeder, B.G., Tyagi, S., Yoshimatsu, T., Scott, C., Ko, C., Carpenter, L., Mehrotra, J., Manabe, Y.C., Fleischmann, R.D., and Bishai, W.R. (2002). Reduced immunopathology and mortality despite tissue persistence in a

Mycobacterium tuberculosis mutant lacking alternative sigma factor, SigH. Proc. Natl. Acad. Sci. USA *99*, 8330-8335.

Kazarian, K.A., Yeremeev, V.V., Kondratieva, T.K., Telkov, M.V., Kaprelyants, A.S., and Apt, A.S. (2003). Proteins of Rpf family as novel TB vaccine candidates. Paper presented at; First International Conference on TB Vaccines for the World (Montreal, Canada).

Kell, D.B., Kaprelyants, A.S., Weichart, D.H., Harwood, C.R., and Barer, M.R. (1998). Viability and activity in readily culturable bacteria: a review and discussion of the practical issues. Antonie Van Leeuwenhoek *73*, 169-187.

Kell, D.B., and Young, M. (2000). Bacterial dormancy and culturability: the role of autocrine growth factors. Curr. Opin. Microbiol. *3*, 238-243.

Kendall, S.L., Movahedzadeh, F., Rison, S.C.G., Wernisch, L., Parish, T., Duncan, K., Betts, J.C., and Stoker, N.G. (2004). The *Mycobacterium tuberculosis dosSR* two-component system is induced by multiple stresses. Tuberculosis *84*, 247-255.

Khomenko, A.G. (1987). The variability of *Mycobacterium tuberculosis* in patients with cavitary pulmonary tuberculosis in the course of chemotherapy. Tubercle Lung Dis. *68*, 243-253.

Khomenko, A.G., Golyshevskaia, V.I., Shaginian, I.A., Safonova, S.G., Nesterenko, L.N., Grishina, T.D., Gintsburg, A.L., and Prozorovsky, S.V. (1994). Is sarcoidosis a chronic persistent infection?. Zh. Mikrobiol. Epidemiol. Immunobiol. *1*, 64-68.

Khomenko, A.G., and Golyshevskaya, V.I. (1984). Filterable forms of *Mycobacterium tuberculosis*. Z. Erkrank Atm-Org. *162*, 147-154.

Khomenko, A.G., Karachunskii, M.A., Dorozhkova, I.R., Chukanov, V.I., and Balta Iu, E. (1980). L-transformation of the mycobacterial population in the process of treating patients with newly detected destructive pulmonary tuberculosis. Probl. Tuberk. *2*, 18-23.

Kuznetsov, B.A., Davydova, M.E., Shleeva, M.O., Shleev, S.V., Kaprelyants, A.S., and Yaropolov, A.I. (2004). Electrochemical investigation of the dynamics of *Mycobacterium smegmatis* cells' transformation to dormant, non-culturable form. Bioelectrochemistry *64*, 125-131.

LeCoeur, H.F., Lagrange, P.H., Truffot-Pernot, C., Gheorgiu, M., and Grosset, J. (1989). Relapses after stopping chemotherapy for experimental tuberculosis in genetically resistant and susceptible strains of mice. Clin. Exp. Immunol. *76*, 458-462.

Lee, B.H., Murugasu-Oei, B., and Dick, T. (1998). Upregulation of a histone-like protein in dormant *Mycobacterium smegmatis*. Mol. Gen. Genet. *260*, 475-479.

Lewis, K. (2000). Programmed death in bacteria. Microbiol. Mol. Biol. Rev. *64*, 503-514.

Lewis, K.N., Liao, R., Guinn, K.M., Hickley, M.J., Smith, S., Behr, M.A., and Sherman, D.R. (2003). Deletion of RD1 from *Mycobacterium tuberculosis* mimics bacille Calmette-Guerin attenuation. J. Infect. Dis. *187*, 117-123.

Li, Z., Kelley, C., Collins, F., Rouse, D., and Morris, S. (1998). Expression of *katG* in *Mycobacterium tuberculosis* is associated with its growth and persistence in mice and guinea pigs. J. Infect. Dis. *177*, 1030-1035.

Lillebaek, T., Dirksen, A., Baess, I., Strunge, B., Thomsen, V.O., and Andersen, A.B. (2002). Molecular evidence of endogenous reactivation of *Mycobacterium tuberculosis* after 33 years of latent infection. J. Infect. Dis. *185*, 401-404.

Lim, A., and Dick, T. (2001). Plate-based dormancy culture system for *Mycobacterium smegmatis* and isolation of metronidazole-resistant mutants. FEMS Microbiol. Lett. *200*, 215-219.

Lim, A., Eleuterio, M., Hutter, B., Murugasu-Oei, B., and Dick, T. (1999). Oxygen depletion-induced dormancy in *Mycobacterium bovis* BCG. J. Bacteriol. *181*, 2252-2256.

Loebel, R.O., Shorr, E., and Richardson, H.B. (1933). The influence of adverse conditions upon the respiratory metabolism and growth of human tubercle bacilli. J. Bacteriol. *26*, 167-200.

Manabe, Y.C., and Bishai, W.R. (2000). Latent *Mycobacterium tuberculosis*-persistence, patience, and winning by waiting. Nat. Med. *6*, 1327-1329.

Manabe, Y.C., Chen, J.M., Ko, C.G., Chen, P., and Bishai, W.R. (1999). Conditional sigma factor expression, using the inducible acetamidase promoter, reveals that the *Mycobacterium tuberculosis sigF* gene modulates expression of the 16-kilodalton alpha-crystallin homologue. J. Bacteriol. *181*, 7629-7633.

Manganelli, R., Voskuil, M.I., Schoolnik, G.K., Dubnau, E., Gomez, M., and Smith, I. (2002). Role of the extracytoplasmic-function sigma factor sigma(H) in *Mycobacterium tuberculosis* global gene expression. Mol. Microbiol. *45*, 365-374.

Manganelli, R., Voskuil, M.I., Schoolnik, G.K., and Smith, I. (2001). The *Mycobacterium tuberculosis* ECF sigma factor σ^E: role in global gene expression and survival in macrophages. Mol. Microbiol. *41*, 423-437.

Mattman, L.H. (1970). Cell wall-deficient forms of mycobacteria. Ann. N.Y. Acad. Sci. *174*, 852-861.

Mazurek, G.H., LoBue, P.A., Daley, C.L., Bernardo, J., Lardizabal, A.A., Bishai, W.R., Iademarco, M.F., and Rothel, J.S. (2001). Comparison of a whole-blood interferon gamma assay with tuberculin skin testing for detecting latent *Mycobacterium tuberculosis* infection. JAMA *286*, 1740-1747.

McCune, R.M., Feldmann, F.M., Lambert, H.P., and McDermott, W. (1966a). Microbial persistence. I. The capacity of tubercle bacilli to survive sterilization in mouse tissues. J. Exp. Med. *123*, 445-468.

McCune, R.M., Feldmann, F.M., and McDermott, W. (1966b). Microbial persistence. II. Characteristics of the sterile state of tubercle bacilli. J. Exp. Med. *123*, 469-486.

McCune, R.M., Tompsett, R., and McDermott, W. (1957). The fate of *Mycobacterium tuberculosis* in mouse tissues as determined by the microbial enumeration technique. II. The conversion of tuberculous infection to the latent state by the administration of pyrazinamide and a companion drug. J. Exp. Med. *104*, 763-802.

McDonough, K.A., Kress, Y., and Bloom, B.R. (1993). Pathogenesis of tuberculosis: interaction of *Mycobacterium tuberculosis* with macrophages. Infect. Immun. *61*, 2763-2773.

McKinney, J.D., zu Bentrup, K.H., Munoz-Elias, E.J., Miczak, A., Chen, B., Chan, W.T., Swenson, D., Sacchettini, J.C., Jacobs, W.R., and Russell, D.G. (2000). Persistence of *Mycobacterium tuberculosis* in macrophages and mice requires the glyoxylate shunt enzyme isocitrate lyase. Nature *406*, 735-738.

Medlar, E.M., Bernstein, S., and Steward, D.M. (1952). A bacteriologic study of resected tuberculous lesions. Am. Rev. Tuberc. *66*, 36-43.

Mitchison, D.A. (1980). Treatment of tuberculosis. The Mitchell lecture, 1979. J. R. Coll. Physicians Lond. *14*, 98-99.

Miyazaki, E., Chaisson, R.E., and Bishai, W.R. (1999). Analysis of rifapentine for preventive therapy in the Cornell mouse model of latent tuberculosis. Antimicrob. Agents Chemother. *43*, 2126-2130.

Morita, R.Y. (1990). The starvation-survival state of microorganisms in nature and its relationship to bioavailable energy. Experientia *46*, 813-817.

Moulder, J.W. (1985). Comparative biology of intracellular parasitism. Microbiol. Rev. *49*, 298-337.

Mukamolova, G.V., Kaprelyants, A.S., Kell, D.B., and Young, M. (2003). Adoption of the transiently non-culturable state - a bacterial survival strategy? Adv. Microb. Physiol. *47*, 65-129.

Mukamolova, G.V., Kaprelyants, A.S., Young, D.I., Young, M., and Kell, D.B. (1998a). A bacterial cytokine. Proc. Natl. Acad. Sci. USA *95*, 8916-8921.

Mukamolova, G.V., Kormer, S.S., Yanopolskaya, N.D., and Kaprelyants, A.S. (1995a). Properties of dormant cells in stationary-phase cultures of *Micrococcus luteus* during prolonged incubation. Microbiology *64*, 284-288.

Mukamolova, G.V., Turapov, O.A., Kazaryan, K., Telkov, M., Kaprelyants, A.S., Kell, D.B., and Young, M. (2002a). The *rpf* gene of *Micrococcus luteus* encodes an essential secreted growth factor. Mol. Microbiol. *46*, 611-621.

Mukamolova, G.V., Turapov, O.A., Young, D.I., Kaprelyants, A.S., Kell, D.B., and Young, M. (2002b). A family of autocrine growth factors in *Mycobacterium tuberculosis*. Mol. Microbiol. *46*, 623-635.

Mukamolova, G.V., Yanopolskaya, N.D., Kell, D.B., and Kaprelyants, A.S. (1998b). On resuscitation from the dormant state of *Micrococcus luteus*. Antonie van Leeuwenhoek *73*, 237-243.

Mukamolova, G.V., Yanopolskaya, N.D., Votyakova, T.V., Popov, V.I., Kaprelyants, A.S., and Kell, D.B. (1995b). Biochemical changes accompanying the long-term starvation of *Micrococcus luteus* cells in spent growth medium. Arch. Microbiol. *163*, 373-379.

Muttucumaru, D.G., Roberts, G., Hinds, J., Stabler, R.A., and Parish, T. (2004) Gene expression profile of *Mycobacterium tuberculosis* in a non-replicating state. Tuberculosis (Edinb.), *84*, 239-246.

Narberhaus, F. (2002). Alpha-crystallin type heat shock proteins: socialising minichaperones in the context of a multichaperone network. Microbiol. Mol. Biol. Rev. *66*, 64-93.

Ng, V.H., Cox, J.S., Sousa, A.O., MacMicking, J.D., and McKinney, J.D. (2004). Role of KatG catalase-peroxidase in mycobacterial pathogenesis: countering the phagocyte oxidative burst. Mol. Microbiol. *52*, 1291-1302.

North, R.J., and Izzo, A.A. (1993). Mycobacterial virulence. Virulent strains of *Mycobacterium tuberculosis* have faster *in vivo* doubling times and are better equipped to resist growth-inhibiting functions of macrophages in the presence and absence of specific immunity. J. Exp. Med. *177*, 1723-1733.

Nystrom, T. (2003). Conditional senescence in bacteria: death of the immortals. Mol. Microbiol. *48*, 17-23.

Opie, E.L., and Aronson, J.D. (1927). Tubercle bacilli in latent tuberculous lesions and in lung tissue without tuberculosis. Arch. Pathol. *4*, 1-21.

Orme, I.M., and Collins, F.M. (1994). Mouse model of tuberculosis. In Tuberculosis: pathogenesis, protection, and control, B. R. Bloom, ed. (Washington, DC, ASM Press).

Pai, S.R., Actor, J.K., Sepulveda, E., Hunter, R.L., Jr., and Jagannath, C. (2000). Identification of viable and non-viable *Mycobacterium tuberculosis* in mouse organs by directed RT-PCR for antigen 85B mRNA. Microb. Pathog. *28*, 335-342.

Parish, T. (2003). Starvation survival response of *Mycobacterium tuberculosis*. J. Bacteriol. *185*, 6702-6706.

Parish, T., Smith, D.A., Kendall, S., Casali, N., Bancroft, G.J., and Stoker, N.G. (2003a). Deletion of two-component regulatory systems increases the virulence of *Mycobacterium tuberculosis*. Infect. Immun. *71*, 1134-1140.

Parish, T., Smith, D.A., Roberts, G., Betts, J., and Stoker, N.G. (2003b). The *senX3-regX3* two-component regulatory system of *Mycobacterium tuberculosis* is required for virulence. Microbiology *149*, 1423-1435.

Park, H.D., Guinn, K.M., Harrell, M.I., Liao, R., Voskuil, M.I., Tompa, M., Schoolnik, G.K., and Sherman, D.R. (2003). Rv3133c/*dosR* is a transcription factor that mediates the hypoxic response of *Mycobacterium tuberculosis*. Mol. Microbiol. *48*, 833-843.

Parrish, N.M., Dick, J.D., and Bishai, W.R. (1998). Mechanisms of latency in *Mycobacterium tuberculosis*. Trends Microbiol. *6*, 107-112.

Paul, S., Laochumroonvorapong, P., and Kaplan, G. (1996). Comparable growth of virulent and avirulent *Mycobacterium tuberculosis* in human macrophages *in vitro*. J. Infect. Dis. *174*, 105-112.

Phyu, S., Mustafa, T., Hofstad, T., Nilsen, R., Fosse, R., and Bjune, G. (1998). A mouse model for latent tuberculosis. Scand. J. Infect. Dis. *30*, 59-68.

Popper, H.H., Winter, E., and Hofler, G. (1994). DNA of *Mycobacterium tuberculosis* in formalin-fixed, paraffin-embedded tissue in tuberculosis and sarcoidosis detected by polymerase chain reaction. Am. J. Clin. Pathol. *101*, 738-741.

Postgate, J.R. (1976). Death in macrobes and microbes. Symp. Soc. Gen. Microbiol. *26*, 1-18.

Primm, T.P., Andersen, S.J., Mizrahi, V., Avarbock, D., Rubin, H., and Barry, C.E., 3rd (2000). The stringent response of *Mycobacterium tuberculosis* is required for long-term survival. J. Bacteriol. *182*, 4889-4898.

Pym, A.S., Brodin, P., Brosch, R., Huerre, M., and Cole, S.T. (2002). Loss of RD1 contributed to the attenuation of the live tuberculosis vaccines *Mycobacterium bovis* BCG and *Mycobacterium microti*. Mol. Microbiol. *46*, 709-717.

Pym, A.S., Domenech, P., Honore, N., Song, J., Deretic, V., and Cole, S.T. (2001). Regulation of catalase-peroxidase (KatG) expression, isoniazid sensitivity and virulence by *furA* of *Mycobacterium tuberculosis*. Mol. Microbiol. *40*, 879-889.

Ramakrishnan, L., Federspiel, N.A., and Falkow, S. (2000). Granuloma-specific expression of *Mycobacterium* virulence proteins from the glycine-rich PE-PGRS family. Science *288*, 1436-1439.

Raman, S., Hazra, R., Dascher, C.C., and Husson, R.N. (2004). Transcription regulation by the *Mycobacterium tuberculosis* alternative sigma factor SigD and its role in virulence. J. Bacteriol. *186*, 6605-6616.

Ravagnani, A., Finan, C.L., and Young, M. (2005). A novel firmicute protein family related to the antibacterial resuscitation-promoting factors by non-orthologous domain displacement. BMC Genomics *6*, e39

Raynaud, C., Etienne, G., Peyron, P., Laneele, M.A., and M.D. (1998). Extracellular enzyme activities potentially involved in the pathogenicity of *Mycobacterium tuberculosis*. Microbiology *144*, 577-587.

Reed, M.B., Domenech, P., Manca, C., Su, H., Barczak, A.K., Kreiswirth, B.N., Kaplan, G., and Barry, C.E., 3rd (2004). A glycolipid of hypervirulent tuberculosis strains that inhibits the innate immune response. Nature *431*, 84-87.

Rhoades, E.R., Frank, A.A., and Orme, I.M. (1997). Progression of chronic pulmonary tuberculosis in mice aerogenically infected with virulent *Mycobacterium tuberculosis*. Tubercle Lung Dis. *78*, 57-66.

Rosenkrands, I., Slayden, R.A., Crawford, J., Aagaard, C., Barry, C.E., 3rd, and Andersen, P. (2002). Hypoxic response of *Mycobacterium tuberculosis* studied by metabolic labeling and proteome analysis of cellular and extracellular proteins. J. Bacteriol. *184*, 3485-3491.

Roszak, D.B., and Colwell, R.R. (1987). Metabolic activity of bacterial cells enumerated by direct viable count. Appl. Environ. Microbiol. *53*, 2889-2893.

Rousseau, C., Turner, O.C., Rush, E., Bordat, Y., Sirakova, T.D., Kolattukudy, P.E., Ritter, S., Orme, I.M., Gicquel, B., and Jackson, M. (2003). Sulfolipid deficiency does not affect the virulence of *Mycobacterium tuberculosis* H37Rv in mice and guinea pigs. Infect. Immun. *71*, 4684-4690.

Russell, D.G. (2001). *Mycobacterium tuberculosis*: here today, and here tomorrow. Nat. Rev. Mol. Cell Biol. *2*, 569-586.

Sander, P., Papavinasasundaram, K.G., Dick, T., Stavropoulos, E., Ellrott, K., Springer, B., Colston, M.J., and Bottger, E.C. (2001). *Mycobacterium bovis* BCG *recA* deletion mutant shows increased susceptibility to DNA-damaging agents but wild-type survival in a mouse infection model. Infect. Immun. *69*, 3562-3568.

Scanga, C.A., Mohan, V.P., Joseph, H., Yu, K., Chan, J., and Flynn, J.L. (1999). Reactivation of latent tuberculosis: variations on the Cornell murine model. Infect. Immun. *67*, 4531-4538.

Schnappinger, D., Ehrt, S., Voskuil, M.I., Liu, Y., Mangan, J.A., Monahan, I.M., Dolganov, G., Efron, B., Butcher, P.D., Nathan, C., and Schoolnik, G.K. (2003). Transcriptional adaptation of *Mycobacterium tuberculosis* within macrophages: insights into the phagosomal environment. J. Exp. Med. *198*, 693-704.

Seiler, P., Ulrichs, T., Bandermann, S., Pradl, L., Jorg, S., Krenn, V., Morawietz, L., Kaufmann, S.H., and Aichele, P. (2003). Cell-wall alterations as an attribute of *Mycobacterium tuberculosis* in latent infection. J. Infect. Dis. *188*, 1326-1331.

Sherman, D.R., Voskuil, M., Schnappinger, D., Liao, R., Harrell, M.I., and Schoolnik, G.K. (2001). Regulation of the *Mycobacterium tuberculosis* hypoxic response gene encoding alpha -crystallin. Proc. Natl. Acad. Sci. USA *98*, 7534-7539.

Shi, L., Jung, Y.J., Tyagi, S., Gennaro, M.L., and North, R.J. (2003). Expression of Th1-mediated immunity in mouse lungs induces a *Mycobacterium tuberculosis* transcription pattern characteristic of nonreplicating persistence. Proc. Natl. Acad. Sci. USA *100*, 241-246.

Shleeva, .M., Mukamolova, G.V., Young, M., Williams, H.D., and Kaprelyants, A.S. (2004). Formation of "non-culturable" cells of *Mycobacterium smegmatis* in stationary phase in response to growth under sub-optimal conditions and their Rpf-mediated resuscitation. Microbiology *150*, 1687-1697.

Shleeva, M.O., Bagramyan, K., Telkov, M.V., Mukamolova, G.V., Young, M., Kell, D.B., and Kaprelyants, A.S. (2002). Formation and resuscitation of "non-culturable" cells of *Rhodococcus rhodochrous* and *Mycobacterium tuberculosis* in prolonged stationary phase. Microbiology *148*, 1581-1591.

Silver, R.F., Li, Q., and Ellner, J.J. (1998). Expression of virulence of *Mycobacterium tuberculosis* within human monocytes: virulence correlates with intracellular growth and induction of tumor necrosis factor α but not with evasion of lymphocyte-dependent monocyte effector functions. Infect. Immun. *66*, 1190-1199.

Sirakova, T.D., Thirumala, A.K., Dubey, V.S., Sprecher, H., and Kolattukudy, P.E. (2001). The *Mycobacterium tuberculosis pks2* gene encodes the synthase for the hepta- and octamethyl-branched fatty acids required for sulfolipid synthesis. J. Biol. Chem. *276*, 16833-16839.

Smeulders, M.J., Keer, J., Speight, R.A., and Williams, H.D. (1999). Adaptation of *Mycobacterium smegmatis* to stationary phase. J. Bacteriol. *181*, 270-283.

Starck, J., Kallenius, G., Marklund, B.I., Andersson, D.I., and Akerlund, T. (2004). Comparative proteome analysis of *Mycobacterium tuberculosis* grown under aerobic and anaerobic conditions. Microbiology *150*, 3821-3829.

Stewart, G.R., Robertson, B.D., and Young, D.B. (2003). Tuberculosis: a problem with persistence. Nat. Rev. Microbiol. *1*, 97-105.

Stewart, G.R., Snewin, V.A., Walzl, G., Hussell, T., Tormay, P., O'Gaora, P., Goyal, M., Betts, J., Brown, I.N., and Young, D.B. (2001). Overexpression of heat-shock proteins reduces survival of *Mycobacterium tuberculosis* in the chronic phase of infection. Nat. Med. *7*, 732-737.

Steyn, A.J., Collins, D.M., Hondalus, M.K., Jacobs Jr, W.R., Kaewakamii, R.P., and Bloom, B. R. (2002). *Mycobacterium tuberculosis* WhiB3 interacts with RpoV to affect host survival but is dispensible for *in vivo* growth. Proc. Natl. Acad. Sci. USA *99*, 3147-3152.

Sun, R., Converse, P.J., Ko, C., Tyagi, S., Morrison, N.E., and Bishai, W.R. (2004). *Mycobacterium tuberculosis* ECF sigma factor *sigC* is required for lethality in mice and for the conditional expression of a defined gene set. Mol. Microbiol. *52*, 25-38.

Sun, Z., and Zhang, Y. (1999). Spent culture supernatant of *Mycobacterium tuberculosis* H37Ra improves viability of aged cultures of this strain and allows small inocula to initiate growth. J. Bacteriol. *181*, 7626-7628.

Tani, T.H., Khodursky, A., Blumenthal, R.M., Brown, P.O., and Matthews, R.G. (2002). Adaptation to famine: a family of stationary-phase genes revealed by microarray analysis. Proc. Natl. Acad. Sci. USA *99*, 13471-13476.

Tufariello, J.M., Chan, J., and Flynn, J.L. (2003). Latent tuberculosis: mechanisms of host and bacillus that contribute to persistent infection. Lancet Infect. Dis. *3*, 578-590.

Tufariello, J.M., Jacobs, W.R.J., and Chan, J. (2004). Individual *Mycobacterium tuberculosis* resuscitation-promoting factor homologues are dispensable for growth *in vitro* and *in vivo*. Infect. Immun. *72*, 515-526.

Turner, D.J., Hoyle, S.L., Snewin, V.A., Gares, M.P., Brown, I.N., and Young, D.B. (2002). An *ex vivo* culture model for screening drug activity against *in vivo* phenotypes of *Mycobacterium tuberculosis*. Microbiology *148*, 2929-2936.

van Pinxteren, L.A.H., Cassidy, J.P., Smedegaard, B.H.C., Agger, E.M., and Andersen, P. (2000). Control of latent *Mycobacterium tuberculosis* infection is dependent on CD8 T cells. Eur. J. Immunol. *30*, 3689-3698.

Vandivière, H.M., Loring, W.E., Melvin, I., and Willis, S. (1956). The treated pulmonary lesion and its tubercle bacillus. II. The death and resurrection. Am. J. Med. Sci. *232*, 30-37.

Voloshin, S.A., Shleeva, M.O., and Kaprelyants, A.S. (2004). Role of intercellular contacts for the initiation of growth and the formation of the non-culturable state for *Rhodococcus rhodochrous* cultivated in a poor medium. Mikrobiologiia (in press).

Voskuil, M. (2004). *Mycobacterium tuberculosis* gene expression during environmental conditions associated with latency. Tuberculosis *84*, 138-143.

Voskuil, M.I., Schnappinger, D., Visconti, K.C., Harrell, M.I., Dolganov, G.M., Sherman, D.R., and Schoolnik, G.K. (2003). Inhibition of respiration by nitric oxide induces a *Mycobacterium tuberculosis* dormancy program. J. Exp. Med. *198*, 705-713.

Voskuil, M.I., Visconti, K.C., and Schoolnik, G.K. (2004). *Mycobacterium tuberculosis* gene expression during adaptation to stationary phase and low-oxygen dormancy. Tuberculosis (Edinb.) *84*, 218-227.

Wayne, L.G. (1960). The bacteriology of resected tuberculosis pulmonary lesions. II. Observation on bacilli which are stainable but which cannot be cultured. Am. Rev. Respir. Dis. *82*, 370-377.

Wayne, L.G. (1976). Dynamics of submerged growth of *Mycobacterium tuberculosis* under aerobic and microaerophilic conditions. Am. Rev. Respir. Dis. *114*, 807-811.

Wayne, L.G. (1994). Dormancy of *Mycobacterium tuberculosis* and latency of disease. Eur. J. Clin. Microbiol. Infect. Dis. *13*, 908-914.

Wayne, L.G., and Hayes, L.G. (1996). An *in vitro* model for sequential study of shiftdown of *Mycobacterium tuberculosis* through 2 stages of nonreplicating persistence. Infect. Immun. *64*, 2062-2069.

Wayne, L.G., and Hayes, L.G. (1998). Nitrate reduction as a marker for hypoxic shiftdown of *Mycobacterium tuberculosis*. Tubercle Lung Dis. *79*, 127-132.

Wayne, L.G., and Lin, K.Y. (1982). Glyoxalate metabolism and adaptation of *Mycobacterium tuberculosis* to survival under anaerobic conditions. Infect. Immun. *37*, 1042-1049.

Wayne, L.G., and Sohaskey, C.D. (2001). Nonreplicating persistence of *Mycobacterium tuberculosis*. Annu. Rev. Microbiol. *55*, 139-163.

Wayne, L.G., and Sramek, H.A. (1994). Metronidazole is bactericidal to dormant cells of *Mycobacterium tuberculosis*. Antimicrob. Agents Chemother. *38*, 2054-2058.

Webb, J.S., Thompson, L.S., James, S., Charlton, T., Tolker-Nielsen, T., Koch, B., Givskov, M., and Kjelleberg, S. (2003). Cell death in *Pseudomonas aeruginosa* biofilm development. J. Bacteriol. *185*, 4585-4592.

Wilson, M., Voskuil, M., Schnappinger, D., and Schoolnik, G.K. (2001). Functional genomics of *Mycobacterium tunberculosis* using DNA microarrays. In Methods in Molecular Medicine, vol 54, *Mycobacterium tuberculosis* protocols, T. Parish, and N.G. Stoker, eds. (Totowa, N. J., Humana Press Inc), pp. 335-357.

Wilson, W., DeRisi, J., Kristensen, H.H., Imboden, P., Rane, S., Brown, P.O., and Schoolnik, G.K. (1999). Exploring drug-induced alterations in gene expression in *Mycobacterium tuberculosis* by microarray hybridization. Proc. Natl. Acad. Sci. USA *96*, 12833-12838.

Yuan, Y., Crane, D.D., and Barry III, C.E. (1996). Stationary phase-associated protein expression in *Mycobacterium tuberculosis*: function of the mycobacterial alpha-crystallin homolog. J. Bacteriol. *178*, 4484-4492.

Yuan, Y., Crane, D.D., Simpson, R.M., Zhu, Y.Q., Hickey, M.J., Sherman, D.R., and Barry, C.E. (1998). The 16-kDa alpha-crystallin (Acr) protein of *Mycobacterium tuberculosis* is required for growth in macrophages. Proc. Natl. Acad. Sci. USA *95*, 9578-9583.

Zahrt, T.C. (2003). Molecular mechanisms regulating persistent *Mycobacterium tuberculosis* infections. Microbes Infect. *5*, 159-167.

Zahrt, T.C., and Deretic, V. (2000). An essential two-component signal transduction system in *Mycobacterium tuberculosis*. J. Bacteriol. *182*, 3832-3838.

Zahrt, T.C., and Deretic, V. (2001). *Mycobacterium tuberculosis* signal transduction system required for persistent infections. Proc. Natl. Acad. Sci. USA *98*, 12706-12711.

Zhang, M., Gong, J.H., Lin, Y.G., and Barnes, P.F. (1998). Growth of virulent and avirulent *Mycobacterium tuberculosis* strains in human macrophages. Infect. Immun. *66*, 794-799.

Zhang, Y. (2004). Persistent and dormant tubercle bacilli and latent tuberculosis. Frontiers Biosci. *9*, 1136-1156.

Zhang, Y., Yang, Y., Woods, A., Cotter, R.J., and Sun, Z. (2001). Resuscitation of dormant *Mycobacterium tuberculosis* by phospholipids or specific peptides. Biochem. Biophys. Res. Commun. *284*, 542-547.

Zhu, W., Plikaytis, B.B., and Shinnick, T.M. (2003). Resuscitation factors from mycobacteria: homologs of *Micrococcus luteus* proteins. Tuberculosis (Edinb.) *83*, 261-269.

Chapter 9

Mycobacterial Resistance to Reactive Oxygen and Nitrogen Intermediates: Recent Views and Progress in *M. tuberculosis*

Kyu Y. Rhee

ABSTRACT

Resistance to the antimicrobial actions of reactive oxygen and reactive nitrogen intermediates (ROI/RNI) is a common property of pathogenic mycobacteria. However, the molecular basis of this resistance in mycobacteria remains poorly understood. This gap in our understanding of mycobacterial ROI/RNI resistance is explained by three factors: (1) the historical absence of a facile system to genetically manipulate mycobacteria, (2) our limited understanding of the free radical stresses encountered by mycobacteria *in vivo*, and (3) apparent redundancies in the ROI/RNI defense mechanisms utilized by mycobacteria. This chapter will review mycobacterial ROI/RNI resistance placed in the framework of our current understanding of the antimicrobial biology of immune-derived ROI/RNI. Reflecting the weight of the published literature, studies of Mtb and murine models of tuberculosis will be emphasized.

INTRODUCTION

In the host-pathogen interaction, pathogen resistance to the host can be explained by two types of mechanisms: (1) those that are '*preventive*' and suppress production of, inhibit delivery of, or detoxify noxious agents, and (2) those that are '*adaptive*' and compensate for and/or repair damage incurred by these agents. In operational terms, resistance determinants can be defined by the following criteria:

(1) The demonstration of a survival advantage for clonal pathogens containing a wild-type allele over knockout counterparts in wild-type hosts
(2) Restoration of this defect in knockout strains complemented with a wild-type allele
(3) Reversal of the attenuation of knockout strains using hosts unable to produce significant amounts of the offending stress.

Resistance to the antimicrobial actions of reactive oxygen (ROI) and reactive nitrogen intermediates (RNI) is a common property of pathogenic mycobacteria. These bacteria establish and maintain long-term intracellular persistence within their hosts facing extensive and prolonged exposures to ROI/RNI (Haas, 2000; Nathan and Ehrt, 2004). Studies of other intracellular pathogens have revealed a wide range of both *preventive* and *adaptive* ROI/RNI resistance strategies (for a review see (Fang, 2004)). In mycobacteria, few such mechanisms have been evaluated and even fewer validated. Thus, the molecular basis of mycobacterial resistance to host-derived ROI/RNI *in vivo* remains poorly understood.

To some degree, the gap in our understanding of ROI/RNI mycobacterial resistance is explained by three factors: (1) the historical absence of a facile system to genetically manipulate mycobacteria, (2) our limited understanding of the free radical stresses encountered by mycobacteria *in vivo*, and (3) apparent redundancies in the ROI/RNI defense mechanisms utilized by mycobacteria.

To date, three general strategies to identifying ROI/RNI resistance genes in mycobacteria have been pursued. ROI/RNI resistance genes were first pursued based on *biochemical and/or genetic hypotheses* originating in experimentally more tractable organisms. This targeted approach leverages the evolutionary conservation of biochemical function and/or genomic structure but neglects important differences in the ecological niches and the associated ROI/RNI stresses from which mycobacteria are taken. Thus, this approach has met varying degrees of success that most likely reflect the physiologic relevance of the conditions tested or the non-orthologous conservation of gene function. Examples illustrating successes of this approach include the *in vivo* validation of the ROI detoxifying activities of the *katG*-encoded catalase-peroxidase-peroxynitritase (Ng et al., 2004) and the Fe-dependent superoxide dismutase, *sodA* (Edwards et al., 2001). Examples illustrating failures of this approach include the discovery of non-functional and/or absent orthologs of the ROI/RNI-responsive *oxyR* and *soxRS* stress regulons (Cole et al., 1998; Deretic et al., 1995; Sherman et al., 1995).

A second approach pursued to identify resistance mechanisms is a *gain-of-function* selection. Here, a library of mycobacterial genes is heterologously expressed in a surrogate host and selected on the ability of individual clones to confer resistance to *in vitro* chemical stresses. This approach also relies on the biochemical and biological transposibility of gene products between organisms and has thus met limited success, emphasizing the species-specific nature of gene function imposed by evolution. For example, while the *Mycobacterium tuberculosis* (Mtb) gene *noxR1* was identified on its ability to confer resistance to acidified nitrite, S-nitrosoglutathione and H_2O_2 in *E. coli*, deletion of *noxR1* in Mtb produced no observable phenotype in infected mice (Ehrt et al., 1997; Ruan et al., 1999; Stewart et al., 2000).

Most recently, completion of the Mtb genome sequencing project and the development of phage-based genetic systems in Mtb have led to the development of *genomically-based, loss-of-function* screening approaches. One particularly powerful approach involved the screening of an ordered library of 10,100 Mtb transposon mutants on the basis of their hypersusceptibility to a low dose of a *physiologic* source of RNI, acidified nitrite (Darwin et al., 2003). This effort identified mutations in seven genes representing six pathways, three of which have been evaluated *in vivo* and whose mutants appear to be attenuated in a nitric oxide synthase-2 (NOS2)-dependent manner. Among these were several otherwise unexpected candidates including two proteasome-asociated proteins (*mpa, paf*), a DNA repair enzyme (*uvrB*), and an enzyme involved in the biosynthesis of the flavin-like cofactor F420 (*fbiC*). The relative success of this approach demonstrated the ability of a genomic approach to bypass the need for preconception and emphasized the value of an autochthonous biologic system. A shortcoming of this approach however, is its insensitivity to genes whose functions are essential for growth *in vitro* and therefore not represented in the transposon mutant pools screened.

Reflecting on these lessons, it is clear that ROI/RNI resistance has evolved in response to specific ecological niches and that further advances in the identification and validation of mycobacterial ROI/RNI resistance mechanisms will require more detailed studies of mycobacterial ROI/RNI biology *in vivo*. In this regard, it is important to emphasize that, while a measure of a gene product's function can be the phenotype of its mutant, the apparent absence of a biological phenotype need not be equated with dispensability. This is because the range of experimental conditions tested is often narrow and fails to discriminate conditionally important functions from those that are either redundant or truly dispensable (Nathan and Shiloh, 2000). This chapter will review mycobacterial ROI/RNI resistance placed in the framework of our current understanding of the antimicrobial biology of immune-derived ROI/RNI. Reflecting the weight of the published literature, studies of Mtb and murine models of tuberculosis will be emphasized.

ROI/RNI PRODUCTION *IN VIVO* AND PREVENTIVE ROUTES TO RESISTANCE

BIOLOGICAL SOURCES AND ROI/RNI PRODUCTION *IN VIVO*

Following immune activation, macrophages produce both ROI and RNI (Nathan and Ehrt, 2004). Phagocytosis or cytokine-mediated immune activation of macrophages stimulates assembly of the phagocyte oxidase complex (phox) and triggers a respiratory burst within minutes that leads to the production of $O_2^{\cdot-}$ through the one electron reduction of molecular oxygen. Cytokine-mediated production of RNI requires *de novo* transcription of NOS2 and occurs more slowly. Once produced however, NOS2 remains catalytically active for days, leading to the sustained production of RNI. Accordingly, *ex vivo* measurements of bone-marrow-derived macrophages have demonstrated that ingestion of Mtb stimulates a brief and limited production of $O_2^{\cdot-}$ followed by the sustained productions of RNI (Piddington et al., 2001).

Phox and NOS2 can also be expressed by polymorphonuclear leukocytes (PMNs). Conversely, enzymes other than phox and NOS2 can produce ROI/RNI. However, the antimicrobial significance of *non*-macrophage, *non*-phox-, *non*-NOS2 sources of ROI/RNI in Mtb infection is unclear. Conflicting evidence surrounds the mycobactericidal properties of PMNs *in vitro* (Denis, 1991; Jones et al., 1990; May and Spagnuolo, 1987) and antimicrobial roles for PMNs in Mtb infection remain undefined as judged by the absence of defects in survival and bacterial burden following infection of PMN-depleted mice (Pedrosa et al., 2000; Seiler et al., 2000). Contributions from *non*-phox sources of ROI, such as myeloperoxidase (MPO), xanthine oxidase (XO), and adventitious electron leak from the mitochondrial electron transport chain, are similarly unclear given the relatively high levels of ROI needed to achieve significant antimicrobial activity, the absence of MPO expression in macrophages, and the relatively minor immunocompromise associated with MPO or XO deficiencies (reviewed in (Fang, 2004)).

While macrophages are the major cell type infected by and capable of killing Mtb, and following infection, produce both ROI/RNI (discussed above) (Nathan and Ehrt, 2004), detailed descriptions of ROI/RNI production *in vivo* remain sparse. MacMicking et al. demonstrated correlations between inducible plasma NO_x, nitrosothiol concentrations and NOS2-mediated immune control of Mtb (MacMicking et al., 1997). Two independent studies have reported the co-localizing immunoreactivity of NOS2 expression and nitrotyrosine staining (a footprint of NO and $O_2^{\cdot-}$ reactivity) in both murine and human Mtb infection (Choi et al., 2002; Flynn et al., 1993; MacMicking et al., 1997; Mogues et al., 2001; Schon et al., 2004). However, quantitative or kinetic descriptions of ROI/RNI production at the tissue or cellular levels remain undescribed. Indeed, the nature, extent, and duration of RNI/ROI production within the macrophage and the granuloma *in vivo* are largely unknown.

ANTIMICROBIAL IMPACT OF PHOX- AND NOS2-DERIVED ROI/RNI *IN VIVO* AND PHOX-, NOS2-SPECIFIC DEFENSES

The functional impact of ROI/RNI in Mtb infection has been addressed using genetically engineered phox- and/or NOS2-deficient mice. These same mice have also been used to evaluate the biological significance of candidate mycobacterial ROI/RNI resistance genes. However, of note these studies have only addressed the roles relative to the acute phase of infection since the gene deficiencies associated with these mice are present from the time of, rather than following, infection. Below is a brief overview of our current understanding of the relative biological impact of phox and NOS2 and the ability of Mtb to detoxify their ROI/RNI products.

PHOX-DERIVED ROI

Neither murine nor human studies of tuberculosis have identified a measurable impact of phox-derived ROI on Mtb *in vivo*. In terms of bacterial burden and survival following infection phox-deficient mice control Mtb as well as their wild-type counterparts (Adams et al., 1997; Cooper et al., 2000; Jung et al., 2002). Human genetic deficiencies in phox, likewise, demonstrate no increased predisposition to infection by Mtb despite a markedly increased susceptibility to a number of life-threatening infections, including those due to saprophytic mycobacteria and the attenuated vaccine strain BCG (Segal et al., 2000).

In contrast, numerous studies have reported a longstanding association between Mtb virulence and catalase activity, such that Mtb strains deficient in catalase activity are significantly attenuated for virulence in animal models of tuberculosis (Cohn et al., 1954; Li et al., 1998; Middlebrook and Cohn, 1953; Pym et al., 2001). Moreover, *in vitro* Mtb is relatively resistant to ROI (Chan et al., 1992). Recent work used a phage-based targeted gene deletion approach to formally evaluate the biochemical ability of the *katG*-encoded catalase-peroxidase-peroxynitritase to detoxify phox-derived ROI *in vivo* (Ng et al., 2004). *katG* encodes the only catalase of Mtb and catalyzes the detoxification of H_2O_2 (that may arise from the spontaneous or catalyzed decomposition of $O_2^{-\cdot}$ produced by phox) to H_2O and O_2 in a heme- and NADH-dependent manner (Cole et al., 1998; Heym et al., 1993; Wengenack et al., 1999; Yamada et al., 2002). This work demonstrated that a *katG* mutant strain of Mtb was significantly attenuated for survival in wild-type, but not phox-deficient, mice and that this phenotype could be complemented by the wild-type *katG* allele in the wild-type mouse. Attenuation of this *katG* mutant was also only observed during a narrow time frame 2-4 weeks after infection, consistent with the early and limited production of ROI observed *in vitro* (Ng et al., 2004; Piddington et al., 2001). Together, these results largely explain the apparent absence of an antimicrobial impact of phox derived ROI on Mtb. The absence of a discernable phenotype in phox-deficient mice further discounts significant antimicrobial roles for *non*-phox derived sources of ROIs.

In addition to *katG*, Mtb also encodes two superoxide dismutases, *sodA*, an Fe-dependent, secreted form, and *sodC*, a Cu, Zn-dependent, membrane-associated form, (Cole et al., 1998). These enzymes catalyze the conversion of $O_2^{-\cdot}$ to H_2O_2 at rates ~100X faster than those occurring spontaneously. Thus, these enzymes may help kinetically shunt $O_2^{-\cdot}$ both away from the more toxic lipid hydroperoxides or peroxynitrite and toward the peroxide catabolizing activity of KatG. Deletion of *sodC* in Mtb increased the susceptibility of Mtb to $O_2^{-\cdot}$ and H_2O_2 *in vitro* and impaired survival in wild-type, but not phox-deficient, bone marrow-derived macrophages (Piddington et al., 2001). However, *sodC*-deficient strains of Mtb demonstrate no observable phenotype when tested in mouse or guinea pig models of infections as measured by bacterial burdens (Dussurget et al., 2001; Piddington et al., 2001). Efforts to delete *sodA* have been unsuccessful suggesting that it is essential for growth *in vitro*. Anti-sense knockdowns of *sodA* in Mtb similarly demonstrated a 5-log attenuation in the bacterial burden of the lungs and spleens of infected mice (Edwards et al., 2001). Nonetheless, despite the imperfect demonstration of either gene as a *bona fide* ROI resistance mechanism, it is likely that, in conjunction with *katG*, these enzymes form a highly efficient, and perhaps partially redundant, ROI detoxification system as well as an indirect RNI defense by preventing the formation of peroxynitrite.

NOS2-DERIVED RNI

Antimicrobial roles for NOS2-derived RNI in acute Mtb infection are more clear cut. *In vitro* NOS2-derived RNI are slowly but potently bactericidal against Mtb (Long et al., 1999). *In vivo*, NOS2-deficient mice fail to control Mtb replication and demonstrate markedly reduced survival rates when compared to their wild-type counterparts (MacMicking et al., 1997; Mogues et al., 2001; Scanga et al., 2001). Expression studies in both humans and mice have also demonstrated the co-localizing presence of NOS2 and nitrotyrosine, a footprint of NOS2 activity, during Mtb infection (Choi et al., 2002; Schon et al., 2004; Venkataprasad et al., 1999). The absence of bacterial burden or survival defects between phox-NOS2 deficient mice and their NOS2-deficient counterparts emphasizes the relative importance of NOS2-derived ROI/RNI over phox-derived ROI (Ng et al., 2004). Together, these data support the production and predominant antimicrobial impact of NOS2-derived RNI in Mtb infection.

Unlike the case for phox-derived ROI however, efforts to identify RNI detoxification and/or suppressive mechanisms have met disappointing results. One extensively studied RNI detoxification candidate is the Mtb-encoded peroxiredoxin, *ahpC*. Mtb *ahpC* was identified by its ability to heterologously transfer RNI resistance to *Salmonella typhimurium* lacking its own *ahpC* (Chen et al., 1998) and exhibits both peroxidase and peroxynitrite reductase activities *in vitro* (Bryk et al., 2000; Bryk et al., 2002). The former reduces alkyl hydroperoxides, generated by the reaction of ROI with mycobacterial lipids and proteins, to alcohols and water. The latter detoxifies the joint ROI/RNI-derived species peroxynitrite ($ONOO^-$). (Peroxynitrite is a distinct RNI species that arises at near diffusion-limited rates under conditions favoring the equimolar production of NO and $O_2^{\cdot-}$ and, at physiologic pH values, demonstrates both hydroxyl radical ($\cdot OH$) and nitrogen dioxide radical ($\cdot NO_2$) oxidant-like properties more potent than either of its precursors; Grisham et al., 1999.) Bryk and colleagues further demonstrated that, in conjunction with three other proteins (Lpd, DlaT, AhpD), AhpC similarly constitutes an *endogenous* peroxidase-peroxynitritase ROI/RNI defense system in Mtb (Bryk et al., 2000; Bryk et al., 2002).

However, functional roles for *ahpC* and its associated PNR system *in vivo* have been difficult to demonstrate. While promoter mutations that increase expression of *ahpC* have been reported in association with loss of KatG activity in isoniazid-resistant strains of Mtb (and increase resistance of Mtb to alkylhydroperoxides), increased levels of AhpC do not correlate with increased resistance to H_2O_2 or restoration of virulence in mouse models of infection using *katG* null strains of Mtb (Heym et al., 1997; Manca et al., 1999). One recent study reported no significant difference in bacterial burden attributable to the disruption of *ahpC* (Springer et al., 2001). Of note however, this gene disruption did not remove the *ahpC* gene but instead, inserted a kanamycin antibiotic resistance cassette into the central third of its coding region while *in vitro* this mutant demonstrated only a modest increase in its susceptibility to chemical ROI (Springer et al., 2001). In contrast, *in vivo* expression profiling of *ahpC* in a mouse model of tuberculosis has demonstrated progressive increases in *ahpC* expression that parallel those observed under conditions of static growth *in vitro* and perhaps mimic those encountered during the chronic phase of infection (S. Ehrt, C. Nathan, unpublished data; Master et al., 2002). Radi et al. further reported the ability of the Mtb atypical peroxiredoxin *tpx* and a thioredoxin-based system to replace *ahpC* and its PNR system *in vitro* (Jaeger et al., 2004). Together these data highlight potential redundancies in ROI/RNI defense that, while perhaps serving to ensure survival *in vivo*, have obscured the ability to experimentally define *bona fide* ROI/RNI resistance mechanisms. Judgment regarding the biological function of *ahpC* and its four-component PNR in Mtb's RNI defense is thus best left suspended.

Mtb also encodes three annotated NO dioxygenase homologues (*hmp*, *glbN*, *glbO*). These enzymes provide potential routes of NO catabolism via conversion to NO_3^- in a heme- and O_2-dependent manner (Cole et al., 1998). *hmp* encodes a flavohemoglobin whose expression in *Escherichia coli* and *S. typhimurium* is induced by RNI and whose deletion results in increased susceptibility to NO and S-nitrosothiols (Crawford and Goldberg, 1998; Hu et al., 1999). In Mtb, *hmp* expression is induced in response to oxygen limitation or ROI or RNI stress during the late exponential phase of growth *in vitro* (Hu et al., 1999). Microarray profiling of Mtb recovered from infected, murine bone marrow-derived macrophages demonstrated that *hmp* expression is also induced intraphagosomally in an interferon γ-and NOS2-independent manner (Schnappinger et al., 2003). However, sequence analysis indicates the absence of highly conserved histidine and phenylalanine residues required for heme coordination and evidence of *in vitro* activity remains unreported (Poole and Hughes, 2000). In contrast, *glbN* and *glbO* encode truncated hemoglobins with demonstrated NO dioxygenase activities *in vitro* (Milani et al., 2003; Ouellet et al., 2002) but little is known about their expression patterns (Couture et al., 1999) and none have been evaluated *in vivo*.

CANDIDATE ROUTES TO RESISTANCE: DETOXIFICATION, EVASION AND SUPPRESSION

To date, the majority of efforts to identify '*preventive*' ROI/RNI resistance mechanisms in Mtb have focused on enzymatic *detoxification* strategies (discussed above). This skew reflects technical challenges associated with cell based screening approaches, the historically limited genetic system of Mtb, and the resulting bias favoring the use of chemical, rather than biological, sources of RNI. However, studies of ROI/RNI resistance in other intracellular pathogens have identified a wide range of '*preventive*' strategies to resist the antimicrobial actions of phox- and/or NOS2-derived ROI/RNI. These include the production of protein and non-protein scavengers of ROI/RNI and/or chemical mediators of their production and/or activity as well as the actions of specialized secretion systems that: (1) interfere with cellular signaling pathways involved in phagocytosis, vesicular trafficking, phagolysosomal maturation, and pro-inflammatory cytokine production, and (2) impair the production and delivery of ROI/RNI (for reviews see (Coombes et al., 2004; Fang, 2004; Koul et al., 2004)).

Early studies of mycobacteria first identified ROI scavenging properties associated with mycobacterial phenolic glycolipids based on their ability to confer cell wall protective properties (Neill and Klebanoff, 1988). Similar protective roles for cyclopropanated lipids from ROI-mediated lipid peroxidation have since been proposed (Chan et al., 1989; Yuan et al., 1998b). More recent studies have suggested a modest protective effect of phthiocerol dimycocerosates from the bactericidal activities of NOS2-derived RNI during acute infection (Rousseau et al., 2004). While additional roles for sulfatides, trehalose and mannitol remain unaddressed, the chemically '*neutralizing*' properties of mycobacterial cell surface-associated molecules commend their potential as non-enzymatic mediators of ROI/RNI resistance.

In addition to meeting physiologic needs, regulation of Fe(II) and Mn(II) accumulation is also an often overlooked, but widespread, enzymatic and non-enzymatic mechanism of ROI/RNI resistance. Tight control of intrabacterial iron availability, afforded by specialized iron acquisition and utilization systems, offers the capacity to limit Fenton-catalyzed production of ·OH. Roles for individual components of the Mtb iron metabolic machinery in ROI defense however remain unaddressed (for a recent review see (Schaible and Kaufmann, 2004)). Mn(II) is similarly a common co-factor for catalases, superoxide dismutases, and peroxidases, and also may indirectly promote enzymatically based ROI resistance. Mn(II) itself can also scavenge both O_2-. and catalyze the detoxification of H_2O_2 to H_2O and O_2. However, while Mtb encodes a Mn transport system (*mntH*) whose orthologs in *Neisseria* sp. afford a catalase-independent ROI protection (Archibald and Fridovich, 1982; Cheton and Archibald, 1988; Horsburgh et al., 2002; Seib et al., 2004; Stadtman et al., 1990), deletion analysis of Mtb *mntH* demonstrated no observable phenotype in a murine model of tuberculosis (Domenech et al., 2002). Roles for metalloregulators of Mn(II) accumulation, such as Fur (*f*erric *u*ptake *r*egulator) and IdeR, and ABC-type and P-type ATPase Mn(II) transporters also await evaluation (Horsburgh et al., 2002).

Beyond chemical detoxification mechanisms, one emerging '*preventive*' mechanism of ROI/RNI resistance may involve the secretory apparatus of Mtb. The Mtb genome does not encode formal orthologs of the Type II and III specialized secretion systems used by many gram-negative pathogens (Coombes et al., 2004). However, increasing evidence supports the existence of its own specialized secretion systems. To date, three secretory systems (*secA1*, *secA2*, *snm*) have been identified in Mtb, two of which (*secA2*, *snm*) appear to serve accessory functions and are important for virulence in animal models of tuberculosis (Braunstein et al., 2001). Of these, one (*secA2*) has been shown to mediate the secretion of the Fe-dependent superoxide dismutase (*sodA*) and may contribute to the ROI defense of Mtb (Braunstein et al., 2003).

A *sec*-independent secretory system in Mtb, the twin-arginine translocase (TAT), whose homologs in *Pseudomonas aeruginosa* and enterhemorrhagic *E. coli* contribute to virulence, was recently identified by sequence homology (*tatA*, *tatC*) (Ochsner et al., 2002; Pradel et al., 2003). Of interest, two of the seven predicted TAT clients (*Rv*0846, *Rv*2577) are annotated to encode proteins potentially involved in ROI/RNI defense. *Rv*0846 encodes a homolog of an *E. coli* multicopper oxidase (Schenk et al., 2000). *Rv*2577 encodes a conserved hypothetical protein with sequence similarity to purple acid phosphatases that have the capacity to inhibit the respiratory burst or function as an extracellular Fenton-type catalyst to prevent ROI induced damage (Schenk et al., 2000). Proof of a functional TAT system however awaits evaluation.

Increasing evidence also supports the ability of Mtb to subvert the production and delivery of ROI/RNI. Numerous studies have demonstrated the ability of Mtb to gain entry into naïve macrophages using a variety of receptors, some of which are unlinked to immune activation. For example, phagocytosis of Mtb through the mannose receptor (MR) does not trigger the production of $O_2^{-\cdot}$ (Astarie-Dequeker et al., 1999). Entry via complement receptor 3 (CR3) is likewise believed to bypass the respiratory burst (Hu et al., 2000). Conversely, recent evidence suggests the presence of an anti-phagocytic capsule on pathogenic, but not saprophytic, mycobacteria that may limit non-opsonic interactions and direct specifically favorable receptor-ligand interactions on certain types of macrophages (Stokes et al., 2004). Limited data also suggest that Mtb actively inhibits recruitment of NOS2 to Mtb-containing phagosomes in an actin-dependent manner (Miller et al., 2004).

Finally, Mtb possesses multiple mechanisms capable of modulating host-signaling pathways involved in phagolysosomal maturation, host-cell apoptosis, and antibacterial responses (for a recent review, see ref. (Koul et al., 2004)). However, Mtb-encoded determinants of these activities remain unidentified and none have been evaluated for their ability to modulate ROI/RNI production. One notable exception is a recent report describing the activity of an Mtb-encoded protein tyrosine phosphatase (*ptpB*) (Singh et al., 2003). PtpB was shown to be a secreted protein that functions optimally at a pH of 5.6, similar to that achieved within the phagolysosome of an activated macrophage (Koul et al., 2000). While the Mtb genome encodes two protein tyrosine phosphatases, Mtb itself encodes no tyrosine kinases, suggesting that these proteins may function specifically to modulate host protein tyrosine phosphorylation signaling (Cole et al., 1998; Singh et al., 2003). Accordingly, *ptpB* mutant strains of Mtb were shown to be attenuated both *in vivo* using guinea pig models of infection and in activated, but not resting, macrophage-like cell lines (Singh et al., 2003). These findings highlight striking similarities to the protein tyrosine phosphatases used by other intracellular pathogens (such as *Salmonella* and *Yersinia*) to evade or suppress ROI/RNI production and thus emphasize a potential role for this protein in ROI/RNI defense.

TARGETS OF ROI/RNI IN MTB AND ADAPTIVE RESISTANCE MECHANISMS

ROI/RNI REACTIVITY AND MACROMOLECULAR DAMAGE

In addition to their primary products, $O_2^{-\cdot}$ and NO, phox and NOS2 give rise to a complex array of RNI and ROI (Fig. 1). These include all NO oxidation states and adducts ranging from NO up to, but excluding, NO_3^- (NO^-, $\cdot NO_2$, NO_2-, N_2O_3, N_2O_4), and O_2 intermediate reduction products from O_2 up to, but excluding, H_2O ($O_2^{-\cdot}$, H_2O_2 and $OH\cdot$). These species arise through both spontaneous and catalyzed reactions of $O_2^{-\cdot}$ and NO. For example, while nitrite accumulates as an auto-oxidation product of NO, phagosomal acidification following infection with Mtb (pH= ~4.5) regenerates NO through the spontaneous formation and disproportionation of nitrous acid. ROI and RNI species can also react with each other to produce the even more reactive species, peroxynitrite ($ONOO^-$) (Nathan and Shiloh, 2000).

The chemical reactivity of phox- and/or NOS2-derived ROI is defined by the production rate, redox potential, stability, lipid solubility and localization of each ROI/RNI species present. For example, while $O_2^{-\cdot}$ demonstrates an unusually strong redox potential, its rates of spontaneous and catalyzed conversion to H_2O_2 predict a steady state concentration of $\sim 10^{-11}$ M that limit its reactivity to a short radius of its site of production (Forman and Torres, 2002). In contrast, H_2O_2 possesses a weaker redox potential and undergoes much slower spontaneous decay but can act more distantly from its site of production, leading to the oxidation of the sulfur atoms of cysteine and methionine residues. While higher concentrations of H_2O_2 are generally required to directly oxidize amino acid residues or induce lipid peroxidation, Fenton-type reactions between H_2O_2 and transition metals can also give rise to more potent oxidants at lower concentrations, such as $OH\cdot$, peroxo- and oxo-metal complexes, and induce significant macromolecular damage, such as oxidation of sugar residues that leads to DNA strand breakage and nucleoside oxidation leading to DNA mutation. Overall, the major chemical reactions of ROIs include oxidation, peroxidation, carbonylation and hydroxylation of proteins, lipids, and nucleic acids (reviewed in Nathan, 2003).

RNI reactivity is defined by its local redox environment and the oxidation state of the RNI species present. For example, while NO and S-nitrosothiols can nitrosylate thiols and amino groups of proteins (forming S-nitrosylated and N-nitrosylated proteins respectively), heme- and non-heme transition metals (forming nitrosyl metal complexes), and amine groups of DNA bases (leading to deamination of guanine, adenosine, and cytosine residues), the stepwise oxidation products $NO_2^{\cdot}$, N_2O_3 and N_2O_4, as well as $ONOO^-$, also possess oxidative properties similar to those

Nathan and Ehrt, 2004; Zahrt and Deretic, 2002). Putting aside current shortcomings in our ability to faithfully model persistence, accumulating evidence suggests that NOS2-mediated immune control of Mtb is ongoing. In contrast to wild-type counterparts, Mtb-infected NOS2-deficient mice treated and apparently cured with chemotherapy uniformly relapse and succumb to infection despite equivalent rates of initial chemotherapeutic cure (J. McKinney, unpublished data). Additional evidence is provided by the observation that Mtb-infected wild-type mice fed a highly specific NOS2-inhibitor during the chronic phase of infection demonstrate a rapid recrudescence of bacillary burden and decreased survival compared to infected counterparts fed an inactive enantiomer (MacMicking et al., 1997). Despite this evidence, little is known about the nature and extent of NOS2 activity during this period of infection. For example, it is known that immune activation by Mtb increases expression of arginase. Thus, it might be predicted that local concentrations of arginine decrease over the course of infection, restricting the production of NOS2-derived RNI to relatively low levels sufficient to inhibit, but not kill, a small, residual bacillary burden (Albina et al., 1989; Ehrt et al., 2001; Gobert et al., 2000). However, under these same conditions of arginine limitation, NOS2 can also produce $O_2^{-\cdot}$ along with, or instead of NO. Thus, it is interesting to note that the 3-4 log recrudescence of *katG*-deficient Mtb observed in the liver and spleen of NOS2-deficient mice at later stages of infection (Ng et al., 2004). Based on the early and limited respiratory burst associated with Mtb infection and the ability of NOS2 to produce $O_2^{-\cdot}$ under conditions of arginine limitation, it is plausible that this recrudescence could, in part, be explained by a late role for NOS2-derived ROI in the chronic and/or persistent phase of Mtb infection. Together, these examples emphasize the outstanding need for more detailed *quantitative* and *kinetic* descriptions of ROI/RNI production at both the tissue and cellular levels *in vivo*.

Significant technical roadblocks also constrain our ability to evaluate genes whose functions may be required for growth *in vitro* or in a conditional manner. Currently available conventional gene disruption approaches preclude the evaluation of 'conditional' functions in genes that serve differing roles at differing times and/or in differing biological contexts (Nathan, 2004). For example, if a gene functions both in establishing an acute infection and persisting in a chronic infection, its role in acute infection might significantly alter, or altogether mask its role in later phases of infection using conventional gene knockout approaches. Thus, 'conditional' gene activation approaches are needed. The development of such approaches is likely essential to ultimately explain how ROI/RNI achieve immune control of Mtb that is often good enough to prevent or significantly delay progression to disease, yet how Mtb, at the same time, is able to resist sterilization and persist successfully, often for the lifetime of the host.

Transposon-mediated genomic approaches have, in the meantime, offered inroads toward the systematic identification of Mtb ROI/RNI resistance genes

(Darwin et al., 2003; Sassetti et al., 2003) while the availability of phox- and/NOS2-deficient mice offer adjunctive opportunities to replace chemically-derived stresses with biological sources of ROI/RNI. One particularly promising approach that was recently initiated involves the use of a variant of signature-tagged transposon mutagenesis (STM) (Camacho et al., 1999; Cox et al., 1999), termed *differential* STM (Hisert et al., 2004). This adaptation applies the same principle of competitive survival among a pool of signature-tagged transposon mutant Mtb clones to identify genes required for growth *in vivo* (which appear under-represented among clones recovered *following* infection) but applies this pool to strains of mice harboring mutations in different host defense pathways. In this manner, genes required for defense against a specific host-derived stress can be identified by their selective appearance in genetically engineered mice unable to generate that stress. Screening-based approaches such as this emphasize the relative infancy of our efforts to understand ROI/RNI resistance in Mtb yet mark equally the abundance of opportunities to study mycobacterial ROI/RNI resistance anew.

ACKNOWLEDGEMENTS

I thank Carl Nathan for his critical review of this work and Sabine Ehrt, Carl Nathan, John McKinney and Heran Darwin for invaluable discussions and sharing unpublished data. I also acknowledge the support of the NIH/NIAID (K08AI061393).

REFERENCES

Adams, L.B., Dinauer, M.C., Morgenstern, D.E., and Krahenbuhl, J. L. (1997). Comparison of the roles of reactive oxygen and nitrogen intermediates in the host response to *Mycobacterium tuberculosis* using transgenic mice. Tuber Lung Dis. *78*, 237-246.

Albina, J.E., Caldwell, M.D., Henry, W.L., Jr., and Mills, C.D. (1989). Regulation of macrophage functions by L-arginine. J. Exp. Med. *169*, 1021-1029.

Archibald, F.S., and Fridovich, I. (1982). The scavenging of superoxide radical by manganous complexes: *in vitro*. Arch. Biochem. Biophys *214*, 452-463.

Astarie-Dequeker, C., N'Diaye, E.N., Le Cabec, V., Rittig, M.G., Prandi, J., and Maridonneau-Parini, I. (1999). The mannose receptor mediates uptake of pathogenic and nonpathogenic mycobacteria and bypasses bactericidal responses in human macrophages. Infect. Immun. *67*, 469-477.

Braunstein, M., Brown, A.M., Kurtz, S., and Jacobs, W.R., Jr. (2001). Two nonredundant SecA homologues function in mycobacteria. J. Bacteriol. *183*, 6979-6990.

Braunstein, M., Espinosa, B.J., Chan, J., Belisle, J.T., and Jacobs, W.R., Jr. (2003). SecA2 functions in the secretion of superoxide dismutase A and in the virulence of *Mycobacterium tuberculosis*. Mol. Microbiol. *48*, 453-464.

Bryk, R., Griffin, P., and Nathan, C. (2000). Peroxynitrite reductase activity of bacterial peroxiredoxins. Nature *407*, 211-215.

Bryk, R., Lima, C.D., Erdjument-Bromage, H., Tempst, P., and Nathan, C. (2002). Metabolic enzymes of mycobacteria linked to antioxidant defense by a thioredoxin-like protein. Science *295*, 1073-1077.

Camacho, L.R., Ensergueix, D., Perez, E., Gicquel, B., and Guilhot, C. (1999). Identification of a virulence gene cluster of *Mycobacterium tuberculosis* by signature-tagged transposon mutagenesis. Mol. Microbiol. *34*, 257-267.

Chan, J., Fujiwara, T., Brennan, P., McNeil, M., Turco, S.J., Sibille, J.C., Snapper, M., Aisen, P., and Bloom, B.R. (1989). Microbial glycolipids: possible virulence factors that scavenge oxygen radicals. Proc. Natl. Acad. Sci. U.S.A. *86*, 2453-2457.

Chan, J., Xing, Y., Magliozzo, R.S., and Bloom, B.R. (1992). Killing of virulent *Mycobacterium tuberculosis* by reactive nitrogen intermediates produced by activated murine macrophages. J. Exp. Med. *175*, 1111-1122.

Chen, L., Xie, Q.W., and Nathan, C. (1998). Alkyl hydroperoxide reductase subunit C (AhpC) protects bacterial and human cells against reactive nitrogen intermediates. Mol. Cell *1*, 795-805.

Cheton, P.L., and Archibald, F.S. (1988). Manganese complexes and the generation and scavenging of hydroxyl free radicals. Free Radic Biol. Med. *5*, 325-333.

Choi, H.S., Rai, P.R., Chu, H.W., Cool, C., and Chan, E.D. (2002). Analysis of nitric oxide synthase and nitrotyrosine expression in human pulmonary tuberculosis. Amer. J. Respir Crit Care Med. *166*, 178-186.

Cohn, M.L., Kovitz, C., Oda, U., and Middlebrook, G. (1954). Studies on isoniazid and tubercle bacilli. II. The growth requirements, catalase activities, and pathogenic properties of isoniazid-resistant mutants. Amer. Rev. Tuberc *70*, 641-664.

Cole, S.T., Brosch, R., Parkhill, J., Garnier, T., Churcher, C., Harris, D., Gordon, S.V., Eiglmeier, K., Gas, S., Barry, C.E., 3rd, et al. (1998). Deciphering the biology of *Mycobacterium tuberculosis* from the complete genome sequence. Nature *393*, 537-544.

Coombes, B.K., Valdez, Y., and Finlay, B B. (2004). Evasive maneuvers by secreted bacterial proteins to avoid innate immune responses. Curr. Biol. *14*, R856-867.

Cooper, A.M., Segal, B.H., Frank, A.A., Holland, S.M., and Orme, I.M. (2000). Transient loss of resistance to pulmonary tuberculosis in p47(phox-/-) mice. Infect. Immun. *68*, 1231-1234.

Couture, M., Yeh, S R., Wittenberg, B A., Wittenberg, J.B., Ouellet, Y., Rousseau, D.L., and Guertin, M. (1999). A cooperative oxygen-binding hemoglobin from *Mycobacterium tuberculosis*. Proc. Natl. Acad. Sci. U.S.A. *96*, 11223-11228.

Cox, J.S., Chen, B., McNeil, M., and Jacobs, W.R., Jr. (1999). Complex lipid determines tissue-specific replication of *Mycobacterium tuberculosis* in mice. Nature *402*, 79-83.

Crawford, M.J., and Goldberg, D.E. (1998). Regulation of the *Salmonella typhimurium* flavohemoglobin gene. A new pathway for bacterial gene expression in response to nitric oxide. J. Biol. Chem. *273*, 34028-34032.

Darwin, K.H., Ehrt, S., Gutierrez-Ramos, J.C., Weich, N., and Nathan, C.F. (2003). The proteasome of *Mycobacterium tuberculosis* is required for resistance to nitric oxide. Science *302*, 1963-1966.

Davies, K.J. (2001). Degradation of oxidized proteins by the 20S proteasome. Biochimie *83*, 301-310.

Denis, M. (1991). Human neutrophils, activated with cytokines or not, do not kill virulent *Mycobacterium tuberculosis*. J. Infect. Dis. *163*, 919-920.

Deretic, V., Philipp, W., Dhandayuthapani, S., Mudd, M.H., Curcic, R., Garbe, T., Heym, B., Via, L.E., and Cole, S.T. (1995). *Mycobacterium tuberculosis* is a natural mutant with an inactivated oxidative-stress regulatory gene: implications for sensitivity to isoniazid. Mol. Microbiol. *17*, 889-900.

Domenech, P., Pym, A.S., Cellier, M., Barry, C.E., 3rd, and Cole, S.T. (2002). Inactivation of the *Mycobacterium tuberculosis* Nramp orthologue (*mntH*) does not affect virulence in a mouse model of tuberculosis. FEMS Microbiol. Lett. *207*, 81-86.

Dussurget, O., Stewart, G., Neyrolles, O., Pescher, P., Young, D., and Marchal, G. (2001). Role of *Mycobacterium tuberculosis* copper-zinc superoxide dismutase. Infect. Immun. *69*, 529-533.

Edwards, K.M., Cynamon, M.H., Voladri, R.K., Hager, C.C., DeStefano, M.S., Tham, K.T., Lakey, D.L., Bochan, M.R., and Kernodle, D. S. (2001). Iron-cofactored superoxide dismutase inhibits host responses to *Mycobacterium tuberculosis*. Amer. J. Respir Crit Care Med. *164*, 2213-2219.

Ehrt, S., Schnappinger, D., Bekiranov, S., Drenkow, J., Shi, S., Gingeras, T.R., Gaasterland, T., Schoolnik, G., and Nathan, C. (2001). Reprogramming of the macrophage transcriptome in response to interferon-gamma and *Mycobacterium tuberculosis*: signaling roles of nitric oxide synthase-2 and phagocyte oxidase. J. Exp. Med. *194*, 1123-1140.

Ehrt, S., Shiloh, M.U., Ruan, J., Choi, M., Gunzburg, S., Nathan, C., Xie, Q., and Riley, L. W. (1997). A novel antioxidant gene from *Mycobacterium tuberculosis*. J. Exp. Med. *186*, 1885-1896.

Fang, F.. (2004). Antimicrobial reactive oxygen and nitrogen species: concepts and controversies. Nat. Rev. Microbiol. *2*, 820-832.

Flynn, J.L., Chan, J., Triebold, K.J., Dalton, D.K., Stewart, T.A., and Bloom, B.R. (1993). An essential role for interferon gamma in resistance to *Mycobacterium tuberculosis* infection. J. Exp. Med. *178*, 2249-2254.

Forman, H.J., and Torres, M. (2002). Reactive oxygen species and cell signaling: respiratory burst in macrophage signaling. Amer. J. Respir Crit Care Med. *166*, S4-8.

Gobert, A.P., Daulouede, S., Lepoivre, M., Boucher, J.L., Bouteille, B., Buguet, A., Cespuglio, R., Veyret, B., and Vincendeau, P. (2000). L-Arginine availability modulates local nitric oxide production and parasite killing in experimental trypanosomiasis. Infect. Immun. *68*, 4653-4657.

Grisham, M.B., Jourd'Heuil, D., and Wink, D.A. (1999). Nitric oxide. I. Physiological chemistry of nitric oxide and its metabolites:implications in inflammation. Amer. J. Physiol *276*, G315-321.

Haas, D.W. (2000). *Mycobacterium tuberculosis*. In Mandell, Douglas and Bennett's Principles and Practice of Infectious Diseases, G.L. Mandell, Bennett, J.E., Dolin, J., ed. (Philadelphia, Churchill-Livingstone).

Heym, B., Stavropoulos, E., Honore, N., Domenech, P., Saint-Joanis, B., Wilson, T.M., Collins, D.M., Colston, M.J., and Cole, S.T. (1997). Effects of overexpression of the alkyl hydroperoxide reductase AhpC on the virulence and isoniazid resistance of *Mycobacterium tuberculosis*. Infect. Immun. *65*, 1395-1401.

Heym, B., Zhang, Y., Poulet, S., Young, D., and Cole, S.T. (1993). Characterization of the *katG* gene encoding a catalase-peroxidase required for the isoniazid susceptibility of *Mycobacterium tuberculosis*. J. Bacteriol. *175*, 4255-4259.

Hisert, K.B., Kirksey, M.A., Gomez, J.E., Sousa, A.O., Cox, J.S., Jacobs, W.R., Jr., Nathan, C.F., and McKinney, J.D. (2004). Identification of *Mycobacterium tuberculosis* counterimmune (cim) mutants in immunodeficient mice by differential screening. Infect. Immun. *72*, 5315-5321.

Horsburgh, M.J., Wharton, S.J., Karavolos, M., and Foster, S.J. (2002). Manganese: elemental defence for a life with oxygen. Trends Microbiol. *10*, 496-501.

Hu, C., Mayadas-Norton, T., Tanaka, K., Chan, J., and Salgame, P. (2000). *Mycobacterium tuberculosis* infection in complement receptor 3-deficient mice. J. Immunol. *165*, 2596-2602.

Hu, Y., Butcher, P.D., Mangan, J.A., Rajandream, M.A., and Coates, A.R. (1999). Regulation of *hmp* gene transcription in *Mycobacterium tuberculosis*: effects of oxygen limitation and nitrosative and oxidative stress. J. Bacteriol. *181*, 3486-3493.

Jaeger, T., Budde, H., Flohe, L., Menge, U., Singh, M., Trujillo, M., and Radi, R. (2004). Multiple thioredoxin-mediated routes to detoxify hydroperoxides in *Mycobacterium tuberculosis*. Arch. Biochem. Biophys *423*, 182-191.

Jones, G.S., Amirault, H.J., and Andersen, B.R. (1990). Killing of *Mycobacterium tuberculosis* by neutrophils: a nonoxidative process. J. Infect. Dis. *162*, 700-704.

Jung, Y.J., LaCourse, R., Ryan, L., and North, R.J. (2002). Virulent but not avirulent *Mycobacterium tuberculosis* can evade the growth inhibitory action of a T helper 1-dependent, nitric oxide synthase 2-independent defense in mice. J. Exp. Med. *196*, 991-998.

Kaushal, D., Schroeder, B.G., Tyagi, S., Yoshimatsu, T., Scott, C., Ko, C., Carpenter, L., Mehrotra, J., Manabe, Y.C., Fleischmann, R.D., and Bishai, W.R. (2002). Reduced immunopathology and mortality despite tissue persistence in a *Mycobacterium tuberculosis* mutant lacking alternative sigma factor, SigH. Proc. Natl. Acad. Sci. U.S.A. *99*, 8330-8335.

Koul, A., Choidas, A., Treder, M., Tyagi, A.K., Drlica, K., Singh, Y., and Ullrich, A. (2000). Cloning and characterization of secretory tyrosine phosphatases of *Mycobacterium tuberculosis*. J. Bacteriol. *182*, 5425-5432.

Koul, A., Herget, T., Klebl, B., and Ullrich, A. (2004). Interplay between mycobacteria and host signalling pathways. Nat. Rev. Microbiol. *2*, 189-202.

Lepoivre, M., Fieschi, F., Coves, J., Thelander, L., and Fontecave, M. (1991). Inactivation of ribonucleotide reductase by nitric oxide. Biochem. Biophys Res. Commun *179*, 442-448.

Li, Z., Kelley, C., Collins, F., Rouse, D., and Morris, S. (1998). Expression of *katG* in *Mycobacterium tuberculosis* is associated with its growth and persistence in mice and guinea pigs. J. Infect. Dis. *177*, 1030-1035.

Liu, L., Hausladen, A., Zeng, M., Que, L., Heitman, J., and Stamler, J. S. (2001). A metabolic enzyme for S-nitrosothiol conserved from bacteria to humans. Nature *410*, 490-494.

Long, R., Light, B., and Talbot, J.A. (1999). Mycobacteriocidal action of exogenous nitric oxide. Antimicrob Agents Chemother *43*, 403-405.

Lundberg, B.E., Wolf, R.E., Jr., Dinauer, M.C., Xu, Y., and Fang, F.C. (1999). Glucose 6-phosphate dehydrogenase is required for *Salmonella typhimurium* virulence and resistance to reactive oxygen and nitrogen intermediates. Infect. Immun. *67*, 436-438.

MacMicking, J.D., North, R.J., LaCourse, R., Mudgett, J.S., Shah, S.K., and Nathan, C.F. (1997). Identification of nitric oxide synthase as a protective locus against tuberculosis. Proc. Natl. Acad. Sci. U.S.A. *94*, 5243-5248.

Malhotra, V., Sharma, D., Ramanathan, V.D., Shakila, H., Saini, D.K., Chakravorty, S., Das, T.K., Li, Q., Silver, R.F., Narayanan, P.R., and Tyagi, J. S. (2004). Disruption of response regulator gene, *devR*, leads to attenuation in virulence of *Mycobacterium tuberculosis*. FEMS Microbiol. Lett. *231*, 237-245.

Manca, C., Paul, S., Barry, C.E., 3rd, Freedman, V.H., and Kaplan, G. (1999). *Mycobacterium tuberculosis* catalase and peroxidase activities and resistance to oxidative killing in human monocytes *in vitro*. Infect. Immun. *67*, 74-79.

Manganelli, R., Provvedi, R., Rodrigue, S., Beaucher, J., Gaudreau, L., Smith, I., and Proveddi, R. (2004). Sigma factors and global gene regulation in *Mycobacterium tuberculosis*. J. Bacteriol. *186*, 895-902.

Maragos, C.M., Andrews, A.W., Keefer, L.K., and Elespuru, R.K. (1993). Mutagenicity of glyceryl trinitrate (nitroglycerin) in *Salmonella typhimurium*. Mutat Res. *298*, 187-195.

Master, S.S., Springer, B., Sander, P., Boettger, E.C., Deretic, V., and Timmins, G.S. (2002). Oxidative stress response genes in *Mycobacterium tuberculosis*: role of *ahpC* in resistance to peroxynitrite and stage-specific survival in macrophages. Microbiology *148*, 3139-3144.

May, M.E., and Spagnuolo, P.J. (1987). Evidence for activation of a respiratory burst in the interaction of human neutrophils with *Mycobacterium tuberculosis*. Infect. Immun. *55*, 2304-2307.

Middlebrook, G., and Cohn, M.L. (1953). Some observations on the pathogenicity of isoniazid-resistant variants of tubercle bacilli. Science *118*, 297-299.

Milani, M., Pesce, A., Ouellet, H., Guertin, M., and Bolognesi, M. (2003). Truncated hemoglobins and nitric oxide action. IUBMB Life *55*, 623-627.

Miller, B.H., Fratti, R.A., Poschet, J.F., Timmins, G.S., Master, S.S., Burgos, M., Marletta, M.A., and Deretic, V. (2004). Mycobacteria inhibit nitric oxide synthase recruitment to phagosomes during macrophage infection. Infect. Immun. *72*, 2872-2878.

Mogues, T., Goodrich, M.E., Ryan, L., LaCourse, R., and North, R.J. (2001). The relative importance of T cell subsets in immunity and immunopathology of airborne *Mycobacterium tuberculosis* infection in mice. J. Exp. Med. *193*, 271-280.

Nathan, C. (2003). Specificity of a third kind: reactive oxygen and nitrogen intermediates in cell signaling. J. Clin. Invest *111*, 769-778.

Nathan, C. (2004). Antibiotics at the crossroads. Nature *431*, 899-902.

Nathan, C., and Shiloh, M.U. (2000). Reactive oxygen and nitrogen intermediates in the relationship between mammalian hosts and microbial pathogens. Proc. Natl. Acad. Sci. U.S.A. *97*, 8841-8848.

Nathan, C.F., and Ehrt, S. (2004). Nitric Oxide in Tuberculosis. In Tuberculosis, Second Edition, W. N. Rom, and S. M. Garay, eds. (Philadelphia, Lippincott Williams and Wilkins), pp. 215-235.

Neill, M.A., and Klebanoff, S. J. (1988). The effect of phenolic glycolipid-1 from *Mycobacterium leprae* on the antimicrobial activity of human macrophages. J. Exp. Med. *167*, 30-42.

Newton, G.L., Arnold, K., Price, M.S., Sherrill, C., Delcardayre, S.B., Aharonowitz, Y., Cohen, G., Davies, J., Fahey, R.C., and Davis, C. (1996). Distribution of thiols in microorganisms: mycothiol is a major thiol in most actinomycetes. J. Bacteriol. *178*, 1990-1995.

Ng, V.H., Cox, J.S., Sousa, A.O., MacMicking, J.D., and McKinney, J.D. (2004). Role of KatG catalase-peroxidase in mycobacterial pathogenesis: countering the phagocyte oxidative burst. Mol. Microbiol. *52*, 1291-1302.

Ochsner, U.A., Snyder, A., Vasil, A.I., and Vasil, M.L. (2002). Effects of the twin-arginine translocase on secretion of virulence factors, stress response, and pathogenesis. Proc. Natl. Acad. Sci. U.S.A. *99*, 8312-8317.

Ohno, H., Zhu, G., Mohan, V.P., Chu, D., Kohno, S., Jacobs, W. R., Jr., and Chan, J. (2003). The effects of reactive nitrogen intermediates on gene expression in *Mycobacterium tuberculosis*. Cell Microbiol. *5*, 637-648.

Ouellet, H., Ouellet, Y., Richard, C., Labarre, M., Wittenberg, B., Wittenberg, J., and Guertin, M. (2002). Truncated hemoglobin HbN protects *Mycobacterium bovis* from nitric oxide. Proc. Natl. Acad. Sci. U.S.A. *99*, 5902-5907.

Parish, T., Smith, D.A., Kendall, S., Casali, N., Bancroft, G.J., and Stoker, N.G. (2003). Deletion of two-component regulatory systems increases the virulence of *Mycobacterium tuberculosis*. Infect. Immun. *71*, 1134-1140.

Park, H.D., Guinn, K.M., Harrell, M.I., Liao, R., Voskuil, M.I., Tompa, M., Schoolnik, G.K., and Sherman, D.R. (2003). Rv3133c/*dosR* is a transcription factor that mediates the hypoxic response of *Mycobacterium tuberculosis*. Mol. Microbiol. *48*, 833-843.

Pedrosa, J., Saunders, B.M., Appelberg, R., Orme, I.M., Silva, M.T., and Cooper, A.M. (2000). Neutrophils play a protective nonphagocytic role in systemic *Mycobacterium tuberculosis* infection of mice. Infect. Immun. *68*, 577-583.

Piddington, D.L., Fang, F.C., Laessig, T., Cooper, A.M., Orme, I.M., and Buchmeier, N. A. (2001). Cu,Zn superoxide dismutase of *Mycobacterium tuberculosis* contributes to survival in activated macrophages that are generating an oxidative burst. Infect. Immun. *69*, 4980-4987.

Poole, R.K., and Hughes, M.N. (2000). New functions for the ancient globin family: bacterial responses to nitric oxide and nitrosative stress. Mol. Microbiol. *36*, 775-783.

Pradel, N., Ye, C., Livrelli, V., Xu, J., Joly, B., and Wu, L.F. (2003). Contribution of the twin arginine translocation system to the virulence of enterohemorrhagic *Escherichia coli* O157:H7. Infect. Immun. *71*, 4908-4916.

Pym, A.S., Domenech, P., Honore, N., Song, J., Deretic, V., and Cole, S.T. (2001). Regulation of catalase-peroxidase (KatG) expression, isoniazid sensitivity and virulence by *furA* of *Mycobacterium tuberculosis*. Mol. Microbiol. *40*, 879-889.

Rand, L., Hinds, J., Springer, B., Sander, P., Buxton, R.S., and Davis, E.O. (2003). The majority of inducible DNA repair genes in *Mycobacterium tuberculosis* are induced independently of RecA. Mol. Microbiol. *50*, 1031-1042.

Ritz, D., and Beckwith, J. (2001). Roles of thiol-redox pathways in bacteria. Annu. Rev. Microbiol. *55*, 21-48.

Roberts, D.M., Liao, R.P., Wisedchaisri, G., Hol, W.G., and Sherman, D.R. (2004). Two sensor kinases contribute to the hypoxic response of *Mycobacterium tuberculosis*. J. Biol. Chem. *279*, 23082-23087.

Rousseau, C., Winter, N., Pivert, E., Bordat, Y., Neyrolles, O., Ave, P., Huerre, M., Gicquel, B., and Jackson, M. (2004). Production of phthiocerol dimycocerosates protects *Mycobacterium tuberculosis* from the cidal activity of reactive nitrogen intermediates produced by macrophages and modulates the early immune response to infection. Cell Microbiol. *6*, 277-287.

Ruan, J., St John, G., Ehrt, S., Riley, L., and Nathan, C. (1999). NoxR3, a novel gene from *Mycobacterium tuberculosis*, protects *Salmonella typhimurium* from nitrosative and oxidative stress. Infect. Immun. *67*, 3276-3283.

Saini, D.K., Malhotra, V., Dey, D., Pant, N., Das, T.K., and Tyagi, J.S. (2004). DevR-DevS is a bona fide two-component system of *Mycobacterium tuberculosis* that is hypoxia-responsive in the absence of the DNA-binding domain of DevR. Microbiology *150*, 865-875.

Sander, P., Papavinasasundaram, K.G., Dick, T., Stavropoulos, E., Ellrott, K., Springer, B., Colston, M.J., and Bottger, E.C. (2001). *Mycobacterium bovis* BCG *recA* deletion mutant shows increased susceptibility to DNA-damaging agents but wild-type survival in a mouse infection model. Infect. Immun. *69*, 3562-3568.

Sassetti, C.M., Boyd, D.H., and Rubin, E.J. (2003). Genes required for mycobacterial growth defined by high density mutagenesis. Mol. Microbiol. *48*, 77-84.

Scanga, C.A., Mohan, V.P., Tanaka, K., Alland, D., Flynn, J.L., and Chan, J. (2001). The inducible nitric oxide synthase locus confers protection against aerogenic challenge of both clinical and laboratory strains of *Mycobacterium tuberculosis* in mice. Infect. Immun. *69*, 7711-7717.

Schaible, U.E., and Kaufmann, S.H. (2004). Iron and microbial infection. Nat. Rev. Microbiol. *2*, 946-953.

Schenk, G., Korsinczky, M.L., Hume, D.A., Hamilton, S., and DeJersey, J. (2000). Purple acid phosphatases from bacteria: similarities to mammalian and plant enzymes. Gene *255*, 419-424.

Schnappinger, D., Ehrt, S., Voskuil, M.I., Liu, Y., Mangan, J.A., Monahan, I.M., Dolganov, G., Efron, B., Butcher, P.D., Nathan, C., and Schoolnik, G.K. (2003). Transcriptional Adaptation of *Mycobacterium tuberculosis* within Macrophages: Insights into the Phagosomal Environment. J. Exp. Med. *198*, 693-704.

Schon, T., Elmberger, G., Negesse, Y., Pando, R.H., Sundqvist, T., and Britton, S. (2004). Local production of nitric oxide in patients with tuberculosis. Int. J. Tuberc Lung Dis. *8*, 1134-1137.

Segal, B.H., Leto, T.L., Gallin, J.I., Malech, H.L., and Holland, S.M. (2000). Genetic, biochemical, and clinical features of chronic granulomatous disease. Medicine (Baltimore) *79*, 170-200.

Seib, K.L., Tseng, H.J., McEwan, A.G., Apicella, M.A., and Jennings, M.P. (2004). Defenses against oxidative stress in *Neisseria gonorrhoeae* and *Neisseria meningitidis*: distinctive systems for different lifestyles. J. Infect. Dis. *190*, 136-147.

Seiler, P., Aichele, P., Raupach, B., Odermatt, B., Steinhoff, U., and Kaufmann, S.H. (2000). Rapid neutrophil response controls fast-replicating intracellular bacteria but not slow-replicating *Mycobacterium tuberculosis*. J. Infect. Dis. *181*, 671-680.

Sherman, D.R., Sabo, P.J., Hickey, M.J., Arain, T.M., Mahairas, G.G., Yuan, Y., Barry, C.E., 3rd, and Stover, C.K. (1995). Disparate responses to oxidative stress in saprophytic and pathogenic mycobacteria. Proc. Natl. Acad. Sci. U.S.A. *92*, 6625-6629.

Singh, R., Rao, V., Shakila, H., Gupta, R., Khera, A., Dhar, N., Singh, A., Koul, A., Singh, Y., Naseema, M., et al. (2003). Disruption of *mptpB* impairs the ability of *Mycobacterium tuberculosis* to survive in guinea pigs. Mol. Microbiol. *50*, 751-762.

Springer, B., Master, S., Sander, P., Zahrt, T., McFalone, M., Song, J., Papavinasasundaram, K.G., Colston, M.J., Boettger, E., and Deretic, V. (2001). Silencing of oxidative stress response in *Mycobacterium tuberculosis*: expression patterns of *ahpC* in virulent and avirulent strains and effect of *ahpC* inactivation. Infect. Immun. *69*, 5967-5973.

St John, G., Brot, N., Ruan, J., Erdjument-Bromage, H., Tempst, P., Weissbach, H., and Nathan, C. (2001). Peptide methionine sulfoxide reductase from *Escherichia coli* and *Mycobacterium tuberculosis* protects bacteria against oxidative damage from reactive nitrogen intermediates. Proc. Natl. Acad. Sci. U.S.A. *98*, 9901-9906.

Stadtman, E.R., Berlett, B.S., and Chock, P.B. (1990). Manganese-dependent disproportionation of hydrogen peroxide in bicarbonate buffer. Proc. Natl. Acad. Sci. U.S.A. *87*, 384-388.

Stewart, G.R., Ehrt, S., Riley, L.W., Dale, J.W., and McFadden, J. (2000). Deletion of the putative antioxidant *noxR1* does not alter the virulence of *Mycobacterium tuberculosis* H37Rv. Tuber Lung Dis. *80*, 237-242.

Stokes, R.W., Norris-Jones, R., Brooks, D.E., Beveridge, T.J., Doxsee, D., and Thorson, L.M. (2004). The glycan-rich outer layer of the cell wall of *Mycobacterium tuberculosis* acts as an antiphagocytic capsule limiting the association of the bacterium with macrophages. Infect. Immun. *72*, 5676-5686.

Tamir, S., Burney, S., and Tannenbaum, S.R. (1996). DNA damage by nitric oxide. Chem. Res. Toxicol *9*, 821-827.

Venkataprasad, N., Riveros-Moreno, V., Sosnowska, D., and Moreno, C. (1999). Nitrotyrosine formation after activation of murine macrophages with mycobacteria and mycobacterial lipoarabinomannan. Clin. Exp. Immunol. *116*, 270-275.

Vogt, R.N., Steenkamp, D.J., Zheng, R., and Blanchard, J.S. (2003). The metabolism of nitrosothiols in the Mycobacteria: identification and characterization of S-nitrosomycothiol reductase. Biochem. J. *374*, 657-666.

Voskuil, M.I., Schnappinger, D., Visconti, K.C., Harrell, M.I., Dolganov, G.M., Sherman, D.R., and Schoolnik, G.K. (2003). Inhibition of respiration by nitric oxide induces a *Mycobacterium tuberculosis* dormancy program. J. Exp. Med. *198*, 705-713.

Weissbach, H., Etienne, F., Hoshi, T., Heinemann, S.H., Lowther, W.T., Matthews, B., St John, G., Nathan, C., and Brot, N. (2002). Peptide methionine sulfoxide reductase: structure, mechanism of action, and biological function. Arch. Biochem. Biophys. *397*, 172-178.

Wengenack, N.L., Jensen, M.P., Rusnak, F., and Stern, M.K. (1999). *Mycobacterium tuberculosis* KatG is a peroxynitritase. Biochem. Biophys Res. Commun *256*, 485-487.

Wink, D.A., Miranda, K.M., Espey, M.G., Pluta, R.M., Hewett, S.J., Colton, C., Vitek, M., Feelisch, M., and Grisham, M.B. (2001). Mechanisms of the antioxidant effects of nitric oxide. Antioxid Redox Signal *3*, 203-213.

Yamada, Y., Fujiwara, T., Sato, T., Igarashi, N., and Tanaka, N. (2002). The 2.0 A crystal structure of catalase-peroxidase from *Haloarcula marismortui*. Nat. Struct Biol. *9*, 691-695.

Yuan, Y., Crane, D.D., Simpson, R.M., Zhu, Y.Q., Hickey, M.J., Sherman, D.R., and Barry, C.E., 3rd (1998a). The 16-kDa alpha-crystallin (Acr) protein of *Mycobacterium tuberculosis* is required for growth in macrophages. Proc. Natl. Acad. Sci. U.S.A. *95*, 9578-9583.

Yuan, Y., Mead, D., Schroeder, B.G., Zhu, Y., and Barry, C.E., 3rd (1998b). The biosynthesis of mycolic acids in *Mycobacterium tuberculosis*. Enzymatic methyl(ene) transfer to acyl carrier protein bound meromycolic acid *in vitro*. J. Biol. Chem. *273*, 21282-21290.

Zahrt, T.C., and Deretic, V. (2002). Reactive nitrogen and oxygen intermediates and bacterial defenses: unusual adaptations in *Mycobacterium tuberculosis*. Antioxid Redox Signal *4*, 141-159.

Index

Books of Related Interest